Numbers and
Proofs

Numbers and Proofs

R. B. J. T. Allenby

School of Mathematics, University of Leeds

OXFORD   AMSTERDAM   BOSTON   LONDON   NEW YORK   PARIS
SAN DIEGO   SAN FRANCISCO   SINGAPORE   SYDNEY   TOKYO

Butterworth-Heinemann
An imprint of Elsevier Science
Linacre House, Jordan Hill, Oxford OX2 8DP
200 Wheeler Road, Burlington, MA 01803

First published 1997
Transferred to digital printing 2003

British Library Cataloguing in Publication Data
A catalogue record for this book is available from the British Library

Library of Congress Cataloguing in Publication Data
A catalogue record for this book is available from the Library of Congress

ISBN 0 340 67653 1

For information on all Butterworth-Heinemann publications
visit our website at www.bh.com

Contents

Preface

In recent years, school/college mathematics has undergone many changes. The most notable of these has been the introduction of a kind of 'investigative' mathematics typified by: 'Given 100 1 cm cubes, find the three-dimensional shape (using all these cubes) for which the surface area is a minimum.' As pointed out by Gardiner (1995) it is frequently the case that an investigation is concluded just as the *real mathematics* underlying the problem is about to enter. In other words, conclusions are reached but there is no *guarantee* that they are correct because they are supported by neither reasoned nor sufficiently precise argument. In the 'good old days', an early course in Euclidean geometry soon taught the student that statements such as 'The angles in a triangle add up to 180 degrees' needed a supporting argument. In fact, *I* can recall finding, using a protractor in a school mathematics class, that the answer is 179 degrees. This result was, if memory serves, greeted with some mirth (or was it despair?) by the master. It was certainly followed by an apparently incontrovertible *proof* that the correct answer is 180 degrees (exactly). Contrast this with the conversation (I hope apocryphal) which I heard reported recently which went:

> QUESTIONER TO TEACHER   Don't you *prove* that the angles of a triangle add up to 180 degrees?
> TEACHER   No! There is no need. I have thirty pupils here each of whom has measured it as 180 degrees ... to within a degree or two.

What would this teacher have made of the fact that, to 12 decimal places, the value of $e^{\pi\sqrt{163}}$ is 262 537 412 640 768 743.999 999 999 999. Presumably it would have been asserted that, 'Clearly $e^{\pi\sqrt{163}}$ is an exact whole number.' Well, it *isn't*. In any case try telling the pilot of a spaceship that his path is accurate *to within a degree or two*. Surely natural inquisitiveness would *demand* that you ask, 'Does this nearness to 180 degrees imply that it actually *is* 180 degrees? If so, how can we be so certain? If not, why are all practical measurements so near to 180 degrees?' In short, the finding of strongly suggestive 'patterns' seems to have replaced *real* mathematical activity.

Now I am *very much in favour* of experimentation involving mathematical ideas – after all, that is what most professional mathematicians spend some of their time doing: it is a fundamental part of research. Gauss, one of the three greatest of all mathematicians, was, in this sense, an experimenter *par excellence*. More often than not, though, all this experimentation can do is to *suggest* certain truths. In order to *guarantee* truth one needs to back up one's claims with logically sound and watertight supporting arguments.

This concern with rigorous argument is one of the major differences between higher education (HE) and pre-HE mathematics (where learning problem-solving techniques succeeds 'investigations') and it seems to come as a surprise to some students, many of whom, because of the trends in school mathematics education, are unaware of its importance.

Without *certainty* mathematics is nothing, so it is the main purpose of this book to introduce to the student beginning a higher education mathematics course (i) the idea that proof is *essential* and (ii) some of the methods by which 'obvious' or controversial or, sometimes, 'unbelievable' claims can be verified.

In those 'good old days', one learnt to construct proofs, from age about 11 onwards, by reading so many of them (mainly in Euclidean geometry) that the methods (and the appropriate style of presentation) were for ever etched on the brain. However, there was never any attempt made by the teacher to give any hint as to the kind of *thoughts* (in particular those ultimately irrelevant ones) which might have passed through the prover's mind on his or her way to the final version of his (her)[1] proof. Yet, as I have long been aware, such 'inside information' would be a boon to the student who, not infrequently, does not know how to begin, especially if he fails to realize that first proceeding up blind alleys may be part of the solution process. (Only fairly recently did I discover that, 50 years ago, Polya, in his book *How to Solve It*, promoted the desirability of giving intuitive sketches of proofs before presenting those logically sound, but occasionally impenetrable, arguments which are given in a style universally favoured since the time of Euclid. Even now, very very few books follow Polya's injunction.)

Because it is futile trying to proceed in a logically sound way if you have no thoughts to proceed with, a relatively large portion of Chapters 3 to 10 of this book is occupied by DISCUSSIONS, the purpose of which is to float such thoughts – even though some of them lead to dead ends. For the (intended) reader of this book, these Discussions are frequently analogous to the preliminary ponderings of the research mathematician. I hope that they, at least occasionally, convey some of the same excitement of the act of discovery. Naturally they are often much longer than the proofs which follow them. The Discussions are, therefore, intended as a chatty guide – a sort of written equivalent of what a teacher may say to the student in ordinary conversation. In particular, I hope they will make the book particularly suitable for self-study.

---

[1] I hope I offend no one if, from now on, I use masculine pronouns. I refuse to use absurd inventions as 'heim' or 'hirs'.

Supporting the Discussions, many proofs are followed by COMMENTS and sections headed WHAT WE HAVE LEARNED. Each chapter begins with a brief account of its content, and each of the first ten chapters ends with a SUMMARY of its main points.

Following a first chapter which stresses the need for proof but does not forget the experimentation through which potential *theorems* arise via *conjectures*, Chapter 2 provides a list of 18 statements whose purpose is twofold: (i) to show the frequent use of the *connectives* NOT, AND, etc. and to make their meaning (often somewhat flexible in ordinary spoken English) exact by the use of *truth tables*; (ii) to act as a sort of test area where a natural line of argument can emphasize universal methods of demonstration. (The 18 statements concern mainly numbers because it is numbers – rather than, say, the properties of geometrical figures or functions – with which the reader will be most familiar.) The testing begins in Chapter 3 and continues in Chapters 5, 6 and 7 after an interlude in Chapter 4 where some elementary set theory is introduced mainly to cope with statements which are more complicated than those in Chapter 3 because of the presence of the words *for all* and *there exists* and their negations.

Once some logically allowable methods of procedure have been identified, they are applied, in Chapters 8, 9 and 10, to determine some of the simpler properties of the rational, real and complex numbers as well as those of the integers.

Chapters 11 and 12 introduce general procedures – *guessing, analogy, transformation, generalization* and *specialization* – of which it is useful to be aware when one is tackling proofs or solving problems. Even though the first ten chapters seem to contain enough material for a first short module, I hope that teachers using this book will find, in these chapters, a number of items which it would be worthwhile their pointing out to their students.

Although, by the end of Chapter 12, occasional examples have been given of 'proofs' which do not stand up to close examination, Chapter 13 is a collection of results which are fallacious or paradoxical or whose proofs are just plain wrong. Some exhibit difficulties experienced by students; to show that the lecturer, too, is not infallible. The reader is asked to judge if each proof is unsatisfactory and, if it is, to say *why*. (I have always believed that one learns more from trying to find an error in a proof of a *false* statement – for example, from a proof that $\pi = 0$ or that $\sqrt{4}$ is not a rational number – than from trying to understand a correct proof of an intuitively believable statement. The former is simply more fascinating and, as in the $\sqrt{4}$ case, does tend to identify the key ideas in corresponding *valid* proofs; in this case the proof that $\sqrt{2}$ *is irrational*.)

Chapter 14, 'A mixed bag', presents a collection of problems challenging the reader to find proofs of 'new' statements and of ones which are analogues or generalizations of some which have appeared earlier. It might be fun to try the problems with a colleague, attempting to solve them as a team or in competition with each other. In this latter case I suggest that you each submit your solution for the other to examine critically to see if your colleague (i) can understand your solution and (ii) believes it. Such activity can help enormously in improving the readability of your work – not an unimportant attribute to offer to a prospective employer. In case a problem causes the reader some pain or frustration, the problems are followed by a very concise list of hints. Should the hints also not suffice, the chapter ends with full (but rather formally presented) solutions to the problems. Of course, I hope that the reader will refrain from delving into the appendix until (as some are wont to say) he has given a problem his *best shot*. Be that as it may, the reader may treat this appendix as a solutions section, as a reading section where he may study *style*, that is, *how* proofs are written out, and as a problem section where he can try to get into the mind of the prover – to see if he can discover what discussions the prover had with himself before setting out the solution in the complete but rather compact form given.

The book ends with a chapter in which answers are given to many of the exercises posed in the text. These solutions are meant to be of more immediate help to the reader than those in the appendix to Chapter 14 so they are written in a more casual style.

It is my hope that this book will both help you to appreciate something of the need for proofs in mathematics and how to go about constructing them. Finding proofs is an exciting research activity. There is no reason why doing problems in undergraduate mathematics should not be similarly stimulating. This means that attempting to solve problems can be exhilarating, even fun and, often, dare I say it, frustrating. Yes, that is part of the price you may have to pay. Most things which are worth doing require some effort (as you recall from learning to ride a bicycle and to swim) and, except for the lucky few, mathematics is no different. You will get fun from chasing a problem or theorem and 'capturing' it. (If you are coming along for an easy ride, *get off now*.) The taste of certainty when a proof has been successfully completed is sweet. As my research student David Doniz rightly said recently in relation to his own thesis, 'The pudding is the proof.'

As with all my other books I have had the good fortune of cajoling colleagues at the University of Leeds in to reading over my text. For the fourth time (and I wish him a long and happy retirement well away from colleagues who prey upon his good nature) I thank Dr Eric Wallace most effusively for his willingness to play the role of student reader. His only failure has been his inability to stop my using ten times the appropriate number of exclamation marks!!! I likewise thank Dr Alan Slomson for pointing out other shortcomings in my original text and Drs John Bowers and Mark Kelmanson for undertaking some fundamental and erudite research for me. Any blemishes which remain (and I urge the reader to tell me of them) can therefore only be put down to my cussedness in wanting to retain them or my deficiencies as a proof-reader.

Also for the fourth time I have benefited from the generosity of Professor Dr Konrad Jacobs of the University of Erlangen-Nurnberg who kindly donated the photographs.

CHAPTER 1

The Need for Proof

What is now proved was, once, only imagined. BLAKE

The main purpose of this introductory chapter is to set the scene. Perhaps the best thing to do is to read it over quickly and then return to it occasionally to refresh your memory. The main points of the chapter are summarized at the end.

Introduction

Mathematics, in particular pure mathematics, is a peculiar subject. It is peculiar in the sense of being special. It is the only subject in which the assertions which are made tend not to be accepted as correct until their truth has been incontrovertibly demonstrated. For example, 2400 years ago the Greeks *believed* that all substances were made out of earth, air, fire and water whilst, at the same time, they *knew* that, in a right-angled triangle, the area of the square constructed on the hypotenuse is equal to the sum of the areas of the squares constructed on the other two sides – because Pythagoras (and later Euclid) had demonstrated it conclusively.

Nowadays our ideas concerning the ultimate structure of matter are somewhat more sophisticated (but are still only conjectured) whereas Pythagoras' theorem is neither more nor less true than it was then.

Subjects other than mathematics frequently progress by trying to fit theories to observational data, making predictions from these theories and testing these predictions against yet more observations. If the new data fit the predictions then the theory is declared sound (for the moment). For (pure) mathematicians this approach is just not good enough.

For example, you will find, if you check it[1] up to 325, that *each positive whole number can be expressed as a sum of four (whole number[2]) squares.* (For example,[3] $196 = 9^2 + 9^2 + 5^2 + 3^2$; $302 = 17^2 + 3^2 + 2^2 + 0^2$.) Are those 325 favourable instances enough evidence, though, for you to claim that the above assertion in italics is true? Of course not – even if you confirm it for the next 100 000 integers too. For all you then know, the assertion may be false for the integer 100 326. *It cannot be emphasized too often that to be certain that an assertion is true for ALL integers, it is not enough to confirm it for a sample of integers, no matter how large the sample.*

Drawing a general inference from limited evidence is known as *induction.* For example, nationwide television viewing figures are inferred from questioning a fairly small sample of the population. Of course, arguments based on induction are insufficent to satisfy the (sceptical) methamatician, for whom confirmation that an assertion is correct is achieved by use of *deduction.*

However, the above remarks do not imply that mathematicians should shun experimentation – far from it. Gauss and others often used to carry out phenomenal amounts of calculation (see Exercise 1(d)) looking for patterns which might *suggest* mathematical assertions or relationships which *ought* to be true. The point is that, in mathematics, experimentation *without subsequent demonstration* is deemed to be of little value.

Conjectures

When the mathematician has the feeling that the result he is interested in is likely to be true – this feeling may come as a result of prolonged experimentation (and the use of induction) or it may just be a case of intuition – he is ready to give this 'likely result' the status of **conjecture.**

To get you experimenting (and then conjecturing), try the following exercise.

EXERCISE 1

(a) For each positive odd integer n (up to 21) find the value of $1 + 3 + \ldots + n$. *Make a general conjecture.*

(b) For the first six integers in the unending list $11, 111, 1111, 11111, \ldots$, determine which, if any, is a perfect square. *Make a general conjecture.*

(c) Determine which of the *even* integers from 4 to 50 inclusive, can be written as a sum of two primes (e.g. $28 = 11 + 17$). *Make a general conjecture.*

(d) For each prime number p from 3 to 17 (excluding 5) write the fraction $1/p$ in decimal form. (For example, $\frac{1}{7} = 0.142\,857\,14\ldots$. Notice that in this case, the repeating 'block', 142 857, is of length 6.) Find the length of the repeating block

---

[1] As Bachet de Meziriac (9 October 1581–26 February 1638) did in 1621.

[2] The positive and negative whole numbers together with zero are usually referred to as the *integers.* We shall tend to use this word from now on.

[3] Notice that we allow 0^2 as a summand. Otherwise, to obtain every positive integer as a sum of squares, we should have to rephrase our assertion as: Every positive integer is a sum of *at most* four (positive) squares.

CARL FRIEDRICH GAUSS
(30 April 1777–23 February 1855)

Carl Friedrich Gauss, generally regarded, along with Archimedes and Newton as one of the three greatest mathematicians of all time, humorously claimed that he could count before he could talk. Born in 1777 into a poor family with an illiterate, but intelligent mother, it is reputed that, aged 3, he corrected a mistake to his father's accounts and that, aged 8, he wrote down immediately the answer to the problem, set in class by a teacher in order to secure an hour's peace: Find the sum of the first 100 integers. (Gauss noticed that the sum is $(1 + 100) + (2 + 99) + \ldots + (50 + 51)$. See Comment 3 to Statement 16 in Chapter 7.)

Before he was 20 he had proved several deep results in the theory of numbers and conjectured the so-called prime number theorem (that the number of primes not exceeding the integer n is, approximately, $n/\log_e n$) the proof of which took another 100 years to find. It is said that it was only after showing how to construct, by straight-edge and compass only, a regular 17-gon, the first new such n-gon for 2000 years, that Gauss decided to study mathematics in preference to studying languages. Then, in his doctoral thesis he gave the first proof of the Fundamental Theorem of Algebra (see Chapter 10), a proof which had eluded such 'giants' as Euler and Lagrange. (He later gave three more proofs.)

In 1801 Gauss published his *Disquisitiones Arithmeticae*. This book begins by introducing the important concept of congruence (see Chapter 8). It also contains the proof, generalizing the above proof about the 17-gon, that, if p is a prime, then a regular p-gon is constructible (by straight-edge and compass) iff p is of the form $p = 2^{2^n} + 1$, that is, iff p is a Fermat prime. In the words of Lagrange, the *Disquisitiones* 'has raised you, at once, to the rank of the first mathematicians'. In the same year he became famous in the public eye when the astronomers, looking where Gauss told them to, rediscovered the planetoid Ceres, after they had 'lost' it and searched for it in vain for nine months.

The 'Prince of Mathematicians', who researched in probability, geodesy, mechanics, optics, actuarial science and electromagnetism, and who expressed the view that 'Mathematics is the queen of science – and number theory is the queen of mathematics', did not appear to relish travel. After his appointment as director of the Göttingen observatory in 1806, he spent only one night in 27 years *not* sleeping under its roof.

for each $1/p$ with $7 \leqslant p \leqslant 17$ and try to relate this length to p in each case. (Gauss did this for $3 \leqslant p \leqslant 1000$.) *Make a general conjecture.*

(e) Find those integers n, from 2 to 17 inclusive, for which $2^n - 2$ is divisible by n. (For example, $2^5 - 2$ *is* divisible by 5; $2^8 - 2$ is *not* divisible by 8.) *Make a general conjecture.* (You will probably arrive at the same conclusion as did Leibniz.[4])

The concept of proof and the main aims of this book

Proofs – and counterexamples

Once you have decided what the various parts of Exercise 1 suggest *might* be true you will, not doubt, be impatient to *confirm* these truths (if truths they be) by supplying incontrovertible and conclusive demonstrations of them. If this can be done, the 'likely results' (i.e. *conjectures*) then assume the status of **theorem**.[5] So how are such demonstations achieved? The answer is: *by means of proofs,* a **proof** being, roughly speaking, *a logically watertight argument comprising sequence of assertions, each deduced from earlier assertions or assumptions, with the last assertion being the conjecture itself.*

The concept of proof is absolutely fundamental to mathematics. In fact one may fairly claim that *without proof (pure) mathematics does not exist.*

It is, of course, just possible that not all the conjectures you make, based on the 'evidence' suggested by the various parts of Exercise 1, will be correct. After some time spent trying to verify a particular conjecture you may reach the conclusion that your conjecture is, after all, wrong. It then becomes appropriate to seek an example which will show that it is wrong; that is, you search for a **counterexample** to the conjecture. Consider, for instance, the assertion: *Every integer in the (unending) list* $31, 331, 3331, \ldots$ *is a prime number.* In fact, whilst it seems pretty remarkable that 31, 331, 3331, 33 331, 333 331, 3 333 331 and 33 333 331 are all primes, it is nevertheless the case that the assertion in italics is *false: not* every integer in the list

---

[4] Gottfried Wilhelm Leibniz (1 July 1646–14 November 1716).

[5] There is a feeling that the word 'theorem' ought to be reserved for true assertions of *some substance or importance.* The statement '5 is the second odd prime' is a true assertion which scarcely seems to deserve the title 'theorem'.

is prime. For example, 333 333 331 is not prime. Thus 333 333 331 is a *counterexample* to the assertion. In fact the next *prime* in the above list is 333 333 333 333 333 331. Notice, however, that the existence of these eight 'extra' (counter)examples immediately following 333 333 331 adds nothing to the fact that the assertion is false. It cannot be made *even more* false by listing these extra examples. So the message is: if a general assertion is false *just one instance* of its falsehood will suffice to show this.

The ability to construct counterexamples can be very useful. Sometimes, failed attempts at constructing a proof of a conjecture can point the way to a counterexample. by narrowing down the areas in which one might be found. Otherwise there is no recipe for developing it other than to acquire a sufficiently large knowledge so that you can more readily intuit when and where to look for one. Of course, for certain types of problem, seeking a counterexample by recourse to the computer may be appropriate ... but beware, just the other day my computer told me that 653 was a counterexample to a conjecture I had made.[6] Unfortunately, due to rounding errors (and bad programming) the computer had made a mistake – and indeed, I now know my conjecture was correct. In any event, seeking out a counterexample to an assertion that you believe to be false can be quite exhilarating (especially if the assertion you are trying to demolish is someone else's).

EXERCISE 2   Find counterexamples to the following assertions.

(a) (Letter to *The Times*) If p is a prime number, then so too is $p! + 1$. (Here $p!$ (**factorial** p) is the product $1 \cdot 2 \cdot 3 \cdot ... \cdot p$ of the first p positive integers.[7])
(b) If n is a positive integer such that n^2 is *palindromic* (i.e. like 9 and 484, n^2 reads the same forwards as backwards) then n must be palindromic.
(c) If a function $f(x)$ takes its least value at $x = 3$ then the value of its derivative, $f'(x)$, at $x = 3$, is 0.

What is needed in constructing proofs?

From our rough definition above, there are two major requirements for constructing a proof of an alleged truth:

1. The need to find a suitable sequence of assertions.
2. The need to employ only logically acceptable schemes of deduction.

How to learn the business

How do students new to the proof construction business learn their trade? One obvious way is to watch the masters at work. By this I mean that one can study proofs which are already in the literature, notice how they are set out and endeavour to understand *why* they do what is claimed of them. In doing this you will build

---

[6] And my eighteenth-century hand-calculator keeps telling me that $(\sqrt{3})^2 = 2.999\,999\,9$.
[7] In this book we will use the symbol $\cdot$ to denote multiplication when doing so makes that multiplication clearer. However, xy will be used rather than $x \cdot y$, for the product of x and y, if it is felt that there is no ambiguity in meaning.

for yourself a useful catalogue of methods, strategies (some used repeatedly) – even clever manoeuvres – and a feeling for what constitutes a logically acceptable procedure which you may be able to bring to bear on problems that you later tackle yourself. (It is perfectly allowable to make use of other people's good ideas. Even Newton[8] said, 'If I have seen further it is by standing on the shoulders of giants.')

Problems with watching the masters

Now it is not too difficult (see Chapter 2) to identify most of the logical concepts which help to guarantee the watertightness of mathematical arguments. However, there are difficulties with the 'watch the masters' approach, one of which is particularly apparent in the case of the first requirement listed above. It arises as follows.

The main task of the theorem prover is to convince the reader of the correctness of his argument. This dictates that the (alleged) proof be presented in a coherent logical order (each step following from earlier ones or assumptions) and in as clear a manner as possible *with all material which is not absolutely necessary to complete the argument being discarded*. All that a proof should contain is a succession of assertions which it is expected that the reader will agree to be valid (sometimes only after much thought).

One consequence of this (well-established) style of presentation is that it often gives the reader very little idea of the *thought processes* which occupied the prover on his way to the final version of his proof.[9]

Having worked through someone's alleged proof, the reader may then well say (with feeling), 'Well, I think I follow the argument and so I believe the result which the writer claims to have proved *but never in a thousand years would I have thought of that line of argument myself.*'

Here are two examples where the reader may experience such feelings.

Example 1

Show that the expression $T = \frac{4}{3} - \frac{1}{2}\cos 2x + \cos 3x + \frac{1}{2}\cos 4x$ is positive for all values of x.

SOLUTION   Let $y = \cos x + \cos 2x$. Then

$$y^2 - y + \tfrac{1}{3} = \cos^2 x + \cos^2 2x + 2\cos x \cdot \cos 2x - \cos x - \cos 2x + \tfrac{1}{3}$$

$$= \tfrac{1}{2}(1 + \cos 2x) + \tfrac{1}{2}(1 + \cos 4x) + \cos 3x + \cos x - \cos x$$

$$- \cos 2x + \tfrac{1}{3} = T$$

But

$$y^2 - y + \tfrac{1}{3} = (y - \tfrac{1}{2})^2 + \tfrac{1}{12} > 0 \quad (\text{since } (y - \tfrac{1}{2})^2 \geqslant 0)$$

[8] Isaac Newton (25 December 1642–20 March 1727), generally regarded, along with Archimedes and Gauss, as one of the three greatest ever mathematicians.

[9] The blame for this state of affairs lies at the door of Euclid. His masterpiece *The Elements* set the style which has persisted for 2000 years.

The reader may think, 'Yes, with some effort I can follow that and I can see that it is correct, but *how did you ever think of beginning* "Let $y = \cos x + \cos 2x$"?'

Possible answers might be: (i) Lucky guess (portrayed, by some, as intuition): (ii) having a good idea of what to try after doing similar problems before (cf. Final Exercise 1(b)); (iii) it was the better of only two useful ideas which emerged from hours of hard slog.

Our second example has a strong historical connection. The standard formula

$$x = \frac{-b \pm \sqrt{(b^2 - 4ac)}}{2a}$$

giving the roots of a quadratic equation was (in essence) known to the Babylonian mathematicians of 1500 BC. On the other hand, similar formulae for the general cubic and quartic equation were not discovered until the first half of the sixteenth century. Here is a way to solve the general cubic equation.

Example 2
Determine the roots of the cubic equation $X^3 + aX^2 + bX + c = 0$.

SOLUTION First put $x = X + \frac{1}{3}a$. The given cubic then reduces to one whose coefficient of x^2 is zero. We may therefore suppose that, when all the tidying up has been done, the equation then takes the simpler form $x^3 + px + q = 0$.

Now (and this is the clever bit) the roots of the cubic equation $x^3 + px + q = 0$ can be obtained as follows. First replace x by $z - (p/3z)$. Tidying up again, we obtain

$$x^3 + px + q = z^3 - \frac{p^3}{27z^3} + q = 0$$

which may be rewritten as

$$z^6 + qz^3 - \frac{p^3}{27} = 0$$

This is a quadratic in z^3 which has solutions given by

$$z^3 = \frac{1}{2}\left\{ -q + \sqrt{\left(q^2 + \frac{4p^3}{27} \right)} \right\}$$

z and x are then easily found.

Reader: 'Again I follow the argument but, again, I must ask: How did the solver think of that nice solution, in particular the neat idea of replacing x by $z - (p/3z)$ which reduces the problem to solving a quadratic?'[10]

Incidentally, the reader may well be itching to know the answer to another question. Because of the $\pm$ sign in the formula for z^3 there appear to be *six* possible values

---

[10] Girolamo Cardano (24 September 1501–21 September 1576) showed how to get rid of the quadratic term in a cubic. François Vieta (? 1540–23 February 1603) introduced the substitution of $z - (p/3z)$ for x.

for z. That is, there seem to be *six roots of a cubic equation*. As we should expect at most three, we cannot allow this to go unchallenged. It needs investigating! However, because of space restrictions here and because it would take us a little too far afield, you are referred to Allenby (1991, p. 181).

A second difficulty with the 'watch the masters' approach is that many of the wonderful proofs we might think of examining are wonderful precisely because they solve difficult problems. As a consequence most of these proofs are quite involved (as well as being compactly written). It therefore seems better for us to begin with somewhat easier problems where we can at least experience the pleasure of joining in and discovering various solutions *for ourselves*.)

Help is at hand

If you come across rather snappy and all too condensed proofs of the above kind – and you *will* – DO NOT DESPAIR! First, *you are not alone* – professional mathematicians frequently meet the same problem, chiefly when they are reading other people's research papers. Second, apart from showing you how to argue logically, one of the *main reasons for writing this book* is that I wanted to take you, so to speak, 'behind the scenes' by revealing some of the preliminaries that a mathematician might go through before producing one of those 'polished' proofs whose genesis has been obscured.

These preliminaries, which we present *before* (almost) every one of the earlier proofs under the heading DISCUSSION are, of course, *not part of the proof*; rather, they are intended to indicate what might be taking place in the mind of the mathematician seeking to find an appropriate proof – or counterexample. (Think how much more readily you would follow – even enjoy – a game of chess between grandmasters if each would share with you his (or her) thinking relating to the problems posed by the opponent.)

The Discussion sections will not reveal a single 'universal' strategy for proof-finding but, if you digest them all, you should be better able to decide which method(s) of attack might be most profitable (and which least) when faced with constructing proofs of your own, simply because you have seen something like it before.

Proofs and crossword puzzles

Constructing a proof is sometimes likened to completing a crossword puzzle. Certainly, in each of these endeavours, a certain amount of 'try it and see' may well be involved. After looking promising initially, some of these attempts may prove fruitless and have to be discarded. The only opening strategy I adopt in solving crossword puzzles is to look for anagrams. Likewise, I know of only one universal opening strategy for problem-solving: *read carefully what is written and make sure you understand what you are told and what you are aiming for*.

Of course, once you have started a crossword puzzle, there is a blindingly obvious strategy you can employ: can I use what I've just discovered to help me with a neighbouring clue? A similar strategy holds in constructing a mathematical argument: can I use what I've just discovered to help me progress? The individual steps in a proof, may, like the answers in the crossword puzzle, arise via use of guesswork, experimentation, ingenuity, intuition, experience ('I've seen that idea – or something very similar – before').

Few people insist on solving a crossword puzzle by taking the clues in numerical order. Likewise it would be foolish to insist upon building a proof straight through 'from beginning to end'. If we picture a proof as being represented by a structure comprising joined up paths, the paths representing deductions (see Figure 1.1 below), it is clear that some of the joins in the middle of a long proof may be completed before those which are nearer to the beginning or to the end. One major difference between crossword puzzle solving and proof construction is that, in proof construction, the last 'clue' may, in some cases of deep mathematical research, take years to solve. Fortunately, readers of this book can be fairly sure that the problems you are given do have answers which will be given to you in due course. Furthermore, the number of steps in each proof is substantially less than the number of clues in all but the most modest crossword puzzle. Nevertheless, the message in brief is: *don't* always expect the various steps in a proof to occur to you in the 'right order' (that is, the order in which you finally present them). In particular, *don't* always expect to know how to begin.[11] Be prepared to fill quite a lot of paper with preliminary thoughts, many of which may end up in the wastepaper bin[12] – or be filed away for possible use on a future occasion.

A pictorial representation

The way in which the chain of assertions used in a proof might finally be fitted together can be indicated pictorially (see Figure 1.1). The diagram actually represents a proof we shall give later to Exercise 27 in Chapter 8. The meeting of the lines leaving G, D and J in a heavy black dot before they reach K indicates that the proof of assertion K makes use of *all three* assertions G, D and J. Similar remarks apply to other lines meeting at a dot. The lines from MI to BT likewise indicates that the proof of BT (the *binomial theorem*) can be made to depend on MI (the *principle of mathematical induction*). The line is dotted because, in the proof that you will see later, BT will not be rederived from MI; rather it will be accepted as a known consequence of MI (as we ask you to show in Problem 32 of Chapter 14). Similar remarks apply to other dotted paths.

Notice that Figure 1.1 is 'redundancy free'. For example, to obain assertion K we really *do* need to call upon *all three* of G, D and J. All paths of enquiry which were eventually seen to make no contribution to the proof have been eliminated. By contrast, Exercise 3 below gives you a diagram *with* redundant paths. You are invited to determine which these are and to cross them out – but only in *your own copy* of this book, not the library's.

---

[11] 'I think I could [shut up like a telescope] if only I knew how to begin,' thought Alice. (From *Alice's Adventures in Wonderland* by Lewis Carroll.)

[12] One famous long-standing conjecture has just been proved after 350 years; another is still unproved after more than 250 years (see the answer to Exercise 1(c) above).

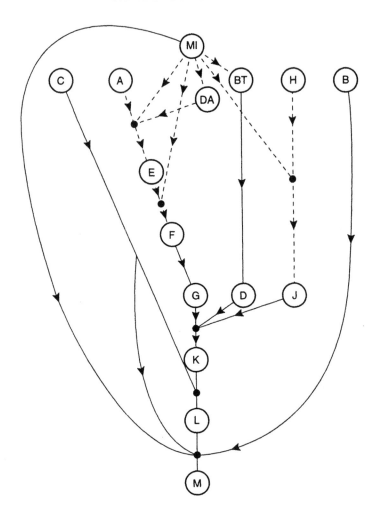

Figure 1.1

All proofs must start somewhere

Figure 1.1 is actually very useful for making two points quite forcefully. First we
ask: do the assertions denoted by A, B, C, H and MI have some kind of universal
quality? Don't they depend on anything at all? Can they not be deduced from
anything more 'basis'? In fact, in Figure 1.1, B represents the assertion that $1^P = 1$,
so one could argue that B is a consequence of the equality $1 \cdot 1 = 1$ (and MI). Then
you may ask: what is $1 \cdot 1 = 1$ a consequence of? What, indeed, is meant by '1'? (As
we are now getting a bit too philosophical we take this no further.)

It is clear that we cannot keep regressing in this way for ever. Accordingly, in
the proofs that we look at later, we shall be happy to accept certain easily believed,
simple assertions as being unquestionably true (for example, that $1^{20} = 1$) and not

requiring any explanation. We shall also accept basic rules of arithmetic such as: *for each pair a, b of integers, $a + b = b + a$* without further ado. As a consequence, whilst all mathematical theorems essentially make claims of the form: *IF ... something* (called the *hypotheses*) are assumed *THEN ... something* (called the *conclusion*) follows, usually only the major hypotheses are prominently displayed. Accordingly, not all mathematical assertions include the words 'if' and 'then'. For example, $2^{1\,257\,787} - 1$ *is a prime number* is not of IF/THEN form, and yet it cannot emerge from nothing. Its truth, or falsity, must depend on the basic rules of arithmetic (which are rarely stated explicitly), the definition of prime number – which in turn depends on the concept of division of one integer by another which depends ...

The second point which Figure 1.1 illustrates is obvious. The validity of a chain of arguments depends totally on that of each of its links. It is, therefore, important to pay equal attention to *all* steps in a proof. Many a proof has foundered because an 'obviously true' linking step has turned out to be false. A good moral is: *when constructing a proof of your own don't allow yourself to use any assertion whose truth you are not (at least) fully convinced of. Be sceptical – even of your own work.*

What to write down

What constitutes an easily believed, simple assertion will vary from reader to reader, and the theorem-prover has to bear in mind the likely knowledge of his or her audience. For example, I would hope that no reader of this book would want to take issue with my assertion that $1^{20} = 1$. However, some assertions are a bit dubious – possibly requiring explanation. Example: *If a and b are integers and each is divisible exactly by the integer c, then $a + b$ is divisible exactly by the integer c.* Yet others are, to say the least, impossible to accept without a fairly detailed and convincing explanation. Example: *For each integer n greater than 1 there is a prime number lying strictly between n and 2n.*

EXERCISE 3   Cross out as many redundant paths as you can find in Figure 1.2. The diagram indicates that both K *and* L are needed to deduce M and that both C *and* M are needed to deduce F.

EXERCISE 4   We wish to prove that assertion B is a consequence of assertion A, a consequence we might denote by $A \rightarrow B$. We can prove the following intermediate short steps: $A \rightarrow C; E \rightarrow J; H \rightarrow L; G \rightarrow N; A \rightarrow E; (D\&E) \rightarrow H$ (that is *both* D and E are needed to deduce H); $C \rightarrow G; (H\&F) \rightarrow K; (L\&M) \rightarrow P; (N\&P) \rightarrow B; A \rightarrow D; (H\&I\&J) \rightarrow M;$ $C \rightarrow F$. Show that we do *not* yet have a proof of $A \rightarrow B$. Which, if any, of (a) $G \rightarrow K$; (b) $L \rightarrow I$; (c) $P \rightarrow I$ will complete the proof?

Some doubtful 'proofs'

Before we get down to business, see if you can put your finger on what is wrong with each of the following arguments. Clearly *something* is wrong with (1), (2), (3) and (6) and it is incumbent upon us to find out what it is so that *we don't unwittingly*

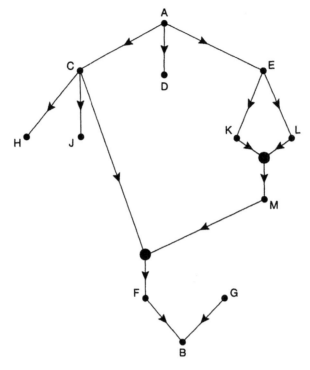

Figure 1.2

make similar mistakes where they may be more difficult to detect.[13] This point is emphasized by arguments (4) and (5), which are also unacceptable – a fact which might escape the reader's notice unless he knows that the claims made are false.

1. CLAIM $0 = \frac{1}{2}$.
    'Proof'

$$0 = \int \sin x \cdot \cos x \, dx - \int \cos x \cdot \sin x \, dx$$

In the first integral use the substitition $u = \sin x$, and in the second use $v = \cos x$. Since

$$\int \sin x \cdot \cos x \, dx = \int u \, du = \tfrac{1}{2} u^2 + c$$

and

$$\int \cos x \cdot \sin x \, dx = - \tfrac{1}{2} v^2 + c$$

we obtain

$$0 = \int \sin x \cdot \cos x \, dx - \int \cos x \cdot \sin x \, dx = \tfrac{1}{2} u^2 + \tfrac{1}{2} v^2 = \tfrac{1}{2}(\sin^2 x + \cos^2 x) = \tfrac{1}{2}$$

2. CLAIM If $(3x + 1)/x > (6x + 1)/(2x + 1)$ then $x > -\frac{1}{4}$.
    'Proof' Multiplying up by x and $2x + 1$ we get $6x^2 + 5x + 1 > 6x^2 + x$. Thus

---

[13] Bridges and dams have collapsed because of insufficient attention to mathematical detail during their construction.

$4x > -1$, that is, $x > -\frac{1}{4}$. (And yet, substituting $x = -\frac{3}{8}$ we get the true inequality $\frac{-1/8}{-3/8} > \frac{-10/8}{2/8}$.)

3. CLAIM $-1 = 1$.
   'Proof'   Assume, for the moment, that $-1 = 1$. Then $(-1)^2 = 1^2$. It follows that $1 = 1$, *which is true*. Hence the original assumption must be true. That is, -1 *is* equal to 1.

4. CLAIM   When two identical coins are tossed, the chances of those coins falling as one head and one tail are 1 in 3.
   'Proof'   The possible outcomes on tossing the coins are: (i) both land as heads; (ii) both land as tails; (iii) there is one head and one tail. There is one chance in three that case (iii) happens.

5. CLAIM   For each real number a we have $\log_e(a) = \log_e(-a)$.
   'Proof' $a^2 = (-a)^2$. Hence $\log_e\{a^2\} = \log_e\{(-a)^2\}$. Consequently $2\log_e(a) = 2\log_e(-a)$. Hence $\log_e(a) = \log_e(-a)$.
   (Johann Bernoulli[14] put forward this argument to counter Leibniz's belief that the logarithms of negative numbers must be complex numbers.)

6. CLAIM $1 - 1 + 1 - 1 + 1 - ... = \frac{1}{2} = 0 = 1$.
   'Proof'   (a) Let $S = 1 - 1 + 1 - 1 + 1 - ...$ Subtracting 1 from each side we obtain, $S - 1 = -1 + 1 - 1 + 1 - ... = -S$. Hence $S = \frac{1}{2}$.
   (b) $S = (1 - 1) + (1 - 1) + (1 - ... = 0 + 0 + 0 + ... = 0$.
   (c) $S = 1 - (1 - 1) - (1 - 1) - ... = 1 - 0 - 0 - ... = 1$.
   So, which is right? (This time Leibniz, Nikolas Bernoulli, Euler[15] and others disputed the answer.)

Style and notation

There is one other item which is, to the best of my knowledge, rarely dealt with in mathematics textbooks – a lack which contributes to the problem that first-year undergraduates have with constructing readable proofs. That item is good mathematical style. The British mathematician G.H. Hardy[16] said that 'There is no place for ugly mathematics.' I think he was referring to the use of ugly, hamfisted and inappropriate mathematical *arguments* ('bringing out the sledgehammer to crack the nut') but it can equally well apply to the *sentences* used to promote the argument. The word 'sentences' is crucial. Too many think of mathematics as a subject in which words are banned, with *all* communications having to be in symbols. Whilst the present author can scarcely claim perfection (!),[17] his experience of many years reading and

---

[14] Johann Bernoulli (6 August 1667–1 January 1748).
[15] Nikolas Bernoulli (21 October 1687–29 November 1759). For Euler, see the biography on p. 89.
[16] Godfrey Harold Hardy (7 February 1877–1 December 1947).
[17] Some, indeed, would say that I am too verbose! But, in a book of this kind, excessive, rather than insufficient, explanation would seem desirable.

writing mathematics might be of some use, at least to the beginner. Accordingly, some comments regarding style will be made from time to time throughout the book. On the whole, however, it is simplest to ask the reader to do two things:

1. To take note of the style in which the 'official' proofs of later chapters are written.
2. To find a colleague with whom you can swap proofs. Each of you should be hypercritical of the other's efforts. In particular, *you should each refuse to be convinced by your colleague's attempts unless you can fully understand (and believe as true) all that he has written.*

The following, just about comprehensible, solution to Exercise 3 in Chapter 6 is, in anyone's language, an example of very poor style because the writer makes no attempt to tell the reader what he is assuming, what he is doing or what he is trying to achieve.

$$\left(M + \frac{1}{n}\right)^2 = m^2 + \frac{2m}{n} + \frac{1}{n^2}$$

$$\frac{1}{n^2} \leqslant \frac{1}{n} = m^2 + \frac{2m+1}{n} < 2$$

$$\frac{2m+1}{n} < 2 - m^2 (\leftarrow + ve)$$

Given M rational $M^2 < 2$

I wrote on the script 'Who says so? Why is this true? If you gave some explanation I might be able to understand what you are trying to do.' Remember, you cannot expect the reader of your work also to read your mind. *Write your solutions/proofs so that in six months' time YOU will be able to understand them.*

The use of good *notation* can also be helpful to both prover and reader. Except in dire circumstances, call your triangle ABC or XYZ (*not* PNu), name the angles at A, B, C by the letters α, β, γ (*not* by *f*, *z*, *Q*); use *p*, *q*, maybe *r* (*not* *M*, *h*) for primes, and so on. That is, try to use letters which, in part, remind you of their role and the relationships between what they represent.

Why re-prove old results?

A tiny handful of mathematics students express the opinion that proofs should be left to those who are interested in examining them and should not be forced on those who 'only wish to use the results'. This attitude almost beggars belief. Others might (more sensibly) ask, Why do we have to prove (say) Pythagoras' theorem again when people have satisfied themselves for over two thousand years that it is true?[18]

Studying and re-proving old theorems helps you remember the results much more easily because you can recall the crucial ideas which show *why* the results are true.

---

[18] The *Guinness Book of Records* reports the existence of a book containing 370 different proofs of Pythagoras' theorem, including one by James Garfield (1831–1881) who became famous in another capacity.

Sometimes we try to re-prove established results simply because our curiosity demands to know why something so remarkable or unexpected is (apparently) true. Such activity can be illuminating. Occasionally, quicker or more elegant proofs and new methods emerge. These may, in turn, generate ideas for dealing with problems which are similar to, but not merely numerically different from, ones you already know how to solve.

Role of the calculator

I have already made a strong point about not placing unquestioning reliance on the output of a calculator. The calculator is a valuable tool *provided you know to what extent you can rely upon its output*. And, of course, certainty requires PROOF.

Final exercises

1. (a) Show that $5 \sin x - 4 \cos x < 9$.
   (b) Show that $\frac{4}{3} + \sin x - \cos x - \sin 2x > 0$ for all values of x. (*Hint:* Look at y^2 where $y = \sin x - \cos x$. This was suggested by the appearance of the $\sin x - \cos x$ and $\sin 2x$ ($= 2 \sin x \cdot \cos x$) terms – and by Example 1 above.[19])
2. Find a counterexample to each of the following false assertions.
   (a) If p is a prime number then $2^p - 1$ is prime. (*Hint:* $2^2 - 1$, $2^3 - 1$, $2^5 - 1$, $2^7 - 1$ are all primes but....)
   (b) Of the two integers on either side of a multiple of 6 (that is, the pairs 5,7; 11,13; 17,19; 23,25; etc.) at least one is a prime. (This was believed by Charles de Bouvelles, 1470–1553.)
   (c) It is impossible to find distinct positive integers a, b, c, d such that $a^2 + d^2 = b^2 + c^2$.
3. Gather evidence which will allow you to conjecture which of the following claims are true and which are false. (You are not asked to try to prove your assertions.)
   (a) Recall (Exercise 2(a)): for each positive integer n, $n!$ is the integer $1 \cdot 2 \cdot 3 \cdot ... \cdot n$.
       CLAIM $n!$ is only a perfect square when $n = 1$.
   (b) CLAIM   There are no integers x and y such that $x^2 - 60y^2 = 1$.
   (c) CLAIM   Dividing up the interior of a circle as shown in Figure 1.3, by

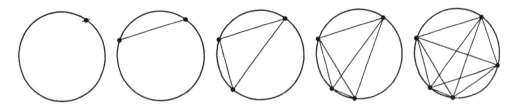

Figure 1.3

---

[19] Do I detect a little circularity in these cross-references?

taking $1, 2, 3, 4, 5, 6, 7, \ldots$ etc. points on the perimeter, splits the interior of the circle into (respectively) $1, 2, 4, 8, 16, 32, 64, \ldots$ etc. regions.

(d) Pairs of primes such as 5,7; 41,43; 107,109; etc. which differ by 2 are called *twin primes*.

CLAIM   The list of twin primes is endless, i.e. *there are infinitely many pairs of twin primes.*

(e) CLAIM $p = 3$ is the only prime for which $p^2 + 2$ is also prime. (An unsolved problem of *number theory* asks if there are infinitely many primes of the form $n^2 + 2$, n being an integer. For example, $1^2 + 2$, $3^2 + 2$, $9^2 + 2$, $15^2 + 2$ are all primes.)

Two that you may wish to try on your computer:

(f) CLAIM   Each *odd* integer greater than 1 is either (i) a prime number or (ii) the sum of a prime number and a power of 2. (As examples: $219 = 211 + 2^3$; $221 = 157 + 2^6$; 223 is prime.)

(g) CLAIM   There are no integers x and y such that $x^2 - 61y^2 = 1$.

4. In Figure 1.4, let the value of x slowly increase. What conjecture can you make?

5. What theorem is suggested by Figure 1.5(i), (ii) if you write the area of the biggest square in two different ways as the sum of the square(s) and triangles contained in it.

6. Given the regular pentagon and its diagonals as in Figure 1.6, do you believe that the two angles marked are equal? If so, why? Can you say more?

7. Try to discover/invent three plausible mathematical assertions which are new to you and whose truth (or falsity), at least initially, is unknown to you.

Summary

Advances in mathematics are frequently made by means of **proof**, often after suggestive (inductive?) evidence has been collected by experimentation. The purpose of proof (using deductive methods) is to allay any *scepticism* that the claim being

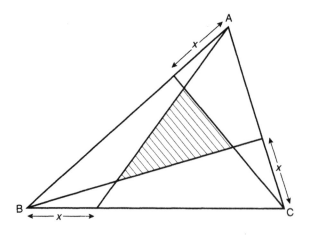

Figure 1.4

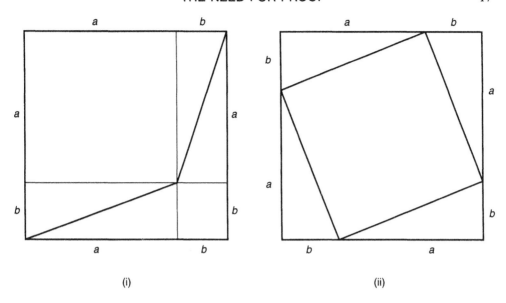

(i)   (ii)

Figure 1.5

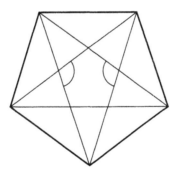

Figure 1.6

made is true. A proof may therefore turn a **conjecture** into a **theorem**. On the other hand, if the scepticism is warranted and a conjecture is wrong, the conjecture's falsity is most easily established by the production of a (single) **counterexample**.

The best way to learn any business is to watch how the experts do it and then to try for yourself. Learning how to prove mathematical theorems by reading the masters can be hard because of the well-intentioned style in which they tend to be written out; few proofs give the reader any clue as to how their authors discovered them. (The chief purpose of the forthcoming Discussions is to offer the reader such clues.) Furthermore, the various sections of a long proof may, because of the need for clarity of presentation, be written down in an order totally different from that in which they were first established. (Pictorial representations, such as Figures 1.1

and 1.2, are rarely used in this context: I included them merely because I thought they might, initially, be helpful.) The level of explanation in a proof should take into account the audience for whom the presentation is intended. In any case, all proofs have to start *somewhere*; one cannot regress for ever.

If you are required to construct a proof or solve a problem, make sure you understand what you may assume as given and what you are required to do. In giving a proof or solution do not use any step which you are not convinced is correct. Try to develop a readable style (test it on your peers), in particular by use of a suitable and advantageous notation. *Be sceptical.*

CHAPTER 2

Statements and Connectives

If ifs and ans were pots and pans, there'd be no trade for tinkers. ANON

The statements below, whose truth or falsity we investigate in Chapters 3, 5, 6 and 7, are used here to show the reliance that mathematical statements place on such words as 'not', 'and', 'or', 'if/then', 'each', 'there exists'. In this chapter we concentrate on the first four of these, noting how in everyday English their use can be very imprecise. We then introduce the idea of truth table in order to give these four *connectives* a precise mathematical meaning. Truth tables can be used to confirm that some intuitively correct methods of argument are logically sound.

Introduction

Probably the best way for you to learn how to construct proofs is for us to do some together, and I'll explain *what I am doing and why* as we go along. Later, having taken note of some of the ideas and methods which recur, you then begin to put a few bits and pieces together, much as an old-time apprentice would.

Although Chapter 1 contains a number of assertions whose truth we could investigate, we list below another eighteen mathematical assertions, or *statements* as we shall now call them, each of which has been specially selected to emphasize a particular form of wording (notice the words in bold capitals) or to demonstrate a particular method of proof, a point of style or some other interesting feature. Some of the statements are not of great mathematical interest, either because the results are not very deep or because they are merely special cases of much more general (and interesting) results. Furthermore, the wording of many is much more formal than one might normally come across, but in a chapter where we begin to aim for some precision, it is best to persevere with such formality. When, later, we prove or give counterexamples to the eighteen statements, we shall often restate them in the more usual and more informal manner.

Most of the statements concern (properties of) numbers of one kind or another. This restriction is deliberate – it is so that we need devote little effort to *understanding* what the statements actually say. Their *meaning* should be fairly transparent. Of course, whether or not you *believe* these statements to be true is another matter.[1]

We first concentrate on the wording. Determining each statement's truth or falsity will begin in Chapter 3. However, as it is possible to be totally unimpressed by even the most wonderful assertions in mathematics *unless one has investigated them for oneself*, I suggest that, when reading the following statements, you give five minutes' thought to each one as to how you might convince a sceptical friend of its truth (or falsity).

The test statements

About specific numbers

STATEMENT 1   The integer $300\,000\,067\,110\,605\,737$ is **NOT** a perfect square.

STATEMENT 2   The ratio (π) of the circumference to the diameter of any given circle is a number lying between 3 and 4. In other words, $\pi > 3$ **AND** $\pi < 4$, or more briefly, $3 < \pi < 4$.

STATEMENT 3(a)   (Either) 1997 is a prime **OR** 1997 is divisible by some prime not exceeding 43.

STATEMENT 3(b)   (Either) there are at least 150 primes which are less than 1000 **OR** the sum of those primes which are less than 1000 is less than 125 000.

STATEMENT 4   **IF** $2^{67} - 1$ is prime **THEN** $2^{67} + 1$ is divisible by 3.

STATEMENT 5

(a) The integer $13\,348\,765\,539$ is divisible by 9 **IF AND ONLY IF** $1 + 3 + 3 + 4 + 8 + 7 + 6 + 5 + 5 + 3 + 9$ is divisible by 9.
(b) $1 + \sqrt{2}$ is a root of the polynomial $37x^7 - (\frac{3}{8})^2 x^5 + 11.623x^4 + 4x - 747.1$ **IF AND ONLY IF** $1 - \sqrt{2}$ is a root of $37x^7 - (\frac{3}{8})^2 x^5 + 11.623x^4 + 4x - 747.1$.

---

[1] The word 'statement' is reserved for sentences which are either true or false (even though you may not know which). Thus 'Play it again, Sam!', 'Whose go is it?', '*x* is a prime' are not statements. 'A succession of fifty consecutive zeros occurs in the decimal expansion of π' is a statement (though its truth/falsity status is unknown (at least to me).

(c) The system of simultaneous equations

$$3\pi x + 4ey = 1$$
$$ex + \pi y = 2$$

where e is the base of natural logarithms, has a unique solution for x and y **IF AND ONLY IF** $3\pi^2 - 4e^2 \neq 0$.

STATEMENT 6 Suppose 6 points in space are connected in (15) pairs by pieces of string, each piece being either red or blue. Then (either) there are 3 points joined only by red string **OR** there are 3 points joined only by blue string.

About collections of numbers

STATEMENT 7 It is **NOT** the case that for **EACH** positive integer n the integer $n^2 - n + 41$ is prime.

STATEMENT 8 For **EACH** positive integer x, $x^3 - x$ is a multiple of 3 **AND** $x^5 - x$ is a multiple of 5.

STATEMENT 9(a) For **EACH** pair x, y of real numbers, if $x^2 = y^2$ then $x = y$ **OR** $x = -y$.

STATEMENT 9(b) For **EACH** real number x, if $x^2 > 9$ then $x > 3$ **OR** $x < -3$.

STATEMENT 10 For **EACH** positive integer n, **IF** $2^n - 1$ is prime **THEN** n is prime.

STATEMENT 11 For **EACH** positive integer x, x^2 is odd **IF AND ONLY IF** x is odd.

About the existence of numbers

STATEMENT 12 **THERE EXISTS NO** rational number a/b such that $(a/b)^2 = 2$.

STATEMENT 13 Let $f(x) = 2x^3 - 9x^2 + 6x + 3$. Then **THERE EXISTS (AT LEAST ONE)** number a such that $1 < a$ and for which $f(a) = 0$.

STATEMENT 14 **THERE EXISTS INFINITELY MANY** primes.

STATEMENT 15 **THERE EXISTS A UNIQUE** positive integer t such that t, $t + 2$ and $t + 4$ are all primes.

Concerning positive integers

STATEMENT 16 **FOR EACH** positive integer n, $1 + 2 + \ldots + n = n(n + 1)/2$.

STATEMENT 17   **FOR EACH** positive integer n

$$\frac{1}{1^2} + \frac{1}{2^2} + \frac{1}{3^2} + \dots + \frac{1}{(n-1)^2} + \frac{1}{n^2} < 2$$

STATEMENT 18   **FOR EACH** positive integer n **THERE EXISTS** a positive integer t such that

$$1 + \frac{1}{2} + \frac{1}{3} + \frac{1}{4} + \dots + \frac{1}{t} > n$$

The connectives NOT, AND etc. – their everyday use

As we have already said, the use of the **connectives** 'not', 'and', etc. in everyday English is not particularly precise. We shall (I hope) have a bit of fun examining this imprecision by means of examples. Then, rather more seriously, I'll tell you what we propose to do about it so that our mathematical work will not suffer likewise.

NOT

I would contend that 'It is NOT that I am unhappy' is not *usually* taken to mean (exactly) 'I am happy'. Further, when introducing the word 'not', you must be careful about its positioning. For example, in ordinary speech the negation of 'Florida is a city in the USA' (Yes, I really did see this statement in one of my daughter's pop-music magazines!) is usually expressed *not* as 'It is NOT the case that Florida is a city in the USA' but rather as 'Florida is NOT a city in the USA'. Sometimes, however, you *have* to use the former style. For example, it is correct to say that the negation of 'Each prime number is odd' is 'it is NOT the case that each prime number is odd' and incorrect to say 'each prime number is NOT odd', since this latter says that every prime number is even. (The trouble here is caused by the presence of the word 'each' – as we shall see in Chapter 4.)

AND

If, with a drink of tea, you take milk but not sugar, and if you are asked 'do you want milk AND sugar?', then the correct answer is surely 'no, thank you' – meaning that you do not want *both*. However, such a reply will almost certainly result in both the milk and the sugar being removed from your presence forthwith. However the above is to be interpreted, it surely means the same as 'do you want sugar AND milk?' Clearly, 'X AND Y' is always the same as 'Y AND X'. But no, *it isn't*. Consider, for example, the statement 'the bus stopped AND I got off'. This is scarcely the same as 'I got off AND the bus stopped'. Another ambiguous example is one mischievously placed on our departmental noticeboard: 'Pat and Reg celebrated 25 years of marriage on August 27th'. So they both did – but they had never been married to each other.

Even in a mathematical setting we have to take care. To say that '13 and 17 are odd' is surely shorthand for '13 is odd and 17 is odd'. However, the expression '13 and 17 are unequal' cannot similarly be expanded and still make sense.

OR

What about the following questions: (i) do you prefer to travel by bus OR train? (ii) do you want tea OR coffee? (iii) are you playing Tina OR Liz at squash? and the statement (iv) you will be hurt if that heavy weight falls on your fingers OR toes.

In (i) the word 'prefer' seems to *insist* that you choose one (but not both) of the possible answers 'bus' and 'train' (even if, given free choice, you would travel by unicycle).

In (ii) it is probably expected that you will ask for *exactly one* type of beverage. Declining both is seen as a possibility but you are certainly not expected to say 'a cup of each, please'.

In (iii) it seems to me that the question allows for the (informal) replies 'Tina' or 'Liz' or 'both' or, even, 'neither'.

Finally (iv) presumably implies that you will also be hurt if the heavy weight traps your fingers *and* your toes.

Here (i) intends the **exclusive** use of the word 'or' (i.e. choose one or the other but not both) whereas (iv) shows we should, perhaps, also admit its **inclusive** use, in the sense of 'one or the other *or both*'. Questions (ii) and (iii) indicate that, sometimes, one can respond to an 'or' question with the answer 'neither'.

IF/THEN

One often hears radio announcements like 'IF you are using the B3143 road [THEN] there are roadworks at Piddletrenthide.' The first part is certainly redundant since the roadworks will still be there whether the listener is using the B3143 or is just lying in bed. So, is the announcement true or false?

Truth tables for NOT, AND, OR, IF/THEN

Since mathematical argument is the very epitome of precision[2] we must make precise the exact meaning we wish to convey when using these words in a mathematical context. One way to do this is via the use of **truth tables**. As we shall see, truth tables are extremely useful in allowing us to be certain that particular forms of argument are (or are not) valid. We begin with the word NOT. Suppose we use the letter A to stand for the statement '$\pi^2 < 9.87$'. We may write this briefly as

$$A: \quad \pi^2 < 9.87$$

Then by NOT A we mean the statement '(it is) NOT (the case that) π^2 is less than

---

[2] The pure mathematician's desire for precision is well expressed by the following well-known story. An artist, seeing one black sheep in a field in Scotland, says joyously, 'Ah, I see there are black sheep in Scotland.' 'Oh, come on,' says his mathematically trained physicist friend. 'All you can say is that, in Scotland there is at least one black sheep.' 'I see why you are not a pure mathematician,' said the third member of the party, a logician. 'All you can say is that there exists, in Scotland, at least one sheep at least half of which is black.'

'9.87', that is, in plainer English, π^2 is NOT less than 9.87, and, in symbols, $\pi^2 \not< 9.87$ (in other words, $\pi^2 \geqslant 9.87$). Now even though, at this moment, you may not know if A is true, it is evident that if A is a true statement then NOT A is false – and vice versa, if A is false than NOT A is surely true.

We could summarize these remarks by means of the following table, where the letters T and F stand for 'true' and 'false' respectively.[3]

A	NOT A
T	F
F	T

However, it is clear that an identical table would be equally appropriate if some other statement were chosen instead of A. Consequently we may summarize our discussion of the word NOT by the following table in which the *statement variable letter P* replaces the *particular statement letter A*, that is:

P	NOT P
T	F
F	T

NOT P is called a *statement form* or *formula* since, like the 'formula' $f(x) = x^2 + x^5\tan(3x - 1) - 4e^x\sin 2x$, which produces a real number whenever a real number is substituted for the variable x, so does NOT P produce a statement whenever a statement is substituted for the statement variable P.

The table above is called the truth table corresponding to **negation**. For brevity we usually write NOT A as $\neg$ A and NOT P as $\neg$ P (though, naturally enough, we shall continue to write the specific negation NOT$\{x = y\}$ as $x \neq y$).

AND

Now let us consider the word AND in the following four statements:

1. $100^{99} < 99^{100}$ AND $\pi^2 + \pi - 13 \neq 0$
2. $100^{99} < 99^{100}$ AND $\pi^2 + \pi - 13 = 0$
3. $100^{99} \geqslant 99^{100}$ AND $\pi^2 + \pi - 13 \neq 0$
4. $100^{99} \geqslant 99^{100}$ AND $\pi^2 + \pi - 13 = 0$

You would surely agree that only statement 1 is true since it is the only one in which the statements (one on each side of the word 'and') are *both* true. (I shall be extremely disappointed if you are prepared to take my word for it. But how can you confirm that $100^{99} < 99^{100}$? Would you opt for your calculator – or the binomial theorem? See Problem 6(a) in Chapter 14.)

These four statements act as models for the following self-explanatory truth table for AND, where we now need to use two statement variables, calling them P and Q. Notice that there are four lines in this table – one for each possible combination of T/F values of P and Q.

---

[3] Some mathematicians use 0 and 1 instead of F and T (since F and T can look alike when written casually). However, since other authors use 1 and 0 instead of F and T, I prefer using the letters since they remind me more readily of the intended interpretation.

P	Q	P AND Q
T	T	T
T	F	F
F	T	F
F	F	F

In our motivating example, we have felt able to ascribe a 'truth value' to each statement (1)–(4) even though there seems little connection between the *content* of the statements on each side of the word 'and'. In everyday speech one might usually expect to find some such connection; but even here there may be ambiguity. For example, 'My uncle went to watch Blackpool FC play football and (my uncle) cried all night.' Was the crying related to the match or not? Who can tell? To avoid such confusion in mathematics, we make the truth value of the **conjunction** of two statements depend only on the truth or falsity of the two statements – and not on any possible connection in meaning *between* the statements, nor, indeed, on the *meanings* of the statements themselves.

For brevity we often write the statement A AND B as A ∧ B and the corresponding statement form P AND Q as P ∧ Q.

EXERCISE 1   Complete the following truth tables for (a) $P \wedge \neg P$; (b) $\neg(P \wedge \neg P)$; (c) $\neg P \wedge \neg Q$; (d) $\neg(P \wedge Q)$

P	¬P	P∧¬P	¬(P∧¬P)
T	F		
F	T		

P	Q	¬P∧¬Q	¬(P∧Q)
T	T		
T	F		
F	T		
F	F		

A statement form which is always true for all T/F values of its component parts is called **valid** or a **tautology**. One which is always false is called **invalid** or a **contradiction**.

OR

Consider the assertion 'if my daughters' ages e and r satisfy the equality $(e - 18)(r - 11) = 0$ then either $e = 18$ OR $r = 11$.' We surely have to admit the possibility that $e = 18$ and $r = 11$ are *both* correct. Because of this need to be able to deduce 'one or the other or possibly both', in other words 'at least one of the two' (as distinct from 'exactly one of the two'), we choose the *mathematical* meaning of OR to be the *inclusive* one so the truth table is as follows:

P	Q	P OR Q
T	T	T
T	F	T
F	T	T
F	F	F

Note that our choice of meaning causes no problem with the assertion 'if $(e - 18)$

$(e - 21) = 0$ then either $e = 18$ or $e = 21$. Here the fact that we clearly cannot have *both* $e = 18$ *and* $e = 21$ *simultaneously* is no reason to deny the truth of '$e = 18$ or $e = 21$ or, *maybe* (*but not on this occasion*) *both simultaneously*.'

In passing, we note a point which sometimes causes difficulty. Is the statement C: $1 \leqslant 2$ true or false? I have heard students mutter that 'we cannot (correctly) say "1 is less than or equal to 2" since 1 is not equal to 2.' However, if statement A is '1 is less than 2', and B is '1 is equal to 2', we see that C '1 is less than or equal to 2', is just the statement A OR B. Since A is true we see that, according to the truth table, A OR B (that is, '$1 \leqslant 2$') is also true (even though B itself is false).

For brevity we often write the **disjunction** A OR B of statements A and B as A ∨ B and the corresponding statement form P OR Q as P ∨ Q.

EXERCISE 2   The ages w, b and n of the brothers Winkin', Blinkin' and Nod are given by the equation $(w - 34)(b - 35)(n - 36) = 0$. Who is the oldest?

EXERCISE 3   Write down the truth table corresponding to the *exclusive* use of 'or', that is, for the form P OR Q but *not both*.[4]

EXERCISE 4   Write down the truth table for $\neg P \vee P$.

EXERCISE 5

(a) Write down the truth table for $\neg(P \vee Q)$ and compare it with that for $\neg P \wedge \neg Q$.
(b) Do the same for $\neg P \vee \neg Q$ and $\neg(P \wedge Q)$.

EXERCISE 6   Is the statement D: $2 \leqslant 2$ true or false?

IF/THEN

Finally let us consider the implication IF/THEN. This is the most important of the connectives because, as already noted in Chapter 1, proofs of theorems tend to proceed in small steps, each of which involves a deduction obtained from one or more previous assumptions or assertions.

We remarked in Chapter 1 that not all statements can be written, naturally, in IF/THEN form, but some can.

EXERCISE 7   Write Statements 3(a) and 3(b) on page 20 in IF/THEN form.

In normal English usage the combination 'If A then B' is often stated as 'A implies B' and is thought of as meaning 'given A (is true) then (the truth of) B necessarily follows'; there is no interest in the implication if A is false. (The father who says to his son, 'If you always drive like that (then) you will surely have an accident', is contemplating only one assumption, namely, that of his son's reckless driving. He is certainly *not* thinking of adding, 'And if you don't always drive like that, you may or may not have an accident.') Nevertheless, consider the following example.

---

[4] Notice how, in ordinary usage, 'but' sometimes plays the role of 'and'.

Example 1

Suppose a mother[5] says to her daughter, 'If you get the grades to study maths at Leeds University I shall buy you a copy of *Numbers and Proofs*'. Under what circumstances will the daughter feel that her mother has not told the truth?

SOLUTION   The four possible outcomes are:

1. The daughter gets the grades and her mother buys the book.
2. The daughter gets the grades but her mother *doesn't* buy the book.
3. The daughter doesn't get the grades but, nevertheless, her mother buys the book. (She mistook it for the latest Barbara Cartland?!!?)
4. The daughter doesn't get the grades and her mother doesn't buy the book.

I suggest that the daughter will only feel that her mother has lied in case 2. In cases (1) and (4) the mother's action probably corresponds to her daughter's expectation. In case (3) the daughter is probably surprised but surely cannot claim that her mother lied.

Using this example as a guide we arrive at the following truth table for the IF/THEN statement form:

P	Q	IF P THEN Q	
T	T	T	(outcome 1)
T	F	F	(outcome 2)
F	T	T	(outcome 3)
F	F	T	(outcome 4)

P is called the **antecedent** and Q the **consequent**. As an alternative we often write the **implication** IF P THEN Q as P→Q.

Even if the above example seems a little artificial, many mathematical contexts demand that we complete the third column of the above table for each of its four rows. Not infrequently in mathematics it is of interest to know that a certain conclusion, B, follows from a given hypothesis, A, *even though we may not, as yet, know the truth or the falsity of A*. In fact, if having proved A→B to be true we can, at some later time, show that B is false, then we shall be able to deduce that A, too, is false. (See the last row of the above table.) Of course, if B is true, we can, from the knowledge that A→B is true, deduce nothing about the truth or falsity of A.

To reinforce our belief that the entries in the third column of the above table are correct, let us consider three examples, the first two of which make statements which are obviously true.

Example 2

(a) If 19 972 001 is prime then 19 972 001 is not divisible by 1999.
(b) If 19 972 001 is prime then 19 972 001 is not divisible by 773.

---

[5] A model parent!

In fact $19\,972\,001 = 7 \cdot 773 \cdot 3691$ so that it appears that we have made the right choice in lines 3 and 4 in the IF/THEN table.

Example 3
Given any two particular statements A and B, consider the assertion '*If* (A and B) *then* B'. Surely we should want this to be true whatever statements A and B are chosen? Giving A and B, in turn, the pairs of truth values T,T; T,F; F,T; F,F and using the above tables for AND and IF/THEN, we obtain the following table in which all truth values for 'If (A and B) then B' are indeed T.

A	B	(A AND B)	B	IF (A AND B) THEN B
T	T	T	T	We want T. Do we get it from 'if...then'? Yes
T	F	F	F	We want T. Do we get it from 'if...then'? Yes
F	T	F	T	We want T. Do we get it from 'if...then'? Yes
F	F	F	F	We want T. Do we get it from 'if...then'? Yes

This example further seems to confirm that we have made the correct choices in the third column of the IF/THEN table especially in the 'troublesome' rows (rows 3 and 4).

EXERCISE 8 Are the following implications true or false?
(a) If $2^{11} - 1$ is prime then $2^{11} + 1$ is divisible by 5. (*Hint: Is $2^{11} - 1$ prime?*)
(b) If $2^{47} - 1$ is prime then $2^{46} + 1$ is divisible by 5.
(c) If $76\,240\,164$ is divisible by 18 then $76\,240\,164^2$ is divisible by 18. (*Hint: Please, don't do the divisions. Think!*)
(d) If $76\,240\,164^2$ is divisible by 18 then so is $76\,240\,164$. (*Hint: If you know a result like that of Statement 5(a) on page 20 but concerning division by 3, it might help you to avoid dividing $76\,240\,164^2$ by 18.*)
(e) If 6π is within 0.16 of the nearest integer then 3π is within 0.08 of the nearest integer.

EXERCISE 9 Confirm, by constructing the appropriate truth table, that the intuitively obvious 'If {(P implies Q) and NOT Q} then NOT P (in symbols $\{(P \rightarrow Q) \wedge \neg Q\} \rightarrow \neg P$) is a tautology.
This statement form is called ***modus tollens***.

EXERCISE 10 Confirm that the intuitively obvious 'If {P and (P implies Q)} then Q' (in symbols $\{P \wedge (P \rightarrow Q)\} \rightarrow Q$) is a tautology.

The statement form in Exercise 10 is called ***modus ponens***. It is *modus ponens* which makes the IF/THEN connective so important in theorem-proving since it allows us to deduce the truth of the statement B if we know the truth of both the statement A and the statement A→B.

EXERCISE 11 Confirm that $\{P \wedge (\neg P \vee Q)\} \rightarrow Q$ is a tautology.

EXERCISE 12 Give examples of specific statements which show that none of the following statement forms is a tautology:

(a) $\{P \vee (P \rightarrow Q)\} \rightarrow Q$
(b) $\{P \rightarrow (P \wedge Q)\} \rightarrow Q$
(c) $(P \vee Q) \rightarrow (P \wedge Q)$

The examples above seem to confirm that we have chosen the correct entries in the final column of the IF/THEN table. Once we accept this, there is the following pleasant consequence.

In order to show (in a mathematical context) that an implication IF A THEN B is true one only needs to check that the pairing A true and B false *cannot hold* (that is, one only needs to assume A true and deduce that B must be true) since all other combinations are true by definition.

EXERCISE 13

(a) Given that $\neg (P \rightarrow \neg Q) \wedge R$ has truth value T, what are the truth values of P, Q and R?
(b) Given that $P \vee Q$, $Q \rightarrow R$ and $(R \vee P) \rightarrow S$ are all true, what is the truth value of S? Is $[\{P \vee Q\} \wedge \{Q \rightarrow R\} \wedge \{(R \vee P) \rightarrow S\}] \rightarrow S$ a tautology?

EXERCISE 14   Find statement forms involving P, Q and using $\rightarrow$, $\wedge$, $\vee$, $\neg$ as required, which give rise to columns (a), (b), (c) of the following table.

P	Q	(a)	(b)	(c)
T	T	T	F	F
T	F	T	T	T
F	T	F	T	F
F	F	T	T	T

(*Hint:* The single F in the third column reminds me of the $P \rightarrow Q$ table – well, almost!)

Equivalent ways of saying 'if/then'

We have already mentioned that the implication 'If A then B' is often stated as 'A implies B' and is denoted briefly by $A \rightarrow B$. There are, however, other commonly used ways of describing the implication.

Since, given the truth of $A \rightarrow B$, the truth of A is (by *modus ponens*) *enough* for us to conclude the truth of B, we say that *A is **sufficient** for B*. Looking at this another way, since, given the truth of $A \rightarrow B$, the truth of B *necessarily* follows from that of A, we say that *B is **necessary** for A*. We also abbreviate 'if A then B' to '*B, if A*' and, somewhat curiously,[6] to '*A only if B*.'

We also ask you to note (making the appropriate grammatical amendments in order to preserve fluency) that an IF/THEN statement will often appear under any one of the pairings of

IF/SUPPOSE/ASSUME/LET/GIVEN ... THEN/DEDUCE/SHOW/IT
FOLLOWS THAT

[6] If you remember it as 'A (can be true) only if B (is true)' then it is *not* so curious.

Thus, for example, 'Suppose (that) A. Show B' is to be understood as meaning the same as 'Prove that, if A then B'.

It seems fairly clear (and it is easily checked by looking at truth tables) that the implication $A \rightarrow B$ and its **converse** $B \rightarrow A$ do not, in general, mean the same thing (see Exercise 15 below). Consequently it is important to be aware of this difference between 'if' and 'only if', especially as the two are easily confused in everyday speech.

EXERCISE 15

(a) Give specific statements A and B such that $A \rightarrow B$ but $B \nrightarrow A$. (I'm afraid I cannot resist repeating an example I was offered in an examination. For A take 'Tottenham Hotspur scores a goal'; for B take 'My boyfriend gets excited'.)

(b) Should the father who is anxious for his son to do well in the forthcoming maths examination say 'You can go to the match *if* you finish your homework by 6 o'clock' or should he have said *only if* instead of *if*?

EXERCISE 16   Let a be the integer $12\,345^{6789} + 6789^{12\,345}$. Of the following conditions, which are (i) sufficient, (ii) necessary, (iii) necessary *and* sufficient for a to be divisible by 12?

(a) a is divisible by 6; (b) a is a multiple of 24; (c) a^2 is divisible by 24; (d) a^2 is divisible by 12. (*Hint:* Please do not attempt to see whether or not a is divisible by 6, 12, 24, etc.)

If and only if: equivalence of statements

Not infrequently in mathematics the two implications $B \rightarrow A$ and $A \rightarrow B$ are true simultaneously. We then have '(A if B) AND (A only if B)' which we write economically as 'A if and only if B'. In symbols this becomes $(A \leftarrow B) \wedge (A \rightarrow B)$ which we write briefly as $A \leftrightarrow B$. Because each of A and B implies the other we then describe the statements A and B as **equivalent statements**. We may also say that A is *necessary and sufficient* for B. The expression *if and only if* is frequently abbreviated, by mathematicians, to **iff** (and so causes confusion to newly appointed secretaries typing mathematical papers who begin to doubt the spelling ability of their mathematical colleagues!!)

EXERCISE 17   Which of the following statements are correct? (No 'formal' proofs are required at this stage, but try to justify your answers to a colleague.)

(a) Let ABC denote a triangle with sides of lengths a, b and c (where $a \leqslant b < c$). Then ABC is a right-angled triangle iff $a^2 + b^2 = c^2$.

(b) The product ab of the integers a and b is positive iff both a and b are positive.

(c) Let x and y be digits in the range from 0 to 9 inclusive. Then the integer $8765x32y$ is divisible by 9 iff $x + y = 5$.

(d) The integer n is divisible by 8 iff the final three digits of n form an integer which is divisible by 8.

Equivalence of statement forms

Generalizing the above, we shall say that two statement forms R and S are **equivalent** precisely when the truth values in corresponding rows of their truth tables are identical.

Example 4
Prove:

(a) The statement forms $P \rightarrow Q$ and $Q \rightarrow P$ are *not* equivalent.
(b) The statement forms $P \rightarrow Q$ and $\neg P \vee Q$ *are* equivalent.

SOLUTION   The tables (where the numbers indicate the order in which the columns were worked out and written down) clearly confirm (a) and (b).

P	Q	$P \rightarrow Q$	$Q \rightarrow P$	$\neg P$	$\vee$	Q
T	T	T	T	F	T	T
T	F	F	T	F	F	F
F	T	T	F	T	T	T
F	F	T	T	T	T	F
(1)	(2)	(3)	(4)	(5)	(7)	(6)

EXERCISE 18   Use truth tables to check that the statement forms $P \rightarrow Q$ and $\neg(P \wedge \neg Q)$ are equivalent.

Example 4(b) and Exercise 18 show that, to prove an implication $A \rightarrow B$ true one may, if it is more convenient, prove *instead* either that $\neg A \vee B$ is true or that $A \wedge \neg B$ is false.

EXERCISE 19   Determine whether or not the following pairs of statement forms are equivalent:

(a) $(P \rightarrow Q) \wedge (P \rightarrow R)$; $P \rightarrow (Q \wedge R)$
(b) $(P \rightarrow R) \wedge (Q \rightarrow R)$; $(P \vee Q) \rightarrow R$
(c) $P \rightarrow \{Q \rightarrow R\}$; $(P \wedge Q) \rightarrow R$

(Note that all truth tables here will have 8 $(=2^3)$ rows corresponding to the 8 possible assignments of T or F to each of P, Q and R.)

EXERCISE 20   Use your intuition to replace each occurrence of '?' below by the appropriate sign ($\rightarrow$, $\wedge$, $\vee$, $\neg$) and check your intuition by constructing truth tables. Compare your conclusions with those of Exercise 19(a), (b).

(a) $(P \rightarrow Q) \vee (P \rightarrow R)$ is equivalent to $P \rightarrow (Q?R)$
(b) $(P \rightarrow R) \vee (Q \rightarrow R)$ is equivalent to $(P?Q) \rightarrow R$

EXERCISE 21   Suppose that R and S are statement forms. Show that R and S are equivalent statement forms iff the statement form $R \longleftrightarrow S$ is a tautology. (*Hint:* This exercise, being 'abstract', may *appear* difficult but the explanation is not much longer than this hint – so, have a go!)

Definitions

Although few mathematicians (including logicians) press the point hard, each *definition* should, for the sake of clarity, be stated in **IFF** form. As an example we shall define the concept of *prime number*. (We have already used this concept many times – but no harm done. If you like, we are now about to find out if you and I have been thinking about the same thing.)

To maintain clarity where there may be room for doubt (see Example 5(e) below) we precede the definition of primeness by one which is even more fundamental.

Definition 1

Let a and b be integers. We say that a **divides** b if (and only if) there is an *integer* c such that $a \cdot c = b$.

When a divides b we write $a|b$; when it doesn't we write $a \nmid b$.

Example 5

(a) $-37|111$; (b) $53 \nmid 1995$; (c) $3|0$; (d) $0 \nmid 3$; (e) $0|0$

(Does Definition 1 *really* allow us to write $0|0$?)

We can now make a second definition.

Definition 2

The positive integer p is said to be **prime** iff (i) $p > 1$ and (ii) the only (positive) integers which divide p are 1 and p.

Example 6

(a) 9973 is the greatest prime less than 10 000; (b) $2^{37} - 1$ is *not* prime (see Exercise 37(a) in Chapter 8); (c) $2^{1\,257\,787} - 1$ is (at present) the largest integer *known for sure* to be prime.

So, did we agree? Do you think that 1 ought to be given prime status? For one reason why it is better *not* to allow 1 to be counted as a prime see the section just before Theorem 9 in Chapter 8.

You can see that one reason for making a definition is to eliminate any possible misunderstanding: in Definition 1 we wished to make it perfectly clear that, for non-zero integers a and b, we write $a|b$ when and only when b/a is an *integer*. A second reason is to give, as in Definition 2, an economical and *precise* description of an (often used) idea. Why do I insist on using **IFF**? Well, if, in Definition 2, we had written 'if' rather than 'iff' we would surely have had to accept 2, 3, 5, 7, 11, 13,... as primes as usual; but then Definition 2 would *not* say that integers such as 1, 4, 999 *must not* be called primes. Indeed it would say *nothing whatever* about what such integers ought to be called. You will often see 'if' in definitions. Beware! It ought, always, to be understood as 'iff'.

Should I commit proofs and definitions to memory?

In later chapters we shall give proofs of (or counterexamples to) many of the mathematical statements we make and we shall see what makes these proofs or counterexamples 'tick'. One of your tasks then will be to try to remember *why* things work as they do. However, *you are not expected to commit the fine details of the proofs to memory:* for most people that would be too awesome a task. Nevertheless, it is quite a good idea to commit many of the (obviously more important) *definitions* to memory. Most are not long and, if you have control over the words, you can apply all your effort to understanding what the words *mean*.

To get you into the habit of playing with definitions and even inventing some of your own, here are some exercises. Perhaps you can pick out the more important ones from the more frivolous.

EXERCISE 22   Integers like 6 ($=2\cdot3$), 34 ($=2\cdot17$), 143 ($=11\cdot13$) might be called *rectangular*. Invent a definition of *rectangular number* so that, according to that definition, 6, 34 and 143 are rectangular numbers whilst 4, 9, 101, 555 are not.

EXERCISE 23   A friend told me that, according to a definition she had seen, 153 and 370 are *autocubes* because $153 = 1^3 + 5^3 + 3^3$ and $370 = 3^3 + 7^3 + 0^3$. Formulate a definition of *autocube* (and find at least one more autocube).

EXERCISE 24   Another friend told me that integers such as 1, 35, 102 (but not 12, 18, 27, 999) are called *square-free*. Formulate a definition of *square-free* from this information.

EXERCISE 25   Some years ago I decided to define a real number to be *delicious* iff its square is less than 9.87. Is π delicious?

EXERCISE 26   Criticize Euclid's definition of a point as 'that which has no part', and a line as 'that which has length but no breadth'.

EXERCISE 27   Define what is meant by the *degree* of a polynomial in x with coefficients coming from the real numbers. What, according to your definition, are the degrees of the polynomials (a) $3x + 2$; (b) $3x^5 - 7x + 1$; (c) 7; (d) 0?

EXERCISE 28   Look up, then write down, the definition of *perfect number*.

To prove something in mathematics we often have to make direct use of a definition. Here are a couple of exercises involving that procedure.

EXERCISE 29

(a) If I tell you that $77|3311$ and that $147|8526$ (both *are* true), then you may correctly deduce that $7|11\,837$. Why? (Do *not* perform any divisions. Just explain *why* the deduction is correct.)

(b) Give three specific integers m, a and b such that $m|(a + b)$ but $m\nmid a$ and $m\nmid b$.

EXERCISE 30   Using Definition 1 it is not difficult to prove, that, if a, b and c are integers such that $a|b$ and $b|c$ then $a|c$. Show, by giving a specific *counterexample*, that it is *not* true that if $a|c$ and $b|c$ then $ab|c$.

Further important statement forms

Example 4(b) and Exercise 18 gave two statement forms equivalent to $P\rightarrow Q$. Here is another very important one. Intuitively it says that 'If the truth of A implies that of B then the falsity of B implies that of A', and vice versa.

Example 7
Show that the statement forms $P\rightarrow Q$ and $\neg Q\rightarrow\neg P$ are equivalent.

SOLUTION   This follows immediately on completing the table:

P	Q	$\neg Q$	$\rightarrow$	$\neg P$	
T	T	F	T	F	(i)
T	F	T	F	F	(ii)
F	T	F	T	T	(iii)
F	F	T	T	T	(iv)
(1)	(2)	(3)	(5)	(4)	

The statement form $\neg Q\rightarrow\neg P$ is called the **contrapositive** of the statement form $P\rightarrow Q$. You will see time and again how a statement of the form $A\rightarrow B$ can be proved much more conveniently and often more easily and naturally by proving, instead, the equivalent implication $\neg B\rightarrow\neg A$.

EXERCISE 31   Complete the following sentence. If $2^{11\,213} - 1$ is prime then 11 213 is prime. Hence, if 11 213 is *not* prime then....

EXERCISE 32   State the converse and the contrapositive of the following statement. If $1729|13^{1728} - 1$ then 1729 is a prime.

The following implications are important because they are sometimes used as **axioms** (starting points) for the logical examination (called the **statement calculus**) of the theory of statements. Perhaps not all of them are intuitively obvious.

1. $\{P\rightarrow(Q\rightarrow R)\}\rightarrow\{(P\rightarrow Q)\rightarrow(P\rightarrow R)\}$
2. $(P\rightarrow Q)\mapsto\{(P\rightarrow\neg Q)\rightarrow\neg P\}$
3. $(P\rightarrow Q)\mapsto\{(P\vee R)\rightarrow(Q\vee R)\}$
4. $\{(P\rightarrow Q)\wedge(Q\rightarrow R)\}\rightarrow(P\rightarrow R)$ (called the **hypothetical syllogism**)

These can be checked, using truth tables, to be tautologies. However, a slightly different approach to making such a check is seen in Example 8.

Example 8
Show that the statement form $\{(P \rightarrow Q) \wedge (Q \rightarrow R)\} \rightarrow (P \rightarrow R)$ is a tautology.

SOLUTION   We check to see if the major implication (i.e. the one immediately after the '}') can possibly be false. Assuming it is, we make deductions, from the truth tables for '$\rightarrow$' and '$\wedge$' as indicated in the successive rows of the following diagram:

$$\{(P \rightarrow Q) \quad \wedge \quad (Q \rightarrow R)\} \quad \rightarrow \quad (P \rightarrow R)$$

$$\begin{array}{cccc}
 & & & F \\
 & T & & \\
 & & & F \\
 T & & T & T \quad F \\
 & & F \quad (F) & \\
 F \quad (F) & & &
\end{array}$$

The (F)s merely repeat what has been determined on the previous line. Note that the assumption that the main implication is false has led, in lines 3 and 5, to the absurdity that P is simultaneously true and false.

EXERCISE 33   Is the statement form $\{(P \vee Q) \rightarrow (P \rightarrow Q)\} \vee (P \wedge \neg Q)$ true for all T/F values of P and Q? (That is, is it tautology?)

Proof format: rules of inference

In Chapter 1 we described a proof (of a conjecture or theorem) as 'a logically watertight argument comprising a sequence of assertions, each deduced from earlier assertions or assumptions, with the last assertion being the conjecture (or theorem) itself'. We also gave, in Figure 1.1, a pictorial representation of how a proof, that a certain *conclusion* M follows from *hypotheses* A, B, C, H and MI, might be developed via a sequence of deductions. Now, how are such deductions made? It is here that various tautologies, of the type obtained above, can help. For example, the tautology $\{P \wedge (P \rightarrow Q)\} \rightarrow Q$ (and the table for '$\rightarrow$') show that, if statements A and $A \rightarrow B$ are both true, then the truth of statement B may be inferred (i.e. deduced). Similarly, from the tautology $\{(P \rightarrow Q) \wedge (Q \rightarrow R)\} \rightarrow (P \rightarrow R)$, we may infer the truth of $A \rightarrow C$ from the truth of both $A \rightarrow B$ and $B \rightarrow C$. Furthermore, in any proof, any statement or statement form may be replaced by an equivalent. A common occurrence is to prove an implication of the form $A \rightarrow B$ by proving, instead, its equivalent $\neg B \rightarrow \neg A$. In our next example we make use of the equivalence of the statement forms $P \rightarrow (Q \rightarrow R)$ and $(P \wedge Q) \rightarrow R$.
    Let us see, by means of a 'theorem' proved in 'skeleton' form, how such inferences are used in practice and what corresponding format the proof of the theorem takes.

'Theorem' 1
Let $A \rightarrow (B \rightarrow C)$, $\neg D \vee A$ and B be given. Show that $D \rightarrow C$.
    The first sentence means that we are to assume that the first three statements are true. The second sentence asks us to deduce that the implication $D \rightarrow C$ must then also be true. In other words, we are being asked to show the truth of the implication

$$[\{A \rightarrow (B \rightarrow C)\} \wedge (\neg D \vee A) \wedge B] \rightarrow (D \rightarrow C) \qquad (I)$$

Now we know from Exercise 19(c) that the statement forms P→(Q→R) and (P∧Q)→R are equivalent. Accordingly, (I) will be true if (and only if) [{A→(B→C)}∧(¬D ∨ A)∧B∧D]→C is true. Translating this back into the form given in the 'theorem', we see that the 'theorem' will be proved if, from the *given* truth of A→(B→C), ¬D ∨ A and B and the *assumed* truth of D we can establish the truth of C.

We do this as follows, giving reasons, and setting it out both pictorially and in a table. As you will see, we shall make much use of *modus ponens* in the form 'If P and (P→Q) are true then so is Q'.

Proof of Theorem 1
1. From (the truth of) D and (of) ¬D∨A we deduce (that of) A (by Exercise 11).
2. From A and A→(B→C) we deduce (B→C) (by *modus ponens*).
3. From B and (B→C) we deduce C (*modus ponens*).

This completes the proof. Figure 2.1 is a pictorial representation and a clearer exposition of the proof.

The inference 'If P and P→Q are true then so is Q' which is derived from the valid implication {P∧(P→Q)}→Q is frequently expressed by writing

$$\frac{P, P→Q}{Q}$$ (We also call this inference *modus ponens*)

Other inferences commonly used in proofs include:

(i) $$\frac{P→Q, ¬Q}{¬P}$$ (*modus tollens*) and its equivalent $$\frac{P ∨ Q, ¬Q}{P}$$

(ii) $$\frac{P→Q, Q→R}{P→R}$$ (*hypothetical syllogism*)

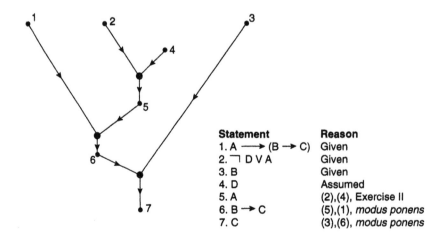

Statement	Reason
1. A ⟶ (B ➞ C)	Given
2. ¬ D V A	Given
3. B	Given
4. D	Assumed
5. A	(2),(4), Exercise II
6. B ➞ C	(5),(1), *modus ponens*
7. C	(3),(6), *modus ponens*

Figure 2.1

(iii) $\dfrac{P, Q}{P}$ and $\dfrac{P, Q}{Q}$ (*conjunctive simplification*)

(iv) $\dfrac{P}{P \vee Q}$, $\dfrac{Q}{P \vee Q}$ and $\dfrac{P, Q}{P \wedge Q}$

(v) $\dfrac{P \to Q}{\neg Q \to \neg P}$ (*contrapositive inference*)

(vi) $\dfrac{P \to R, \; Q \to R}{(P \vee Q) \to R}$ (*inference by cases*)

(vii) $\dfrac{P \to (Q \vee R)}{(P \wedge \neg Q) \to R}$ (viii) $\dfrac{P \to (Q \to R)}{(P \wedge Q) \to R}$

The statement forms above and below the line in all these inferences are equivalent, and as most of them are intuitively obvious, they take little remembering.

COMMENT   We *could* have proved the above theorem by applying truth table techniques directly to the implication (I). Note that a full truth table would have 16 ($= 2^4$) lines. The 'inference' method of proof can, therefore, often save us a fair amount of work.

Here are two for you to try.

EXERCISE 34   Show, in the above manner, that from $A \to B$, $A \vee C$ and $C \to \neg D$ that you can deduce $D \to B$.

EXERCISE 35   Show, in the same manner, that the hypotheses $A \to B$, $B \to C$, $D \to \neg C$ and $A \wedge D$ are inconsistent by demonstrating that they permit you to deduce $C \wedge \neg C$.

A political statement

As an amusing interlude let us check the validity of the following statement which was, of course, *not* made by any politician *I* know of. Some words have been placed in brackets so as to ease the analysis.

'If the government (is) returned to power then we shall all be better off financially. If we are *not* better off financially then there will be unrest. Hence if there is not (to be) unrest (then) the government (must be) returned to power.'

This seems to be of the form 'If $A \to B$ and $\neg B \to C$ then $\neg C \to A$'. The question is: is this a logically sound argument?

EXERCISE 36   Is the following argument sound?

'If the government reduces taxes and increases spending, inflation will get worse. If the government does not increase spending, unemployment will increase. If inflation gets worse or unemployment increases, the government will lose the next election. Therefore, either the government will reduce taxes or lose the next election.'

Another example of poor style

We begin with a final exercise.

EXERCISE 37   Show that the statement forms (a) $(P \rightarrow Q) \rightarrow R$ and (b) $P \rightarrow (Q \rightarrow R)$
are not equivalent. In the light of your findings discuss what meaning, if any, can
be ascribed to the implication $A \rightarrow B \rightarrow C$.

Setting out a proof as a succession of implications $(\cdots \rightarrow \cdots \rightarrow \cdots \rightarrow \cdots)$ (a common
occurrence amongst first-year undergraduates) may be regarded as, at best, poor
style. Try to avoid it by thinking, carefully, about the meaning of what you have written.
    Can you see why, on several grounds, the solution to the following example is
absolutely ghastly?

Example 9
Show that the two roots of the quadratic equation $x^2 + 7x + 11$ differ by $\sqrt{5}$.

SOLUTION $x^2 + 7x + 11 = 0 \rightarrow x = \frac{-7 \pm \sqrt{(49 - 4 \|1\| \|11)}}{2} \rightarrow \frac{(-7 + \sqrt{5}) - (-7 - \sqrt{5})}{2} \rightarrow$
 $\sqrt{5} \rightarrow$ true.

(Aargh! I know what is *meant* but what is written comes nowhere near to saying it.
The expression ' $\rightarrow$ true' is a great favourite of many undergraduates, but as it is
meaningless, DON'T USE IT!)

Summary

Since the **connectives**, NOT, AND, OR, IF/THEN, and IFF occur frequently in mathematics
it is important to be certain about how to interpret them (in everyday English this
is often unclear). This certainty is achieved by spelling out their intended interpretation
by means of **truth tables**. Once this has been done it is possible to confirm,
mechanically, what, admittedly in many cases, is intuitively clear, namely that certain
statement forms are **equivalent**. For example, the statement form $P \rightarrow Q$ is equivalent
to $\neg P \vee Q$, to $\neg(P \wedge \neg Q)$ and to its **contrapositive** $\neg Q \rightarrow \neg P$. Likewise one may
confirm that the statement forms $(P \wedge \{P \rightarrow Q\}) \rightarrow Q$ (called *modus ponens*) and
 $\{(P \rightarrow Q) \wedge \neg Q\} \rightarrow \neg P$ (called *modus tollens*) are **tautologies**, that is, they take only
the truth value T. (We also say that such statement forms are **valid**.)
    Tautologies give rise naturally to various universally valid inferences and
equivalences which we can employ in proof construction.
    If $A \rightarrow B$ we say that A is **sufficient** for B and that B is **necessary** for A. Examples
show that $A \rightarrow B$ is, in general, not equivalent to its **converse** $B \rightarrow A$. However, both
may hold, in which case we write $A \longleftrightarrow B$ which we read as 'A **if and only if** B'. A
principal use of IFF is in framing definitions where we need to know *precisely* which
objects satisfy the definition and which do not.
    Think carefully before writing $A \rightarrow B \rightarrow C$ since the two statement forms
 $(P \rightarrow Q) \rightarrow R$ and $P \rightarrow (Q \rightarrow R)$ are not equivalent.

CHAPTER 3

True or False?

Be sure of it – give me ocular proof. SHAKESPEARE

In this chapter we try to find proofs (or counterexamples) for the first six of the eighteen test statements listed in Chapter 2. Since students often find difficulty knowing where to begin both in thinking about a problem *and* in writing out its solution, we try to show both phases in action by first discussing, often in detail, how one might (or might *not*) approach the problem at hand – and only then giving what we term a formal proof. Certain methods of attack tend to reappear. In order to be sure that our methods are logically sound we shall occasionally need to call upon the equivalence of certain statement forms, as established in Chapter 2.

Introduction

Few proofs occur to us instantaneously: to find a proof of any particular claim we may have to fish around guessing, trying and sometimes rejecting potential lines of enquiry, even hoping for a burst of inspiration, for quite some time before we find an acceptable line of argument. (Those proofs which appear to come from nowhere are more likely a result of a certain amount of subconscious mulling over, perhaps after you have stepped aside from the problem for a while.) Sometimes these preliminary thoughts – which we place under the heading DISCUSSION – will take us along what prove to be dead ends. Indeed, one aim of the Discussion section is to present the material with the spirit and excitement (and even frustration) of mathematical research. Sometimes we shall suddenly realize that our fumblings can be shortened or improved. You must accept that this trial and error approach is *part of the process of building a proof.* It does not necessarily indicate a lack of ability on your part. *There is no quick fix for finding proofs.* In particular, *there is no universal method available for you to follow which will invariably lead you to the production of a desired proof.*

Once you have an idea of how a proof might proceed, you should attempt to write it in a clear and logically developed way *which will best help the reader to verify that your line of argument is correct.* However, it is not part of your task to tell the reader *how* you discovered the proof – even less to tell of all the false trails you may have followed before you realized your goal. Consequently the proofs we give are often substantially shorter than the Discussions which precede them: our Discussions will be chatty – formal proofs should not be. (Of course, each proof you construct has to pass the validity test. I feel that a good idea here would be to test your argument on a colleague to see if he will accept your argument as a valid proof of your claim. Perhaps you could offer a sum of money for each error detected – this should concentrate your mind wonderfully.)

From our Discussions – which are by no means unique – certain strategies and methods will emerge. We shall list these under the heading WHAT WE HAVE LEARNED at the end of most investigations and we shall summarize the most important of them at the end of the chapter. It is useful to try to remember something of the *methods* employed without committing to memory *too* much of the fine detail.

Where it seems appropriate, I shall do away with the rather formal wording employed when the statements were first listed. This will give you an idea of how mathematicians are happy to be fairly informal *provided they know that they can back up their informality with rigour,* and it will exhibit the translation between formality and informality which will help you to recognize exactly what is meant when you meet informality in other books and courses. Please compare the 'new' statements with the old.

I remind you: the Discussions are not part of final proofs. Accordingly, some readers may prefer to proceed as follows:

1. Reread the statement.
2. Try to prove or disprove it.
3. Read the formal solution – and only then turn to the Discussion so see what *my* thoughts were about solving the problem.

In any case the discussions are meant only as a chatty backdrop, to be read casually, rather than to be studied deeply.

The proofs

STATEMENT 1   300 000 067 110 605 737 is not a perfect square.

DISCUSSION
We adopt our sole opening strategy. Do we understand fully what Statement 1 says and what we are expected to do? I think so. There seems no problem here. For

convenience let us call the given integer z. Perhaps the first move you might make is to try to find $\sqrt{z}$. If you have a sufficiently powerful piece of computing equipment to hand, you will be able to determine straight away whether or not Statement 1 is true. How might you proceed if you only had pencil and paper?[1] Rather than trying to find $\sqrt{z}$ you *could* try to find, by making a succession of increasingly accurate approximate *guesses*, some integer s, say, such that $s^2 < z$ whilst $(s + 1)^2 > z$. (Since z is just greater than 30×10^{16} a reasonable first guess for s would be in the region of 5.4×10^8, since $(5\frac{1}{2})^2 = 30\frac{1}{4}$. Finding such an s would prove that z is not a perfect square. Maybe z *is* a perfect square though? If it *were*, then it would be the square of some integer, t, say. Now t would certainly have to be odd – and hence it would have to end in a 1, 3, 5, 7 or 9. But then, t^2 would have to end in a 1, 9, 5, 9 or 1 (wouldn't it?). If that's right then z *can't* be a perfect square – since *it* ends in a 7.

This looks convincing. Can we tidy it up – perhaps being a bit less woolly about t^2 ending in a 1 or 5 or 9? In fact every odd integer t is surely of the form $(10 \times m) + r$ where $r = 1$ or 3 or 5 or 7 or 9, and where m is some suitable integer. So then

$$t^2 = (10m + r)^2 = 100m^2 + 20mr + r^2$$

Looking at the units digit we see that it is the same as that of r^2 (since the $100m^2 + 20mr$ part is an integer ending in a 0). So we only have to check the five values of r^2.

I think we are now ready to give our proof.

In fact, what we, in this book will call a *formal* proof[2] (to emphasize its difference from the very informal Discussion) may well say:

Proof of Statement 1   Suppose that the given integer, z, *is* the square of some (odd) integer t, say. Now $t = 10m + r$ for some suitable integer m and for some suitable value for r from the list 1, 3, 5, 7 and 9. But then

$$t^2 = (10m + r)^2 = 100m^2 + 20mr + r^2$$

and so the last digit of $z\,(=t^2)$ is the same as the last digit of r^2, which is, therefore, 1 or 9 or 5. But z's final digit is a 7. Hence z cannot be a perfect square.

Another proof of Statement 1   An even shorter proof might say: the given integer cannot be a square since no (integer) square can end in a 7.

COMMENTS

1. The most obvious difference between the above Discussion and the proofs is the brevity of style of the latter. For example, the first proof assumes that the reader can see for himself why the last digits of t^2 and r^2 are the same. The second proof, whilst highlighting the key observation which prevents z from being a square, asks even more of the reader. Accordingly you may feel that the second proof is *too* short.

---

[1] For an interesting method, now rarely taught, see, for example, Durrell (1950, p. 274).
[2] This is *not* what logicians would mean by the term 'formal proof'. What we have given is, to them, no more than an abbreviated demonstration of a proof.

2. Our first proof of Statement 1 certainly 'feels' valid. Let us examine the argument. Without knowing whether or not z is a perfect square we said 'Suppose it were ...'. That is, we assumed the contrary of what is claimed. (One advantage in this approach is that it gives us something (some supposition) to work from.) We then deduced that, *if* it were a square, then z would have to end in a 1, a 5 or a 9 – thereby contradicting the fact that z ends in a 7. This contradition led to our having to admit that our supposition was wrong.

   This kind of argument – assuming false that which you wish to prove true and deducing a contradiction from this assumed falsity – is commonly employed in mathematical proof. It is called **proof by contradiction** or *reductio ad absurdum*.

3. I have tried to impress on you the idea of remembering the method – not the fine detail. Here, if you will only recall 'Squaring up $10m + r$ shows that no square can end in a 7' the details of how to show this will not be difficult to recall. Moreover, in remembering the key idea, you will find it equally easy to believe – or even to *work out for yourself* – that, likewise, no odd square can end in a 3.

4. Some readers might be interested in seeing what logical principles we have used in the above argument. These principles – which help confirm that our reasoning is sound – are presented in (a), (b) below. The uninterested reader should pass to item 5.

   (a) Let us denote by B the statement B: z is not a perfect square. We first *supposed*, not knowing whether we were right or wrong, that z *is* a perfect square, that is, that $\neg B$ is the case. We then deduced that z must end in a 1 or 9 or 5 or 9 or 1 by considering five separate cases – see (b) below. Thus, if we let A be the (true) statement A: 'z has final digit 7', we can claim to have proved that $\neg B \rightarrow \neg A$, which is, of course, equivalent to $A \rightarrow B$ (see page 34, Chapter 2). So, at first glance, our proof seems to be of the **contrapositive** type. However, our proof continues 'But z's final digit *is* a 7', that is, we are adding the remark that $\neg A$ is definitely false. Thus, assuming the truth of $\neg B$ is seen to imply the *false* statement $\neg A$. We therefore have our proof by contradiction. Because our proof showed the falsity of $\neg B$ (rather than the truth of B) we say we have used an **indirect** rather than a **direct** proof of B.

   (b) The second method of proof which we employed above (namely, in deducing $\neg A$ from $\neg B$), is that of **proof by cases**. (We showed that no odd integer, t, when squared, can end in a 7 by considering all *five possible cases* for t, namely, t's last digit being 1 or 3 or 5 or 7 or 9.)

   The logical principle behind the 'cases' form of proof is described (cf. Exercise 19(b) in Chapter 2) by the equivalence.

$$\{(C_1 \rightarrow D) \wedge (C_2 \rightarrow D) \wedge (C_3 \rightarrow D) \wedge (C_4 \rightarrow D) \wedge (C_5 \rightarrow D)\}$$
$$\longleftrightarrow \{(C_1 \vee C_2 \vee C_3 \vee C_4 \vee C_5) \rightarrow D\}$$

In our proof, C_1 is 'The integer t has final digit 1', C_2 is 'The integer t has final digit 3', C_3 is 'The integer t has final digit 5', etc., whilst D is 'The integer t^2 doesn't end in a 7.' Notice that $C_1 \vee C_2 \vee C_3 \vee C_4 \vee C_5$ is then 'The integer t has final digit 1 or the integer t has final digit 3 or ... or final digit 9'. In other words $C_1 \vee C_2 \vee C_3 \vee C_4 \vee C_5$ says that t ends in a 1, a 3, a 5, a 7 or a 9 and, hence, covers all possibilities.

5. We could have shown z not to be a square by employing a sufficiently power-ful calculating device. However, what would we then have known? Could we then have said immediately and *with certainty* that, for example, neither 7 390 846 724 536 489 097 nor 93 008 139 562 448 576 309 093 is a perfect square? I suggest we could not. The calculator only gives us an answer – it doesn't tell us *why*. On the other hand, the *theoretical* investigation we undertook shows, with no extra work, that *no odd integer whose final digit is a 3 or a 7 can be a perfect square*. Theory can tell you something your computer cannot.

WHAT WE HAVE LEARNED

- Do not rush into involving the computer/calculator.[3]
- It may pay to take an educated guess or, even better, to do some *thinking*.
- Given any statement, it may be possible to show that assuming the statement's negation leads to another statement which is blatantly false. Then the given statement must be true, the proof being indirect.
- Sometimes a proof by (covering all) cases is appropriate.
- Infinitely many cases may be covered by grouping them into finitely many subcases of the same type. (Here every odd integer is of one of the five forms $10k + r$ where $r = 1, 3, 5, 7$ or 9.)

EXERCISE 1   Show that 187 198 574 197 820 121 945 181 952 is not a perfect square.

EXERCISE 2

(a) Which digits can be the final digit of the cube of an integer?
(b) With this information you can decide, instantly, whether or not 300 000 067 110 605 737 is a cube?

EXERCISE 3

(a) Using the fact that each integer n may be written in one of the forms $100m + r$ where $0 \leqslant r \leqslant 99$, find all possible pairs of *last two digits* of a perfect square. (*Hint:* You need only to consider r in the range $0 \leqslant r \leqslant 50$. Why? In fact you need only consider $0 \leqslant r \leqslant 25$. Why?)
(b) Show that 174 429 696 251 511 866 227 734 759 is not a perfect square.

The information you obtained in Exercise 3(a) was employed by Pierre de Fermat in a clever scheme he invented to help factorize fairly large integers quickly by hand. (Allenby (1989, pp. 44–6).)

---

[3] I am reminded of the student who wrote, in an examination, that $7^{20} = 79\,792\,266\,280\,000\,000$. No need to guess where this absurd assertion came from!

PIERRE DE FERMAT
(20 August 1601–12 January 1665)

Pierre de Fermat, who was born near Toulouse in 1601, came from a prosperous family, and so his choice of law as a profession was a natural one. Obtaining a law degree in 1631, he became a magistrate, entitled to include the 'de' in his name. That same year he married. Of his five children, three entered the church.

Our knowledge of Fermat comes mainly from letters he wrote from 1636 onwards. It seems that he only got seriously involved in mathematics in his late twenties when, in reconstructing some theorems of Apollonius, he invented, simultaneously with, but independent of, Descartes, the subject of analytic geometry. In 1643 Fermat established what we today know as the second derivative conditions $f''(x) > 0$, $f''(x) < 0$ for determining minimum and maximum values of functions. He later used this to good effect in confirming Descartes' Law of Diffraction. Around 1654 he and Pascal laid the foundations of the theory of probability when they answered a question posed to Pascal by a gambler: How should the spoils in a game of chance be fairly divided if the game has to end early?

Nowadays Fermat is best remembered as 'The father of modern number theory'. It was he who insisted that the proper objects of study were the integers. In his early days Fermat was much interested in *perfect numbers n*, such as 6 ($= 1 + 2 + 3$) and 28 ($= 1 + 2 + 4 + 7 + 14$), where the sum of the proper divisors of n is equal to n. This led to his proving Statement 10 and, later, to his Little Theorem (see Exercise 27 in Chapter 8) which lies at the root of much recent work devising tests for more rapid investigation of the possible primeness of a given integer, itself of interest in connection with the so-called public key encryption systems.

> Despite his name now being on everyone's lips due to Wiles's recent proof of Fermat's Last Theorem, concerning the equation $x^n + y^n = z^n$, Fermat's number theory received little recognition in his lifetime and faded into obscurity until resurrected by Euler and others in the eighteenth and nineteenth centuries.

As we have already said, it is worthwhile remembering general methods of procedure: they may prove useful in different contexts. Here is an example where one method of procedure tried above turns out to be of little use whilst the other is just what we need.

STATEMENT 1.1   There is no integer n for which $n^2 + 4$ is exactly divisible by 7.

DISCUSSION

Once again we start by asking, do we understand what we have to prove? For example, it is not sufficient to say (as was offered in a recent examination answer), 'If I put $n = 1$ then $n^2 + 4 = 5$ and $7 \nmid 5$. This completes the proof.' In fact we have to make sure that for *all* integers n, $7 \nmid (n^2 + 4)$. Proving that $7 \nmid (n^2 + 4)$ for the single case of $n = 1$ is, therefore, not enough.

In this problem there seems little chance of using a computer directly since a computer is no more use than a human for checking an infinite list one by one. Maybe we should adopt the useful ploy of the previous proof of assuming that the result stated is false and try to deduce a contradiction? Let us try. We assume that there is some integer n for which $n^2 + 4$ *is* divisible by 7. That is, we assume that there are integers n and k such that $n^2 + 4 = 7k$. Some of you might follow this with: 'Hence $n^2 = 7k - 4$. Consequently $n = \sqrt{(7k - 4)}$.' Now this doesn't seem to help *me* very much.[4] So what else can we do? One possibility is to *try to get some feeling* as to whether or not the stated result is true by evaluating $n^2 + 4$ for various values of n. Trying $n = 1$ to, say, 20, we find none is divisible by 7 because successive remainders, on division of $n^2 + 4$ by 7 are: 5, 1, 6, 6, 1, 5, 4, 5, 1, 6, 6, 1, 5, 4, 5, 1, 6, 6, 1, 5,.... These 20 remainders *do not*, of course, even remotely begin to prove that Statement 1.1 is true but they do, at least, exhibit a repeating pattern of length 7: Does this pattern persist for ever? *Does this (apparent) pattern remind us of anything?* In Statement 1 we dealt with all odd integers by considering five cases of the form $10m + r$. (There r was one of the 'remainders' 1, 3, 5, 7 or 9.) Maybe here we can deal with *all* integers n via just seven *cases* by observing that each integer is of one of the forms $7m + r$ for suitable integer m and suitable 'remainder' $r = 0, 1, 2, 3, 4,$ 5 or 6. Writing n in the form $7m + r$ shows that *every* value of $n^2 + 4$ is of one of the forms $(7m + r)^2 + 4 = 49m^2 + 14mr + r^2 = 7(7m^2 + 2mr) + r^2 + 4$. Consequently $n^2 + 4$ will be divisible by 7 if and only if $r^2 + 4$ is – since $7(7m^2 + 2mr)$ clearly is. Thus there are only seven values to check for divisibility by 7, namely the seven values of $r^2 + 4$ where $r = 0, 1, 2, 3, 4, 5$ or 6. Since none of these is a multiple of 7 we have confirmed the truth of Statement 1.1. A formal proof may well go as follows.

Proof of Statement 1.1   Every integer is of the form $7m + r$ where $r = 0, 1, 2, 3, 4,$ 5 or 6. Since $(7m + r)^2 + 4 = 7K + r^2 + 4$ (for suitable integer K) we see that

---

[4] In a recent examination some students did try this approach. See Exercise 6 below.

$(7m + r)^2 + 4$ is divisible by 7 if and only if $r^2 + 4$ is divisible by 7. But the possible values of $r^2 + 4$ (if $0 \leqslant r \leqslant 6$) are 4, 5, 8, 13, 20, 29 and 40 – and none of these is divisible by 7.

COMMENTS

1. The above proof exhibits another instance of *proof by cases*. The 'obvious' remark that 'Every integer is of one of the forms $7m + r$ where $r = 0, 1, 2, 3, 4, 5$ or 6' is one instance of the *Division Theorem* which will play a prominent role in Chapter 8.
2. Notice, again, how the formal proof is a logically coherent but compressed account of the relevant parts of our Discussion. In particular, without the preliminary Discussion, the reason why the formal proof begins 'Every integer is of the form' would perhaps not be completely obvious until one had read the entire proof, and had seen how it all fitted together.
3. Notice, too, that, after all, our formal proof did *not* assume that $7 | n^2 + 4$ and endeavour to obtain a contradiction. Rather we proved that if X is an integer of the form $n^2 + 4$ then $7 \nmid X$. Hence this is an example of a **direct** proof.

WHAT WE HAVE LEARNED

- Even with apparently straightforward statements always make sure, by careful reading, that you know what is being claimed before you start a proof.
- Some results can be proved by a 'direct' approach even though, initially, one doesn't see any alternative to an indirect attack.
- It can be useful to try some instances to get a feel for the problem. In particular, a recognizable pattern (or a counterexample) may appear. Of course, a proof *for all n* cannot be deduced from the correctness of the statement for the first 20 values of *n*.
- Ask yourself if you have come across a similar problem, proof or idea before. If so, can you use the same or similar methods now? Here the key move is a reduction of infinitely many cases to 7 cases.
- Sometimes a 'hopeful' method (here writing $n = \sqrt{(7k - 4)}$) can appear to lead nowhere. Then there may be no alternative but to start afresh.

EXERCISE 4   Show that 15 never divides any integer of the form $n^2 - 2$.

EXERCISE 5   Determine for precisely which integers n we have $15 | (n^2 + 11)$.

EXERCISE 6   Why can we not prove Statement 1.1 by saying 'if $n^2 + 4 = 7k$ then $n = \sqrt{(7k - 4)}$. But this is impossible since n is an integer and the square root clearly is not'?

STATEMENT 2 $3 < \pi < 4$.

DISCUSSION

(Do we understand the problem?) π is, of course, the ratio c/d of the length, c, of the circumference of a circle to the length of the circle's diameter d. To confirm the inequality $3 < c/d < 4$ we need to show that $3 < c/d$ *and* that $c/d < 4$. Perhaps we can confirm these inequalities by drawing a very accurate circle (of diameter d) and carefully measuring its circumference? Well, not really. Even with a pristine picture, how sure are you that your measurement will be anywhere near accurate? Furthermore, geometry really deals with idealized figures in which lines have no thickness. (See Exercise 26 in Chapter 2.) Therefore the pictures we draw on paper can be, at best, rough approximations whose only purpose is to assist our reasoning. In addition you have all seen examples of optical illusions – and, perhaps, the splendid drawings of M. C. Escher. So it can be dangerous to place too much reliance on diagrams (see Exercise 10 below).

What we need are two lengths, m and n, say, which can be determined *exactly* and for which $3 \leqslant m < c/d < n \leqslant 4$. What about the perimeters of squares or hexagons, etc. lying just inside and just outside the circle? Here just a rough picture of our circle may help (see Figure 3.1).

Now it is easy to calculate the perimeters of the squares we have drawn. The outer square has perimeter of length $4d$ whilst the inner one has, by Pythagoras' theorem, $4(d/\sqrt{2})$. Consequently c lies between $2\sqrt{2}d$ and $4d$. Oh dear! $2\sqrt{2}d < 3d$ and so we have not fully achieved the result claimed. How we might improve on the lower quantity $2\sqrt{2}d$? Presumably an internal (regular) polygon P with more than 4 sides would sit 'closer' to the circumference and P's perimeter would be a better approximation (still on the lower side) to c than the value $2\sqrt{2}d$.

We leave you to check that the perimeter of the inscribed hexagon of Figure 3.2 is exactly $3d$. (Why is it?) Having done this you may care to feel admiration for Archimedes[5] who, taking regular 96-sided figures inside and out, arrived at the conclusion that $3\frac{10}{71} < \pi < 3\frac{1}{7}$ (Heath, 1897). You may also be amused by the fact

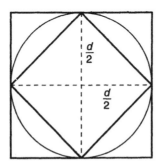

Figure 3.1

[5] Archimedes of Samos (287–212 BC) is generally agreed to be, with Newton and Gauss, one of the three greatest mathematicians of all time.

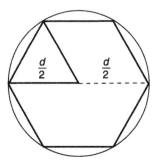

Figure 3.2

that, in 1897, a bill was enacted in Indiana stating that π was *de jure* 4. A formal proof of Statement 2 may therefore take the following form.

Proof of Statement 2   Let C be a circle of diameter d units. Then the perimeters of the inscribed regular hexagon and the escribed square are respectively $3d$ and $4d$. The required ratio thus lies between $3d/d$ and $4d/d$.[6]

COMMENTS

1. Since this proof goes directly from what is given (the circle) to the desired conclusion (that $3 < c/d < 4$) we have another example of a *direct proof*.
2. The above proof is a bit on the curt side – and there isn't even a picture! There is no particular merit in paring one's proofs right to the bone. You could add a little more by way of explanation – including a picture if you wished – especially if you think it would help the reader.

WHAT WE HAVE LEARNED

- Sometimes an initial failure to achieve our aim can be overcome if we are prepared to *modify* our first attempts rather than abandon them and start afresh.
- Some proofs can be aided by drawing pictures.

EXERCISE 7   Is the following a proof of Statement 2? 'π is approximately 3.1415926..., so, clearly, $3 < \pi < 4$'.

---

[6] This last sentence is optional, but it 'rounds off' the proof nicely.

EXERCISE 8 By choosing inscribed and escribed (regular) 12-gons (and using Pythagoras' theorem) show that $6\sqrt{(2 - \sqrt{3})} < \pi < 12{\cdot}(2 - \sqrt{3})$ and hence that $3\frac{1}{10} < \pi < 3\frac{11}{50}$.

EXERCISE 9 The above proof *does not* show that the ratio of the length of the circumference and the diameter is the same for *every* circle. Why not? Can you therefore think how you might demonstrate that π is the same for *all* circles? Eudoxus (408–355 BC) could – see Eves (1976, p. 307).

The following statement is surely wrong.

STATEMENT 2.1 Every triangle is isosceles (indeed equilateral).

Proof of Statement 2.1 In Figure 3.3 take OA as the bisector of the *angle* BAC and OX as the perpendicular bisector of the *side* BC. Let Z and Y be the feet of the perpendiculars from O onto the sides AB and AC respectively. Then triangles AZO and AYO are congruent since they have (three pairs of) corresponding angles equal *and* corresponding sides(s) AO coincident. In particular, AZ and AY are equal in length as are OZ and OY. Next, OB and OC are equal in length (why?). Consequently ZB and YC are equal in length (why?). This implies that AB and AC are equal in length so that triangle ABC is isosceles. (Likewise one may prove that AB and BC are equal in length, thus showing that triangle ABC is equilateral.)

EXERCISE 10 Identify where the error in the above proof lies. (It is not enough to say, 'Well, it must be wrong since, plainly, not all triangles *are* isosceles.' If it is wrong we want to know *why* and *where* so that, in more subtle situations, we don't allow a similar mistake to creep past unnoticed.)

EXERCISE 11 Prove (or find a proof in a book) that, in any triangle, the perpendicular bisectors of the three sides meet in a common point. Draw pictures to cover all possible cases.

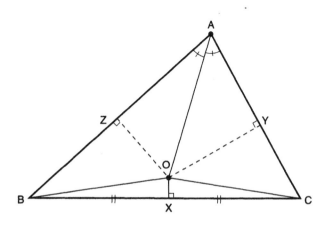

Figure 3.3

EXERCISE 12   Do Figures 3.4 and 3.5 suggest theorems? Discuss them with your tutor.

EXERCISE 13   Using the same set of axes draw rough sketches of the graphs of
(a) $y = \cos x$ and (b) $y = x$. From your graph do you believe that the equation
$\cos t = t$ has a solution or not? Is it clear (pictorially) whether or not $\cos t - \sin t = \frac{1}{2}$
has a solution with $0 < t < \pi/4$?

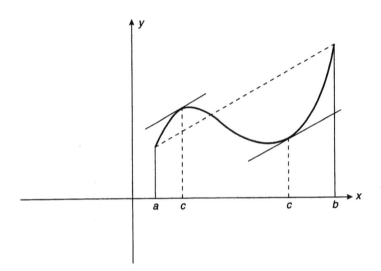

Figure 3.4

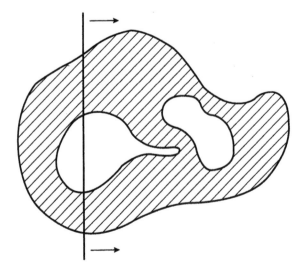

Figure 3.5

WHAT WE HAVE LEARNED

● Drawing pictures can be helpful in explaining (Statement 2, Exercise 11) and/or suggesting (Exercises 12, 13) possible results and even in *remembering* them (Exercises 11–13) but you must be careful not to attribute to a picture more than it warrants (Statement 2.1). Some picture 'proofs' (for example that in Figure 3.5) will seem convincing; others (Figure 3.4) may leave you feeling, at best, unsure.

STATEMENT 3(a)   Either 1997 is a prime or it is divisible by some prime not exceeding 43.

DISCUSSION

This time it is quite appropriate that we begin by asking if you fully understand what you are required to prove. The words 'prime' and 'divisible' presumably cause no trouble, so the only problem is, what does the statement claim? You will *not* be being stupid if you feel the need to read the statement several times. The statement does *not* claim that 1997 is prime *nor* that 1997 is divisible by a prime not exceeding 43 – only that at least one of these two alternatives holds. (It will become apparent that we have here an 'exclusive or' statement.) Now, if Statement A: '1997 is a prime' is true then Statement 3(a) is true. So, to be *sure* that Statement 3(a) is true we are only left to check that *if* A is false (that is, if 1997 is *not* prime) *then* B: '1997 is ... 43' is true. So let us suppose that 1997 is *not* prime. Then 1997 is expressible as a product: $1997 = mn$, where m and n are integers each greater than 1. Notice that we cannot have both m and n being greater than $\sqrt{1997}$ (which is approximately 44.7) since, if they were, their product would exceed 1997). Thus at least one of m, n is no greater than 44. Then, either the smaller of these two factors is itself prime or this smaller factor is divisible by some (even smaller) prime number which is certainly no bigger than 44. A formal proof of Statement 3(a) may therefore take the following form.

Proof of Statement 3(a)   Either 1997 is prime or it is not. In the former case there is nothing more to prove. In the latter case there are integers m and n such that $1997 = mn$ where either $1 < m \leqslant \sqrt{1997}$ or $1 < n \leqslant \sqrt{1997}$. Since m and n can be written as products of primes there is a prime no bigger than $\sqrt{1997}$ which must divide 1997. Since $\sqrt{1997} < 44.7$ our proof is complete.

COMMENTS

1. The above proof is still quite wordy. The first three sentences could be replaced by 'If 1997 is *not* prime then there are integers m, n such that $1997 = mn$....'
2. We still do not know whether or not 1997 is prime. Rather we have established a watertight (and fairly quick) criterion for *determining* whether or not it is, namely: try dividing it by all primes up to its square root.

3. We have sneaked in here a couple of results without proof. Did you spot them? They were: (i) that m could be expressed as a product of primes and (ii) that, when m was assumed to divide 1997, each divisor of m must also divide 1997. We perhaps ought to prove both. In fact one is Theorem 1 of Chapter 8 and the other is Exercise 30 of Chapter 2. When you see the proof of Theorem 1 or, rather, its extension to Theorem 8 of Chapter 8 you will agree that it would be dangerous to assume (i) as being 'obvious'. The trouble here is that you have 'known' (i) for so long that you may give it no more thought than you give to whether or not the moon will still be orbiting the earth tomorrow.

4. As Comment 1 emphasizes, our formal proof of $A \vee B$ is, in essence, a proof of $\neg A \rightarrow B$ (A and B being as in the Discussion). The equivalence of $A \vee B$ and $\neg A \rightarrow B$ was noted in Example 4(b) in Chapter 2. (In that example take P to be $\neg A$ and Q to be B.)

5. Statement 3(a) is an 'exclusive or' statement: clearly there is no possibility that *both* A and B can be true simultaneously. (Surely it is also an iff statement? Surely 1997 is a prime *if and only if* 1997 is divisible by no prime less than 44? So it looks as if $A \veebar B$ ($\veebar$ denoting the 'exclusive or') is equivalent to $A \longleftrightarrow \neg B$. Can this be true? See Exercise 16 below.

EXERCISE 14 Use the idea of looking for divisors not exceeding the square root to determine (using only pencil and paper) which of the following integers are primes: 101, 103, 107, 109, 221, 289, 421, 667, 899 and (why not?) 1997.

EXERCISE 15 In 1588, Cataldi[7] proved that $2^{17} - 1$ and $2^{19} - 1$ are primes by making a list of all primes up to ... what?

EXERCISE 16 Use the truth table technique to determine whether or not the statement forms $P \veebar Q$ and $P \longleftrightarrow Q$ are equivalent.

STATEMENT 3(b) Either there are at least 150 primes less than 1000 or the sum of these primes is less than 125 000.

DISCUSSION
Do you fully understand what is claimed? If not read and reread the statement until you are completely happy that you *do* understand it – and that you can recall it without looking at the book again. Note that you are not called upon to find the number of primes less than 1000 nor their sum. Is the statement true or false? Let us first (for no particularly good reason) see if we can prove Statement 3(b) to be true.

   Since this is another 'or' statement we shall recall the method which we have just used and try the ploy '*if* there are more than 149 primes less than 1000 then Statement 3(b) will be true – and if there *aren't* then Statement 3(b) will *still* be true *if their sum* is less than 125 000.' So, can we prove, on the assumption that the first part of Statement 3(b) is *false*, that the second part must be true?

[7] Pietro Antonio Cataldi (15 April 1552–11 February 1626).

Now, rather trivially, if there are at most 149 primes less than 1000 then their sum is surely less than $149 \cdot 999$ ($= 148\,851$). Oh dear! We have not got below our target of 125 000. What can we do? It is conceivable that, as a last resort, we might have to find *all* the primes in question and, if there are fewer than 150, add them up. (In particular, Statement 3(b) is a *finite problem*, which, failing everything else, can be checked by *exhaustion* of all possible cases.[8] But I'd rather not do that. Let us *think* first.) Clearly, saying that every prime in the given range is less than 999 is rather 'wasteful' – most are much less than that. Furthermore, one might suspect that, in the range from 1 to 1000, there are more primes nearer to 1 than to 1000 (since the larger numbers at the top end of the range have more 'chance' of being divisible by an earlier integer). One might therefore suspect that the primes below 1000 can be paired (e.g. 2–997, 3–993, 5–991,...) so that each pair has a sum of less than 1000. If this could be carried out their sum would be less than 75×1000 – well below our target. Of course I'm not *certain* that the estimate I've just made is valid but the difference between it and the target of 125 000 is so great that I am hopeful that I shall be able to beat the desired target by taking into account the smallness of the earlier primes. Now it is easy to check that there are 25 primes which are less than 100. (I chose 100 since I happen to know, without having to list them, that there are 25 primes less than 100.) On each of these primes our overestimate of 999 is 'costing' us more than 900. Thus we may certainly reduce our estimate of 148 851 by (at least) $25 \cdot 900$ ($= 22\,500$). If we then note that the primes 101, 103 and 107 are also 'costing' us more than 800 each we can lower our total by a further 2400. This achieves our aim. A formal proof of Statement 3(b) may therefore take the following form.

Proof of Statement 3(b)   Either there are at least 150 primes less than 1000 or there are at most 121 in the range from 108 to 1000 (since there are 25 primes less than 100 and 101, 103 and 107 are primes). In this case the total contribution from these primes is no more than $121 \cdot 999 + 25 \cdot 100 + 3 \cdot 107$ which is less than 125 000.

COMMENTS

1. This is another $A \vee B$ assertion which is, in essence, being proved in the form $\neg A \rightarrow B$; *if* there are fewer than 150 primes less than 1000 *then* their sum is less than 125 000. Our proof of this implication is direct.
2. Even now you are probably not *certain* whether or not there are at least 150 primes less than 1000 – nor whether the sum of those primes less than 1000 really is less than 125 000. Yet you know that Statement 3(b) is correct.
3. Note that here is an 'inclusive or' – since it is conceivable that each possibility is true.
4. Because $A \vee B$ and $B \vee A$ are surely equivalent it seems that to prove $A \vee B$ true we might, instead, try to prove that $\neg B \rightarrow A$ is true. Which of the two statements should we start from: $\neg A$ or $\neg B$? Perhaps the simplest answer is to take whichever of $\neg A$ and $\neg B$ seems to give you more to work with. (See Exercises 17–19 below.)
5. Sometimes proofs may call upon other statements about whose truth you have

[8] And probably of the prover!

some doubt. In that case the validity of the statement under consideration must remain questionable.

WHAT WE HAVE LEARNED

- Because it is especially important here, we repeat: *always make sure you understand what is being sought before embarking on an attempt at a solution or proof.*
- We also remind you not to make an immediate grab for your calculator but to *think*; and to try to modify your thinking if your first attempt isn't quite good enough.
- To prove the truth of A ∨ B one possible ploy is to prove that ¬A → B is true.

EXERCISE 17 Prove Statements 3(a) and 3(b) by negating the statements *not* negated in the above proofs. (That is, prove that '¬B implies A' rather than '¬A implies B' as carried out above.)

EXERCISE 18 Let $f(x) = 2x^3 - 27x^2 + 126x - 203$ and let n be some positive integer. Show that either $f(n) \leqslant 0$ or $n \geqslant 4$. (*Hint: Don't* assume $f(n) > 0$.)

EXERCISE 19 Show that either

$$\frac{d^2y}{d^2x} + 4\frac{dy}{dx} + 3y = 0$$

$$\text{or } y \neq e^{-3x}$$

(*Hint: Don't* assume that

$$\frac{d^2y}{d^2x} + 4\frac{dy}{dx} + 3y \neq 0)$$

STATEMENT 4 If $2^{67} - 1$ is prime then $2^{67} + 1$ is divisible by 3.

DISCUSSION
Do we understand this claim? We do *not* have to *prove* that $2^{67} - 1$ is prime nor that $2^{67} + 1$ is divisible by 3. Rather we must show that the *assumption* that $2^{67} - 1$ is a prime number forces us to *conclude* that $2^{67} + 1$ is a multiple of 3. Of course, *if* $2^{67} + 1$ *is* divisible by 3 then Statement 4 will be true even if its antecendent ('$2^{67} - 1$ is prime') is false (see the IF/THEN table on page 27 in Chapter 2). Now it is not a *very* great nuisance to discover whether or not $2^{67} + 1$ is a multiple of 3. (I'm fairly confident what the answer is and I haven't even bothered to get my abacus off the shelf! Can you think how I might be so confident? What divides $2^1 + 1, 2^3 + 1, 2^5 + 1, 2^7 + 1, ...?$) Nevertheless, we don't want to work even that hard.

Recall that our task is to prove (or disprove) the implication, not to determine whether or not each piece of the implication is true. [So we] suppose [whether it is true or not!] that $2^{67} - 1$ is prime. Then certainly $2^{67} - 1$ is not divisible by 3 [why not?]. Neither will 2^{67} be divisible by 3 since *it is divisible only by powers of* 2. Consequently [under the supposition that $2^{67} - 1$ is prime] we deduce that $2^{67} + 1$ must be divisible by 3 since *every third integer is a multiple of 3.*

For a formal proof of Statement 4 we may take that part of the above Discussion from '[So we]' onwards after deleting the words in square brackets as they are a trifle redundant.

COMMENTS

1. Once again we have used a couple of results, namely those highlighted by italics, without any proof. Do you think any proof is needed? Are the italicized results 'obvious'? In proving Statement 3(a) we used, without proof, the claim that every integer can be written as a product of primes. Here, the first italicized claim is that 2^{67} cannot be factorized into a product of primes in any way other than as a product of 67 twos. Is that obvious? I hope that you don't think so, for, in Chapter 8, it will take us some effort to prove that factorization of integers into primes is *unique* – and we shall give a very similar example where the corresponding claim is *false.*

   The other italicized statement – concerning multiples of 3 – is a little easier to accept. You might think how you would convince someone who wasn't too sure.
2. Setting A to be '$2^{67} - 1$ is prime', B to be '$3 \nmid 2^{67}$' and C to be '$3 | (2^{67} + 1)$' we claim to have proved A → C by showing that (A ∧ B) → C. Is this fair? (Remember that we are assuming that we know that B is true.)

Interlude: a lecture without words

The choice, above, of the integer $2^{67} - 1$ is not accidental. It gives me the chance to tell you a story.

Following investigations by Fermat, a French minimite friar called Marin Mersenne,[9] who held scientific meetings in his rooms, claimed that, in considering primes $p \leqslant 257$, the integer $2^p - 1$ is prime iff $p = 2, 3, 5, 7, 13, 17, 19, 31, 67, 127, 257$. This amazing claim ($2^{257} - 1$ is an integer of some 78 digits) is quite close to the truth but not totally correct. In 1903, at a research meeting, Professor F. N. Cole[10] calculated, by hand, the value $2^{67} - 1$ on one half of a blackboard and wrote down the integers 193 707 721 and 761 838 257 287 on the other. He then, still in longhand, calculated the product of these integers to be ... $2^{67} - 1$. He sat down to prolonged applause having said nothing. (He later revealed that the determination of the factors of $2^{67} - 1$ had taken him '20 years of Sunday afternoons.')

We can learn a few things from this story. (i) Do not *uncritically* believe everything that 'higher authorities' (including authors of books) tell you. (ii) You may solve your problem if you persevere. (iii) Cole might have better spent his time inventing a decent computer! (Several readily available computer packages will factorize $2^{67} - 1$ in a few seconds.)

---

[9] Marin Mersenne (8 September 1588–1 September 1648).
[10] Frank Nelson Cole (20 September 1861–26 May 1926).

WHAT WE HAVE LEARNED

- In proving the truth of an implication A→B one is not concerned with the truth or falsity of A and B themselves but only with whether or not B can be deduced logically from A.
- Follow the example of Professor Cole: persistence can pay.

EXERCISE 20   Rewrite Statement 4 in contrapositive form. Prove it in that form. Is your proof simpler than that in the text?

EXERCISE 21   Earlier we saw that $A \to B$ is equivalent to $\neg A \vee B$. Use this to write Statement 4 in disjunctive form. Prove it in that form.

EXERCISE 22   On 1 January 2001 I am having a party to celebrate the start of the third millennium. Show that, *if* at least 23 people come, *then* it will be more likely than not that there will be two people present who share the same birthday in 2001. (*Hint:* The probability that the second person to arrive at my party will have a birthday different from the first is $\frac{364}{365}$. The probability that the third has a birthday different from those of the first two is $\frac{364}{365} \cdot \frac{363}{365} \cdots$). Show that the converse assertion about birthdays is false.

STATEMENT 5

(a) $1\,334\,876\,565\,539$ is divisible by 9 if and only if its digit sum $1 + 3 + 3 + 4 + 8 + 7 + 6 + 5 + 6 + 5 + 5 + 3 + 9$ is divisible by 9.
(b) $1 + \sqrt{2}$ is a root of the polynomial $37x^7 - (\frac{3}{8})^2 x^5 + 11.623x^4 + 4x - 747.1$ if and only if $1 - \sqrt{2}$ is a root (of this polynomial).
(c) The simultaneous equations

$$3\pi x + 4ey = 1$$

$$ex + \pi y = 2$$

have a unique solution if and only if $3\pi^2 - 4e^2 \neq 0$.

As you will recognize, Statements 5(a), (b) and (c), all of which happen to be true, are clearly, like Statements 1, 3(a) and 4, special cases of much more general statements. (For example, as we shall prove in Chapter 8, a statement similar to Statement 5(a) can be made for *every* integer.) Accordingly we shall not prove any of 5(a), (b) or (c) here but we shall restrict ourselves to making some comments.

COMMENTS

1. One might have thought that, in any true statement of the form A iff B, it ought to be equally easy (or equally difficult) to establish the truth or falsity of A and B. Statement 5(a) should dispel that feeling. Indeed, the whole point of Statement 5(a) is that it tells us that testing $1\,334\,876\,565\,539$ for divisibility by 9 is equivalent to the obviously easier problem of testing the sum of its digits for divisibility by

9. Statement 5(c) is in essence of this same type. On the other hand, the two component statements of 5(b) *do* appear equally trivial (or difficult) to establish. The point of Statement 5(b) is that, if you determine that $1 + \sqrt{2}$ is (is not) a root of the given equation then, *without extra work*, you will know that $1 - \sqrt{2}$ is (is not) a root of the equation. We shall see a similar result involving complex numbers in Chapter 10 (Theorem 1, page 163). But TAKE CARE! For results such as 5(b) to be true, we have to place certain restrictions on the coefficients of the polynomials concerned. (See Comment 2 following Theorem 3 in Chapter 11, page 189, for an exact statement.)

2. Of course, if you wish to confirm that Statements 5(a), (b) and (c) are true you merely have to check that, in each case, the simple statements on either side of the 'iff' are either both true or both false.

WHAT WE HAVE LEARNED

● Iff statements are valuable for several reasons. They sometimes tell you that something complicated is the case when and only when something much simpler to check is true; witness 5(a), 5(c) above or Exercises 23, 24 below. Sometimes, like Statement 5(b), iff statements tell you, 'for free' that if one thing is a fact then so is another.

EXERCISE 23   Use the fact that $4|100$ to make the following statement true. $4|19\,901\,991\,199\,219\,931\,994$ iff $4|●●$ (Replace each ● by one digit.)

EXERCISE 24   Use the fact that $8|1000$ to make the following statement true. $8|135\,792\,468\,012\,345\,678$ iff $8|●●●$ (Replace each ● by one digit.)

EXERCISE 25   Let a, b, c, d, e, f be numbers. Is it true that the system of equations

$$ax + by = e$$

$$cx + dy = f$$

has a solution iff $ad - bd \neq 0$? If it is not true, give an example which shows that it is not true.

STATEMENT 6   Suppose that 6 points in space are connected in pairs by 15 pieces of string, each piece of string being either red or blue. Then either (i) there are 3 points which are joined in pairs only by red string, *or* (ii) there are 3 points which are joined in pairs only by blue string.

DISCUSSION

Here we have to prove that the hypothesis H: 'there are 6 points ... connected by ... strings' implies R: 'There is a red "triangle"' *or* B: 'There is a blue "triangle"'. So, assuming H to be true we must prove that '(R or B)' is true. Now if R is true then the conclusion (R or B) holds no matter whether B is true or not. Thus we may as well suppose that R is false – but then we *must* prove B true. This, then, is our method. A diagram may help but one surely cannot hope to prove the result by enumerating all possible *cases*? Even a picture of 6 points and *all* pieces of string is a bit messy, so let us draw just a bit of it. Figure 3.6 shows one point, A, joined to each of the other five. Does this picture help? Recall that we may suppose that there is *no red triangle*.

Proof of Statement 6 Let A be any one of the six given points and label the remaining five points B, C, D, E, F. Now either *at least* three of the strings AB, AC, AD, AE, AF are red *or at most* two are red and so *at least* three are blue. In the first case suppose that the strings AX, AY, AZ are red (X, Y, Z being three of the points B, C, D, E, F). Then strings XY, YZ and ZX must all be blue (since we are assuming that there is no red triangle). Consequently, in this case, there does exist a blue triangle, XYZ. In the second case suppose that the three strings, AR, AS and AT are blue. Now *at least* one of the strings RS, ST and TR must be biue (why?). Consequently, in this case too, there must be a blue triangle.

COMMENTS

1. Notice the method of proof. We showed that H⟶(R or B) by showing that (H ∧ ¬R)⟶ B. Intuitively it seems reasonable that these implications are equivalent. You should confirm this assumption using truth tables. (This type of proof is also used in Theorem 6 of Chapter 8.)
2. Perhaps the most important observation concerning the above proof is that it exploits a method we are often *obliged* to use where *infinitely many* possibilities are under consideration, not just finitely many, as here. In the proof we made A the centre point of our argument. A could have been any one of the six points. Someone else may have chosen a different point. In fact our proof will cover all

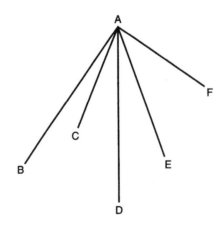

Figure 3.6

possible cases provided we take care not to attribute to A any property not shared by *all* of the six points. We do this by saying (as we have): 'Let A be (any) one of the six given points'. This idea of allowing one 'general proof' to deal with all the different individual cases in one go is logically acceptable. Its next use is in Theorem 1 of Chapter 4.

WHAT WE HAVE LEARNED   Two methods of proof:

- To prove that A → (B ∨ C) we may prove instead either that (A ∧ ¬B) → C or, of course, (A ∧ ¬C) → B.
- In proofs requiring consideration of a large (possibly infinite) number of indistinguishable cases, we may, with care, deal with such cases by proving what is required in one 'general' setting.

EXERCISE 26   Show that it is possible to join 5 points in space by (10) bits of red or blue string so that *no* triple of points is joined in pairs by only red string or only blue string. (Thus 6 is the minimum number of points required to *ensure* that either a red or a blue 'string triangle' (or both) appears.)

Summary

In what we have done above a few *strategies* have emerged.

1. First, always read the claim being made and make sure that you understand it.
2. Ask yourself if you believe the claim. Sometimes you may have no idea. In that case try some *instances* of the claim (cf. Statement 1.1).
3. Look for a *pattern*. Can you prove that the pattern persists?
4. As with any problem or proof, ask yourself if you have seen anything similar before. If so, try the same method or procedure. (A method – proof by cases – used in Statement 1 was helpful in proving Statement 1.1.) Will a picture help (cf. Statement 2)? However, make sure that you recognize its limitations (cf. Exercise 10).
5. Don't always blunder in with the calculator. Calculators don't tell you *why* the answer is what it is and they are not always as accurate as you would wish. Furthermore they only deal with 'one-off' cases.
6. Intelligent guessing may be helpful in showing you the way forward.
7. Be prepared to modify an argument which 'nearly' works (Statements 2, 3(b)) or to start afresh when one goes wrong (Statement 1.1).
8. Recognize when you have used unproven results (Statements 3(a) and 4) even if you think they are obviously true. Remember proof outlines, not details.

As a bit of extra advice: after proving a result look over your proof. Try to criticize it. Are the steps valid? Is it clearly presented?

What about *methods*? We have had several examples of **proof by cases** (Statements 1, 1.1) and of proofs which are **direct** (Statements 2, 4 and, eventually, 1.1) and

indirect, that is, involving proof by contraposition or by contradiction and involving the words 'Suppose it were (not)' (Statement 1) and, to some extent, Statement 6). We also noted how we could prove statements of the form $A \vee B$ by using the equivalence of $A \vee B$ with $\neg A \rightarrow B$ (Statements 3(a), (b)) and of $A \rightarrow (B \vee C)$ with $(A \wedge \neg B) \rightarrow C$ (Statement 6). Intuition said that these equivalences were acceptable but, if we have the slightest doubt, we can employ truth tables as a check. Statement 6 also introduced the 'general' method where all cases are dealt with in one go by considering just one sufficiently general case. (The proof of Statement 1.1 used a variant of this method), dealing with all integers by splitting them into *five* 'general' cases (corresponding to the remainders 1, 3, 5, 7 and 9).

We have admitted that Statements 1–6 are special in that most are obviously particular cases of very much more general results. (Even Statement 6 is, in essence, the simplest example in that branch of mathematics known as Ramsey Theory.) Nevertheless their inclusion had some value, in particular in regard to the role played in determining their validity by use of truth tables.

Statements 7–18 make claims about infinite collections of numbers. Consider Statement 10. Informally this says 'if $2^n - 1$ is prime then n is prime'. In this form it looks like one of our earlier IF/THEN statements – but it isn't. Indeed, neither '$2^n - 1$ is prime' nor 'n is prime' is a statement. Each is, in fact, an example of a *predicate*: something which becomes a statement (and is therefore true or false) only when some specific integer is substituted for the symbol n.[11] Consequently it looks as if we have to abandon our reliance on truth tables as a final arbiter of logical reasoning. Fortunately the logical theory underlying Statements 7–18 (the predicate calculus) tells us that we can (in essence) continue to argue as before. For example, if we wish to prove that for all integers n, if $2^n - 1$ is prime then n is prime, then we can, if we find it helpful, prove instead the contrapositive version: 'for all integers n, if n is *not* prime then $2^n - 1$ is *not* prime'.

In more symbolic form this says that 'for all integers n: $\{A(n) \rightarrow B(n)\}$' is equivalent to 'for all integers n: $\{\neg B(n) \rightarrow \neg A(n)\}$'. As another example, consider Theorem 6 of Chapter 8, which is of the form 'for all integers n: $[P(n) \rightarrow \{A(n)$ or $B(n)\}]$'. This will be proved in the form 'for all integers n: $[\{P(n) \wedge \neg A(n)\} \rightarrow B(n)]$', using the principle adopted to prove Statement 6.

---

[11] Technically the predicate is (as in English) the expression ' is prime'. This makes it even clearer that a predicate cannot be either true or false. The 'n' is merely a place-marker.

CHAPTER 4

Sets, Negations, Notations and Functions

I don't want to belong to any club that will accept me as a member.
GROUCHO MARX

From Chapter 5 onwards, nearly all statements we make will concern collections of items (usually numbers). The mathematical theory of collections is called the *theory of sets*. Here we introduce the basic notions and notations of this theory. We also introduce a symbolic notation for expressing the concepts of 'for all' and 'there exists' which, amongst other things, allows us to determine more easily the negations of complicated statements. We also show how the function concept can be defined in terms of set notation.

Introduction

Many of the more interesting mathematical statements are concerned with *collections* of entities of one kind or another rather with individual ones. For example, Statement 1 is a fairly uninteresting fact about a single integer, whereas its generalization to *No integer ending in a 2, a 3, a 7 or an 8 can be a perfect square* is much more interesting (if not exactly profound), especially when a proof of the generalization is no more difficult than that for the particular case. It is, therefore, appropriate that we spend a few pages discussing this idea of collections before continuing on our journey proper.

The introduction of sets

When dealing with collections of objects it is often convenient to think of each collection as a single, complete entity. In mathematics the overworked word 'set' has been chosen, instead of 'collection', as the name for such an entity. We therefore have the following definitions.

Definition 1

A **set** is just a collection A, say, of objects each of which is called an **element** or **member** of the set. If an object, denoted by m, is an element of A we write $m \in A$ (and say 'm belong to A'). If m is *not* an element of A we write $m \notin A$ (and say 'm does not belong to A').

In a modern mathematical setting we might want to talk about the set of all roots of a given equation, the set of all differentiable functions, the set of all prime numbers. The theory of sets entered mathematics in the late nineteenth century via Cantor's study of the representation of functions by trigonometric series, and it has so infiltrated mathematics that I have had to struggle to avoid mentioning the word 'set' in the first three chapters.

There are many problems in the theory of sets which are of a very deep nature. However, we shall use sets mainly as a convenient language in which to express some mathematical notions succinctly.

The author would like to think that amongst his readers there is at least a handful who, suffering from the obligatory bout of healthy scepticism, are mumbling to themselves, 'If a set is defined to be a collection, how is one to define "collection"?' Here we recall that we cannot keep on defining each new concept in terms of 'simpler' ones, so I shall risk disappointing some readers, and possibly infuriating others, by not defining the word 'collection'.

Some special sets

Certain sets occur so frequently that we denote them by special symbols. Thus the sets of all integers, of all rational numbers, of all real numbers and of all complex numbers are denoted, respectively, by $\mathbb{Z}$, $\mathbb{Q}$, $\mathbb{R}$ and $\mathbb{C}$.[1] We may therefore (correctly) write $-3 \in \mathbb{Z}$, $31/117 \notin \mathbb{Z}$, $39/23 \in \mathbb{Q}$, $\sqrt{2} \notin \mathbb{Q}$, $\pi \in \mathbb{R}$, $3 + 4i \notin \mathbb{R}$ and $3\pi + \sqrt{12}i \in \mathbb{C}$.[2] It also proves convenient to denote the sets of positive integers, of positive rationals and of positive reals by $\mathbb{Z}^+$, $\mathbb{Q}^+$ and $\mathbb{R}^+$ respectively.

EXERCISE 1   Let S be the set consisting of the two roots s_1, s_2 of the quadratic equation $1 + x + x^2 = 0$, and let T denote the set of the three roots (all real) of the cubic equation $2x^3 + 9x^2 + 4x - 3 = 0$. Are the following statements true or false?

(a) $s_1 \in \mathbb{R}$ or $s_2 \in \mathbb{R}$
(b) If $t \in T$ then $t \in \mathbb{R}^+$.

EXERCISE 2   Does $x = \sqrt{(17 + 4\sqrt{13})} - \sqrt{(17 - 4\sqrt{13})} \in \mathbb{Z}$? (*Hint:* First find x^2.)

---

[1] $\mathbb{Z}$ for German *Zahl*, $\mathbb{Q}$ for quotient.
[2] $\sqrt{2} \notin \mathbb{Q}$ is Statement 12. It will be proved correct in Chapter 6.

GEORG CANTOR
(3 March 1845–6 January 1918)

Georg Cantor, who was born in St Petersburg in 1845, studied at the University of Berlin where he came in to contact with the prominent mathematicians Weierstrass, Kummer and Kronecker.

Although Cantor made significant contributions to mathematical analysis, he is principally known as the founder of set theory, a theory which arose naturally out of his investigations into trigonometric series.

According to the *Dictionary of Scientific Biography*, the theory of sets was born on 7 December 1873 when Cantor proved that the members of the sets $\mathbb{R}$ and $\mathbb{Z}$ could not be paired off in a one-to-one fashion (see Theorem 4 in Chapter 9), thus demonstrating the existence of different sizes of infinity. Subsequently he introduced and developed the arithmetic of infinite numbers.

For quite some time Cantor's ideas were not well received. Many mathematicians were against accepting infinite sets as 'completed wholes'. Even worse, certain paradoxes kept on appearing. Cantor himself was suspicious of his 'proof' (now readily accepted) that the points in the real line could be placed in one-to-one correspondence with the points in the plane as this seemed to undermine the intuitively clear idea of 'dimension'. Furthermore, Cantor had shown that for no set S is it possible to match the elements of S in a one-to-one manner with the collection of S's subsets. Thus, if X denotes the (harmless?) collection of all sets then the collection of X's subsets must contain some sets not already in X.

Perhaps in part because of the strong criticism that his work received, especially at the hands of Kronecker, the 'father of set theory' suffered frequently from severe depression.

He died in the university at Halle where he had taught since 1869 never hearing

> the words of Hilbert in 1925, 'No-one will expel us from this paradise which Cantor has created for us.'

Set notation: the empty set

If a set has only a small (finite) number of members we may fully describe it by l isting all its elements between curly brackets. For example, $\{101, 103, 107, 109\}$ is the set of all primes between 100 and 110. If we wish to give this set a name, B say, then we write $B = \{101, 103, 107, 109\}$. (We may also write $B = \{107, 101, 109, 103\}$ or even $B = \{107, 103, 103, 109, 101, 103, 101, 107\}$ if, as is usual, we agree to take no notice of the *order* in which the elements are written down or of *repetitions* of members of the set.)

Where it is not easy to list all the elements of a **finite set** (that is, a set with a finite number of members) we can often describe the set by indicating the *property* shared by those *and only those* members of the set. For example, $\{x: x \in \mathbb{Z} \text{ AND } 1 < x < 9999 \text{ AND } x \text{ is prime}\}$ is 'the set of (all those) x such that x is an integer AND $1 < x < 9999$ AND x is prime.'[3] This is certainly a finite set – with 1229 members. An alternative notation, somewhat more 'user-friendly' is $\{x \in \mathbb{Z}: 1 < x < 9999 \text{ AND } x \text{ is prime}\}$ which would be read as 'the set of integers such that $1 < x < 9999$ AND x is prime'. Of course a more natural description of this set is 'the set of all primes between 1 and 9999'.

We may readily generalize this idea. We let $\{x: P(x)\}$ denote 'the set of (precisely) those x which satisfy the property P'. For example, if, for $P(x)$, we take '$x \in \mathbb{Z}$ AND $(0 < x < 2)$' we obtain the set $\{1\}$ whose only element is the integer 1.[4] Alternatively, we may write this set as $\{x \in \mathbb{Z}: Q(x)\}$ where $Q(x)$ is now $0 < x < 2$.

Identifying a set by a common property of its elements is especially useful when the set has infinitely many members. We shall prove later that the set $\{x \in \mathbb{Z}: x \text{ is a prime number}\}$ is an *infinite* set. (Curiously, no-one knows whether the set $\{x \in \mathbb{Z}: x \text{ and } x + 2 \text{ are } both \text{ primes}\}$ – the set of *twin primes* – is infinite or finite.)

A third notation which is very convenient is often used. For example, instead of $\{x: x \in \mathbb{Z} \text{ and there exists } n \in \mathbb{Z} \text{ such that } x = n^2\}$ we could write $\{n^2: n \in \mathbb{Z}\}$. This is the set of all squares of integers.

A fourth notation sometimes used describes a set by listing several elements and a 'general' element. Thus $\{2, 4, 6, \ldots, 2n, \ldots\}$ indicates the set of all positive even integers.

At the opposite extreme to infinite sets let us consider the sets $\{x: P(x)\}$ in the following cases:

1. $P(x)$ is $x \in \mathbb{R}$ AND $x^2 = -1$
2. $P(x)$ is x is a live human being AND x is 1000 years old

In each case the set $\{x: P(x)\}$ has no member. Such a set is said to be *empty*. The **empty set** is denoted by the symbol $\emptyset$. Note that neither $\{0\}$ nor $\{\emptyset\}$ is the empty set: each of these sets has exactly one element. In particular, $0 \in \{0\}$ and $\emptyset \in \{\emptyset\}$.

---

[3] We use the colon ':' to mean 'such that'. So authors use '|', instead.

[4] Note that x here is a 'dummy variable': $\{x: P(x)\}$ can equally well be written $\{y: P(y)\}$ or, even, $\{\blacksquare: P(\blacksquare)\}$. As in Chapter 3, $P(\blacksquare)$ is called a 'predicate'. $P(\blacksquare)$ becomes a statement when an appropriate object replaces the 'variable' $\blacksquare$.

EXERCISE 3 Write down the elements of the following sets:

(a) $\{1, 2, 3, \{4, 5\}, \{6, \{7, 8\}\}\}$
(b) $\{x \in \mathbb{Z}^+ : x^2 = 3 \text{ or } x^2 = 4\}$
(c) $\{n + (1/n - 6) : n \in \mathbb{Z} \text{ and } -1 \leqslant n \leqslant 5\}$
(d) $\{x \in \mathbb{Z}^+ : x^2 = 3 \text{ and } x^2 = 4\}$

EXERCISE 4 Which set has the most elements $\{x : x \in \mathbb{Z} : 0 < x^3 < 12\,345\}$, $\{x : x \in \mathbb{Z} \text{ AND } 1 < x < 100 \text{ AND } x \text{ is prime}\}$ or $\{x^3 : x \in \mathbb{Z} \text{ AND } -12 < x \leqslant 12\}$?

EXERCISE 5 Find three $P(x)$, different from that above, for which $\{x : P(x)\}$ is the set $\{1\}$ and three different $Q(x)$ such that $\{x \in \mathbb{Z} : Q(x)\} = \varnothing$.

EXERCISE 6 Find suitable predicates $P(x)$, $Q(x)$ such that $\{x : P(x)\}$ has exactly 41 members all being odd integers greater than 100 whilst $\{x : Q(x)\}$ has infinitely many members, none of which is a number of any kind.

If we take our discussion a little further we shall see that there is an intimate connection between the theory of sets and the sentential connectives we introduced in Chapter 2. We begin with the fundamental concept of equality of two sets.

Equality and containment of sets: subsets

Two sets, say A and B, are said to be **equal** when they have precisely the same elements. We indicate this, rather unsurprisingly, by writing A = B. As an example, the sets U = $\{113, 127\}$ and V = $\{x \in \mathbb{Z} : x \text{ is prime AND } 111 \leqslant x \leqslant 129\}$ are equal, so we may write U = V. Likewise $\{1, 2, 3\} = \{1, 3, 2, 3\}$.

When every element of one set A belongs to another set B we say that A is a **subset** of B (also that A is *contained in* B[5]). We then write A $\subseteq$ B. As a consequence, one way to show that two sets, for example U and V, are equal, is to show that U $\subseteq$ V and that V $\subseteq$ U. (This method is used in Theorem 1 below.)

If A $\subseteq$ B and there are elements in B which are *not* in A we sometimes write A $\subset$ B if we wish to draw attention to this fact. If A $\subset$ B then we refer to A as a **proper subset** of B. Thus the use of $\subset$ and $\subseteq$ in relating sets corresponds to the use of $<$ and $\leqslant$ in relating real numbers.

Note that the empty set $\varnothing$ is a subset of *every* set A, for if $\varnothing$ were not contained in A, there would have to be an element x, say, such that $x \in \varnothing$ but $x \notin A$. This is impossible since *there are no elements in* $\varnothing$. (Notice the indirect proof.)

If A is *not* a subset of B we write A $\nsubseteq$ B or A $\not\subset$ B, as appropriate.

Example 1

(a) $\{x \in \mathbb{R} : x^2 = 1\} = \{-1, 1\}; \quad \{x \in \mathbb{R} : x^2 = -1\} = \varnothing.$
(b) $\mathbb{Z} \subset \mathbb{Q}$ and $\mathbb{Q} \subset \mathbb{R}$ and $\mathbb{R} \subset \mathbb{C}$; more briefly, $\mathbb{Z} \subset \mathbb{Q} \subset \mathbb{R} \subset \mathbb{C}$.

---

[5] 'Contained in' involves a relationship between two sets; 'belongs to' involves the relationship between an element and a set. It is *very* easy to get these confused – try not to (by doing all of Exercise 11).

(c) If $A = \{\alpha, \beta, \gamma\}$ then its subsets are $\{\alpha\}$, $\{\beta\}$, $\{\gamma\}$, $\{\alpha, \beta\}$, $\{\beta, \gamma\}$, $\{\gamma, \alpha\}$, $\{\alpha, \beta, \gamma\}$ and $\emptyset$.

(d) If $A = \{a, \{b\}, \{c, d, e\}\}$ then A has three elements a, $\{b\}$, $\{c, d, e\}$.

Notice that none of b, c, d, e is an *element* of A. Amongst A's eight subsets are $S = \{a, \{b\}\}$, $T = \{a, \{c, d, e\}\}$ and $A = \{a, \{b\}, \{c, d, e\}\}$ itself. We thus have $a \in S$, $a \in T$, $a \in A$, $b \notin S$, $\{b\} \notin S$, $\{b\} \notin T$, $\{b\} \in A$, $\{c, d, e\} \in A$, $e \notin A$, $\emptyset \notin A$, $a \not\subset A$, $\{a\} \subset A$, $\{b\} \notin S$, $\{\{b\}\} \subset S$, $S \subset A$, $T \subset A$, $S \not\subseteq T$, $A \subseteq A$, $\{c, d, e\} \notin A$, $\{\{c, d, e\}\} \subset A$, $S \notin T$, $\emptyset \subset S$, $\emptyset \subset T$, $\emptyset \subset A$.

EXERCISE 7 Let $A = \{1/n: n \in \mathbb{Z}^+\}$. An examination answer concluded, 'Therefore $0 < A \in \mathbb{R}$'. Criticize this conclusion (on several counts). What do you think the writer meant?

EXERCISE 8 Let $A = \{x \in \mathbb{Z}: x^3 - x = 0\}$, $B = \{x \in \mathbb{Z}: x^4 = 1\}$, $C = \{-1, 1\}$, $D = \{-1, 0, 1\}$. Which pairs of A, B, C, D (if any) are equal?

EXERCISE 9 Is the set $\{x \in \mathbb{R}: x^2 + x + 1 = 0\}$ equal to any of 0, $\emptyset$, $\{0\}$, $\{\emptyset\}$?

EXERCISE 10

(a) List all the subsets of (i) $\{1, 2\}$; (ii) $\{1, 2, 3\}$; (iii) $\{1, 2, 3, 4\}$.

(b) Make a conjecture about the number of subsets of $\{1, 2, ..., n\}$.

EXERCISE 11

(a) List the subsets of $S = \{a, b, \{c, d\}, 47\}$

(b) Is $c \in S$? $\{c, d\} \in S$? $\emptyset \in S$? $S \in S$? Is $\{c, d\} \subset S$? $\{\{c, d\}\} \subset S$? $\{b, 47\} \subset S$? $\{c, d, 47\} \subset S$? $S \subseteq S$?

EXERCISE 12

(a) Show that no two of the sets $\emptyset$, $\{\emptyset\}$, $\{\{\emptyset\}\}$ are equal.

(b) Do the same for the sets $\emptyset$, $\{\emptyset\}$, $\{\emptyset, \{\emptyset\}\}$.

EXERCISE 13 One of my students claimed that $\emptyset \subset \{\emptyset\} \subset \{\{\emptyset\}\}$. Was he right?

We now do a little more set theory in order to show a connection with the sentential connectives introduced in Chapter 2.

Set operations: intersection and union

The set of elements common to two sets A and B is called the **intersection** of A and B: in symbols $A \cap B$ $(= \{x: (x \in A) \text{ AND } (x \in B)\})$. $A \cap B$ is a subset of both A and B. The totality of elements which are in A or in B (or in both) is called the **union** of A and B: in symbols $A \cup B$ $(= \{x: (x \in A) \text{ OR } (x \in B)\})$. A and B are subsets of $A \cup B$. If[6]

---

[6] That the elements of these four sets are listed in the order shown is, of course, pure chance!

$A = \{s, t, r, a, b, d, y, l\}$ and $B = \{b, l, i, n, d, a, g, e\}$ then $A \cap B = \{b, a, l, d\}$ whilst $A \cup B = \{s, t, a, l, y, b, r, i, d, g, e\}.$[7]

EXERCISE 14

(a) Describe, in the form $\{\cdots: k \in \mathbb{Z}\}$, the set $\{4m + 1: m \in \mathbb{Z}\} \cap \{5n + 2: n \in \mathbb{Z}\}$.
(b) Find an *infinite arithmetic progression* $a, a + d, a + 2d, a + 3d, \ldots$ which has no elements in $\{2m: m \in \mathbb{Z}\} \cup \{3n: n \in \mathbb{Z}\}$.

EXERCISE 15   Find, if possible:

(a) Infinite sets A and B such that $A \cap B = \{1\}$ whilst $A \cup B = \mathbb{Z}$.
(b) Sets C and D such that $C \cup D = \{t, h, i, c, k\}$ whilst $C \cap D = \{t, h, i, n\}$.

Usefulness of Venn diagrams

Given three sets A, B and C what would you say is the relationship, if any, between the sets $A \cup (B \cap C)$ and $(A \cup B) \cap (A \cup C)$? Questions like this can be 'answered' (always remembering the danger of 'proving' things by pictures) by means of **Venn diagrams**.[8]

If $A = \{2m: m \in \mathbb{Z}\}$ and $B = \{3n: n \in \mathbb{Z}\}$ we can picture $A \cap B$ and $A \cup B$ as the shaded regions in the Figures 4.1 and 4.2. Here the surrounding box, denoted by $\mathscr{U}$, represents a convenient *universe of discourse* – in this example we might take $\mathscr{U}$ to be $\mathbb{R}$ or maybe $\mathbb{Q}$ or perhaps $\mathbb{Z}$.

The shaded region in Figure 4.3 represents all those objects (in $\mathscr{U}$) which are NOT in A. We denote this **complement of A (with respect to the universe $\mathscr{U}$)** by A^c. Figure 4.4 appears to indicate that IF $(x$ is in$)$ A THEN $(x$ is in$)$ B has exactly the same meaning as IF $(y$ is$)$ NOT (in) B THEN $(y$ is$)$ NOT (in) A thus mirroring the fact that the implications $A \rightarrow B$ and $\neg B \rightarrow \neg A$ are equivalent.

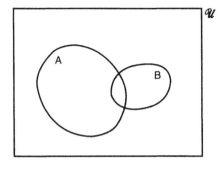

Figure 4.1 $A \cap B$

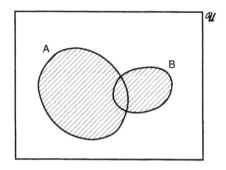

Figure 4.2 $A \cup B$

[7] Stalybridge, the name of a town in England, uses 11 distinct letters. Write to tell me of any longer place name using no repeated letter!
[8] John Venn (4 August 1834–4 April 1923).

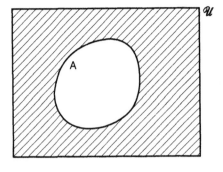

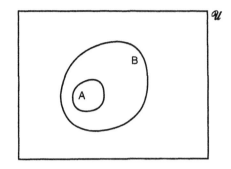

Figure 4.3 A^c **Figure 4.4** $A \subseteq B \rightarrow B^c \subseteq A^c$

With three sets A, B and C things look a bit more complicated. Nevertheless we can see from Figures 4.5 and 4.6 that the equality of $A \cup (B \cap C)$ and $(A \cup B) \cap (A \cup C)$ looks likely to hold.

Let us prove the above equality conclusively for *every* triple of sets (and not just three whose intersections resemble those in the pictures).

Theorem 1   Let A, B and C be three sets. Then $A \cup (B \cap C) = (A \cup B) \cap (A \cup C)$.

DISCUSSION
Recall that to prove $U = V$ we may try to prove $U \subseteq V$ and $V \subseteq U$. So, here, we attempt to show that *each* element of $A \cup (B \cap C)$ belongs to $(A \cup B) \cap (A \cup C)$ and, conversely, that *each* element of $(A \cup B) \cap (A \cup C)$ belongs to $A \cup (B \cap C)$. Because there may be infinitely many elements in these sets, it seems reasonable to proceed by arguing about a 'typical' element x, say, in $A \cup (B \cap C)$ (typical in the sense that *we assume nothing of x other than it belongs to $A \cup (B \cap C)$*). We then show that this typical, but unspecified, element belongs to $(A \cup B) \cap (A \cup C)$). This procedure demonstrates that *every* element of $A \cup (B \cap C)$ belongs to $(A \cup B) \cap (A \cup C)$, in other words that $A \cup (B \cap C) \subseteq (A \cup B) \cap (A \cup C)$. That this method of argument is valid can be supported by logical analysis of the situation.

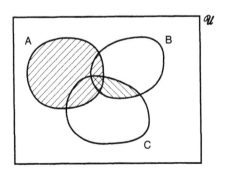

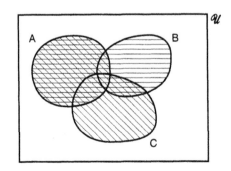

Figure 4.5 $A \cup (B \cap C)$ **Figure 4.6** $(A \cup B) \cap (A \cup C)$

When employing this 'typical element' method it is usual to begin with the words 'Let x'; so we write out the formal proof as follows.

Proof of Theorem 1   Let $x \in A \cup (B \cap C)$. Then (i) $x \in A$ or (ii) $x \in B \cap C$. If (i) is the case then $x \in A \cup B$ AND $x \in A \cup C$ (since each of these sets contains A). If (ii) is the case then $x \in B$ (so that, consequently, $x \in A \cup B$) AND $x \in C$ (so that, consequently $x \in A \cup C$). Thus, in each of the two cases, $x \in (A \cup B) \cap (A \cup C)$. This proves that

$$A \cup (B \cap C) \subseteq (A \cup B) \cap (A \cup C) \qquad \text{(I)}$$

We now have to prove the inequality the other way round. One can almost feel that our argument ought to go just as easily backwards. Here it is, much more briefly.

Let $x \in (A \cup B) \cap (A \cup C)$. Then $x \in (A \cup B)$ AND $x \in (A \cup C)$.
If $x \in A$ then certainly $x \in A \cup (B \cap C)$.
If $x \notin A$ then $x \in B$ (since $x \in A \cup B$) AND $x \in C$ (since $x \in A \cup C$).
Thus, if $x \notin A$ then $x \in B$ AND $x \in C$, i.e. $x \in B \cap C$. Thus, whether $x \in A$ or not, $x \in A \cup (B \cap C)$. We have therefore shown that

$$(A \cup B) \cap (A \cup C) \subseteq A \cup (B \cap C) \qquad \text{(II)}$$

The inequalities (I) and (II) establish the *equality*

$$A \cup (B \cap C) = (A \cup B) \cap (A \cup C)$$

COMMENT   Alternatives to beginning 'Let $x \in A \cup (B \cap C)$' include 'Suppose $x \in A \cup (B \cap C)$', 'Take an arbitrary $x \in A \cup (B \cap C)$'.

WHAT WE HAVE LEARNED

● To prove two sets equal we may show that every element in each set belongs to the other. For each of these steps, especially for infinite sets, we argue about a 'typical' element from one set and prove it lies in the other. When proving X = Y by showing X ⊆ Y and Y ⊆ X, the proof of the second inequality is often similar to that of the first – in reverse.

CHALLENGE   Close the book and try to write out the above proof in your own words. (Do *not* aim to be word perfect.) If you fail, reread the whole proof and try again.

EXERCISE 16   Use a Venn diagram to decide if it is always true that $A \cap (B \cup C) = (A \cap B) \cup C$. Then either prove it correct (as in Theorem 1) or give three specific sets showing the equality failing.

EXERCISE 17   Let A, B and C be three sets. Prove that $(A \cap B) \cap C = A \cap (B \cap C)$. (It is likewise true, for all sets A, B and C, that $(A \cup B) \cup C = A \cup (B \cup C)$.)

EXERCISE 18   Use a Venn diagram to illustrate the following

(a) $A \subseteq B \cup C$
(b) (i) $(A \cup B)^c = A^c \cap B^c$
     (ii) $(A \cap B)^c = A^c \cup B^c$

The equalities (i) and (ii) are called **De Morgan's laws.**[9] Explain the connection between (i) and the statement form $\neg(A \vee B) \longleftrightarrow (\neg A) \wedge (\neg B)$ and between (ii) and the statement form $\neg(A \wedge B) \longleftrightarrow (\neg A) \vee (\neg B)$. Because of this we shall now use $\wedge$ and $\vee$ instead of AND and OR in describing sets.)

EXERCISE 19   Prove De Morgan's laws in the manner of Theorem 1.

EXERCISE 20   Use Figure 4.7 to help you decide whether or not the equality

$$(A \cup B) \cap (C \cup D) = (A \cap C) \cup (B \cap D)$$

holds for every four sets A, B, C, D. Prove that it does – or give a counterexample.

EXERCISE 21   Let A, B and C be sets. Does it necessarily follow that $A = B = C$

(a) if $A \cap B \subseteq C$, $B \cap C \subseteq A$ and $C \cap A \subseteq B$
(b) if $A \cup B \subseteq C$, $B \cup C \subseteq A$ and $C \cup A \subseteq B$?

EXERCISE 22   For sets A, B and C in a universe $\mathcal{U}$ is it necessarily true that $(A \cup B \cup C)^c = A^c \cap B^c \cap C^c$?

Venn diagrams are useful in suggesting an important result known as the *inclusion–exclusion principle* which we meet later in the chapter.

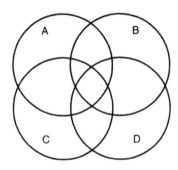

Figure 4.7

[9] Augustus de Morgan (June 1806–18 March 1871).

There exists ... some notation ... for all

We have seen examples of mathematical statements which involve the words 'for each' and 'there exist(s)'. Some statements involve several occurrences of these expressions. Consider, for example, the true statement 'For each positive real number h there exists a positive integer N_h (depending on h) such that, for each integer $n > N_h$

$$\frac{\pi^2}{6} - h < 1 + \frac{1}{2^2} + \frac{1}{3^2} + \dots + \frac{1}{n^2} < \frac{\pi^2}{6},$$

Informally this says that, provided you take n to be large enough, the sum of the reciprocals of the squares of the first n positive integers can be made as near to $\pi^2/6$ as you wish (whilst remaining less than $\pi^2/6$).

In order more easily to understand statements and, in particular, their negations, it is useful to introduce the abbreviations $\forall x$ for 'for all x (it is the case that)' and $\exists x$ for 'there exists x (such that)'. $\forall$ and $\exists$ are called, respectively, the **universal quantifier** and the **existential quantifier**.

Example 2
The statement 'for each pair of positive real numbers r and s there exists a real number t such that $rs = t^2$' may be written

$$(\forall r)(\forall s)(\exists t)[rs = t^2] \tag{III}$$

which we read as 'For each r (and) for each s there exists t such that r times s equals t squared.'[10]

In fact we may need to take the symbolism a short step further to indicate, in some manner, that r and s are positive real numbers and that t is a real number, since, for example, (III) is *false* if we let r, s and t range over *all* real numbers and also if we restrict r, s and t to being integers. Thus it is better to write (III) as:

$$(\forall r \in \mathbb{R}^+)(\forall s \in \mathbb{R}^+)(\exists t \in \mathbb{R})[rs = t^2]$$

although, in practice, the context will usually determine exactly to which sets r, s, t, etc. belong.

Example 3
In symbols, the (true) assertion 'For each pair x and y of integers we have $x + y = y + x$' may be written

$$(\forall x \in \mathbb{Z})(\forall y \in \mathbb{Z})[x + y = y + x]$$

Example 4
In symbols, the (true) assertion 'For all (pairs of) integers x and y for which $x < y$ there exists a positive integer z such that $x + z = y$' may be written

$$(\forall x \in \mathbb{Z})(\forall y \in \mathbb{Z})(\exists z \in \mathbb{Z}^+)[\{x < y\} \rightarrow \{x + z = y\}]$$

---

[10] The statement of Example 2 may also begin (somewhat closer to its symbolic form) 'for all (pairs of) positive real numbers r and s'. I just feel that 'for each' is a bit more emphatic than 'for all'. Some will disagree.

Example 5

The assertion '1999 is prime' may be written symbolically as

$$(\forall x \in \mathbb{Z}^+)(\forall y \in \mathbb{Z}^+)[\{1999 = xy\} \rightarrow \{(x = 1) \vee (y = 1)\}]$$

EXERCISE 23   Write in words the statements:

(a) $(\forall x \in \mathbb{R})[x^2 - 1 = (x - 1)(x + 1)]$
(b) $(\exists x \in \mathbb{R})[x^2 - 5x + 6 = 0]$

(Leave the symbols within [ ] unchanged.)

EXERCISE 24   Write in symbols the (true) assertion 'For each integer x there exists an integer y such that $x < y$'.

EXERCISE 25   Write in words

$$(\forall m \in \mathbb{Z}^+)(\forall n \in \mathbb{Z}^+)\left[\frac{m}{n}^2 \neq 2\right]$$

EXERCISE 26   Write in words

$$(\exists y \in \mathbb{Z})(\forall x \in \mathbb{Z})[x < y]$$

Is this assertion true? Compare the symbolism with that of Exercise 24. What do you conclude?

EXERCISE 27   Write in symbols using $\forall$, $\exists$, $\in$, $\rightarrow$, etc., (a) $A \subseteq B$, (b) $A \subset B$.

EXERCISE 28   Find a predicate $P(x, y)$ such that $(\forall x \in \mathbb{Z})(\exists y \in \mathbb{Z})[P(x, y)]$ and $(\exists y \in \mathbb{Z})(\forall x \in \mathbb{Z})[P(x, y)]$ *do* mean the same thing.

EXERCISE 29   What result (stated easily in words) is given in symbolic form by

$$(\forall x \in \mathbb{Z})[(\exists m \in \mathbb{Z})[x = 2m + 1] \rightarrow (\exists n \in \mathbb{Z})[x^2 = 2n + 1]]]$$

(Use as few words as you can to translate this, but don't just translate it verbatim.)

EXERCISE 30   Which of the following are true?

(a) $(\forall x \in \mathbb{R})(\exists y \in \mathbb{R})[x + y = 0]$
(b) $(\exists y \in \mathbb{R})(\forall x \in \mathbb{R})[x + y = 0]$
(c) $(\forall x \in \mathbb{R})(\exists y \in \mathbb{R})[x + y = 0] \rightarrow (\exists y \in \mathbb{R})(\forall x \in \mathbb{R})[x + y = 0]$

EXERCISE 31   Write in symbols 'No integer of the form $4k + 3$ is a sum of two integer squares'.

EXERCISE 32   Write in words

$$(\forall r \in \mathbb{R}^+)(\exists N \in \mathbb{Z}^+)(\forall n \in \mathbb{Z}^+)\left[(n > N) \rightarrow \left(\frac{1}{n} < r\right)\right]$$

Is the assertion true?

EXERCISE 33 Let $\mathbb{Z}_2$ temporarily denote the set $\{x \in \mathbb{Z}: x > 1\}$. Write in words, as informally as possible

$$(\exists x \in \mathbb{Z})[\{x > 2^{1\,257\,787} - 1\} \wedge \{(\forall y \in \mathbb{Z}_2)(\forall z \in \mathbb{Z}_2)[x \neq yz]\}]$$

EXERCISE 34 What does the following say? Is it true?

$$(\forall x \in \mathbb{Z}^+)(\forall y \in \mathbb{Z}^+)\left[\left\{(x \geqslant 3 \wedge 1 < y < x) \to \frac{x}{y} \notin \mathbb{Z}\right\} \to x \text{ is prime}\right]$$

(*Hint:* $x = 6, y = 4$.)

EXERCISE 35 What does the following say? Is it true?

$$(\forall x \in \mathbb{Z}^+)\left[\left\{x \geqslant 3 \wedge (\forall y \in \mathbb{Z}^+)\right\}\left[1 < y < x \to \frac{x}{y} \notin \mathbb{Z}\right]\right\} \to x \text{ is prime}\right]$$

Our main reason for introducing this new notation concerns *negation*.

Negation

In many areas of mathematics, in particular in mathematical analysis, it is important to deal with the negations of statements involving quantifiers. Once we have sorted out the simplest cases, the more involved ones cause no extra problems (especially when dealt with symbolically).

Not for all ...

Consider the statement 'For each positive integer n the integer $2^{2^n} + 1$ is prime.' This may be written partially symbolically as

$$(\forall n \in \mathbb{Z}^+)[2^{2^n} + 1 \text{ is prime}]$$

Fermat knew that $2^{2^n} + 1$ *is* prime for $n = 1, 2, 3, 4$ (and $n = 0$) but was unable to decide if $2^{2^5} + 1$ ($= 4\,294\,967\,297$) is prime. Curiously enough, using a method invented by Fermat, Euler, in 1732 was able to discover that $2^{2^5} + 1$ is divisible by 641. As a consequence, it is *not* true that $(\forall n \in \mathbb{Z}^+)[2^{2^n} + 1$ is prime] because there exists at least one n, namely $n = 5$ for which $2^{2^n} + 1$ is *not* prime. In symbols

$$(\exists n \in \mathbb{Z}^+)[2^{2^n} + 1 \text{ is not prime}]$$

(How did Euler think of trying 641 as a factor? We leave you in a state of suspense until Chapter 8.)

More generally, for each set S and each predicate $P(x)$, we get the negation of $(\forall x \in S)[(P(x)]$ by changing $\forall$ to $\exists$ and negating the condition $P(x)$. That is, the negation of $(\forall x \in S)[P(x)]$ is

$$(\exists x \in S)[\neg P(x)]$$

Furthermore, just as truth tables tell us is the case for statements, so we have, for predicates:

1. If $P(x)$ is the disjunction $A(x)$ OR $B(x)$ then $\neg P(x)$ is $\neg A(x)$ AND $\neg B(x)$.
2. If $P(x)$ is the conjunction $A(x)$ AND $B(x)$ then $\neg P(x)$ is $\neg A(x)$ OR $\neg B(x)$.

Example 6

The negation of the *false* statement

$$(\forall k \in \mathbb{Z}^+) [6k - 1 \text{ is prime OR } 6k + 1 \text{ is prime}]$$

is the *true* statement

$$(\exists k \in \mathbb{Z}^+) [6k - 1 \text{ is composite AND } 6k + 1 \text{ is composite}^{11}]$$

For example, $k = 20$ shows this.

Note that the negation of $(\forall x \in S) [A(x) \rightarrow B(x)]$ is

$$(\exists x \in S) [A(x) \nrightarrow B(x)]$$

EXERCISE 36   Write the negations of the following statements in $\exists$ form. (You are not asked to check their truth or falsity.)

(a) $(\forall x \in \mathbb{Z}) [x^2 - 10x + 26 > 0 \wedge x^4 - x + 2 \text{ is even}]$
(b) $(\forall x \in \mathbb{Z}^+) [\text{that last digit of } x \text{ is a 3} \vee \text{the last digit of } x^5 \text{ is not a 3}]$
(c) $(\forall x \in \mathbb{Z}^+) [x \text{ is prime} \rightarrow 4^x - 3 \text{ is prime}]$

EXERCISE 37   Convince yourself by means of a specific example that the negation of $(\forall x \in S) [P(x)]$ is *not* $(\forall x \notin S) [P(x)]$.

Not there exists ...

The negation of 'The polynomial $x^4 - 8x^3 + 22x^2 - 24x + 40$ has a (real) root' – in symbols

$$(\exists r \in \mathbb{R}) [r^4 - 8r^3 + 22r^2 - 24r + 40 = 0]$$

is 'The polynomial $x^4 - 8x^3 + 22x^2 - 24x + 40$ has *no* real root'. In other words, each real number is *not* a root of $x^4 - 8x^3 + 22x^2 - 24x + 40$. In symbols

$$(\forall r \in \mathbb{R}) [r^4 - 8r^3 + 22r^2 - 24r + 40 \neq 0]$$

Note, again, how the condition $r^4 - 8r^3 + 22r^2 - 24r + 40 = 0$ has changed to its negative $r^4 - 8r^3 + 22r^2 - 24r + 40 \neq 0$ and how, here, $\exists x$ has changed to $\forall x$.

Example 7

(See Statement 7.1 in Chapter 5.) The negation of the statement: 'There exists an integer x greater than 1 such that $x^4 + 4$ is prime', in symbols

$$(\exists x \in \mathbb{Z}^+) [\{x > 1\} \wedge \{x^4 + 4 \text{ is prime}\}]$$

is the statement $(\forall x \in \mathbb{Z}^+) [x = 1 \vee x^4 + 4 \text{ is composite}]$.

---

[11] An integer n is composite iff $n = r \cdot s$ where r, s are integers each greater than 1.

EXERCISE 38   Write the following in words and its negation in symbols and in words:

$$(\exists x \in \mathbb{Z}^+)[\{x \text{ is prime}\} \wedge \{x^2 + 200 \text{ is prime}\}]$$

(You might like to ponder its truth or falsity.)

Negating multiple quantifiers

The same principle holds for obtaining the negatives of statements when more than one quantifier is involved. For example, the negation of $(\forall x)(\exists y)(\forall z)[P(x, y, z)]$ is seen, successively, to be equivalent to each of

1. $(\exists x)\{\neg(\exists y)(\forall z)[P(x, y, z)]\}$
2. $(\exists x)(\forall y)\{\neg(\forall z)[P(x, y, z)]\}$
3. $(\exists x)(\forall y)(\exists z)[\neg P(x, y, z)]$

That is, we change all the quantifiers and negate the predicate $P(x, y, z)$.

Example 8
The negation of the (true) statement

$$(\forall y \in \mathbb{Z})(\exists x \in \mathbb{Z})[x \geqslant y]$$

is the statement

$$(\exists y \in \mathbb{Z})(\forall x \in \mathbb{Z})[x < y]$$

which you should have found (above) to be false.

Example 9
$(\forall x \in \mathbb{R})(\exists y \in \mathbb{R})[y^2 = x]$ is false. (Why? Give a counterexample!) Its negation is $(\exists x \in \mathbb{R})(\forall y \in \mathbb{R})[y^2 \neq x]$.

EXERCISE 39   Show by means of a specific example that the assertion

$$(\forall x \in \mathbb{Z})(\forall y \in \mathbb{Z})(\exists z \in \mathbb{Z})[x^2 + y^2 = z^2]$$

is false. Write down the symbolic form of its negation (using $\forall$ and $\exists$ but not $\neg$).

Functions

Introduction

Just as it is difficult to avoid using the word 'set', so it is equally difficult to do much mathematics without the concept of **function**. Our attitude in this book is to treat this concept informally. Nevertheless, it is of some interest to see how functions can be defined formally in terms of sets. We do this at the end of the section.

You will all be familiar with examples of functions from $\mathbb{R}$ to $\mathbb{R}$. Most of these will have been given by a formula, e.g. $y = 4x^2 + 3$, $y = \sin 2x + 3e^x$, which may be plotted on a graph. In your early schooldays you were rather more sophisticated if you considered functions which associated with each member of the class that person's height in centimetres, etc. This latter example better epitomizes the present-day informal definition of function which is given here.

Definition 2

Let A and B be non-empty sets. A **function** f (also called a *map*, a *mapping* **from A to B**) is any rule which associates with *each* element of A some *unique* element of B. (There are problems with this 'definition': (i) what is a rule; (ii) should different rules yield different functions? The definition we give in terms of sets gets round these problems.)

Notation

If f is a function from A to B we write $f: A \to B$. If $a \in A$, the (unique) element of B to which a is sent by f is denoted by $f(a)$ which we read as 'f of a'. It is called the **value of f at a** or the **image of a under f**. *Note that the function is denoted by f whereas $f(x)$ denotes the value of f at x.*

We give some specific examples in a moment. First it is important to notice the lopsided (unsymmetrical) nature of the definition. Figures 4.8 and 4.9 illustrate what a function can and cannot do.

In particular, from Figure 4.8 we see that:

1. f sends *every* element of A to *some* element in B.
2. *Many* (even infinitely many) elements of A may be 'sent' to the same element of B.
3. *Not* every element of B need be 'visited'.
4. Figure 4.9 tells us that: *No* element of A can 'visit' more than one element of B. (In particular, we do not allow as a function that rule which associates with each positive real number both its square roots.)

Example 10

(a) Each function $f: \mathbb{Z}^+ \to \mathbb{R}$ is called a **sequence**. Thus the function $f: \mathbb{Z}^+ \to \mathbb{R}$ given by $f(n) = 1 + (1/\sqrt{n})$ is an example of a sequence.
(b) The function given, for each $n \in \mathbb{Z}^+$, by $g(n) = \{p: p$ is prime AND $p|n\}$ is a

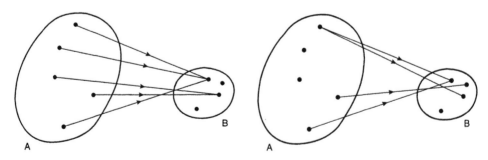

Figure 4.8 What a function *can* look like   **Figure 4.9** What a function *can't* look like

function from $\mathbb{Z}^+$ to the set of all subsets of $\mathbb{Z}^+$. (Note that $g(1) = \emptyset$, $g(60) = \{2, 3, 5\}$, $g\{11\,213\} = \{11\,213\}$, and that, for each $n \in \mathbb{Z}^+$, $g(n^2) = g(n)$.
(c) For each set S we may define the **identity function** ι on S by $\iota(s) = s$ for all $s \in S$.
(d) If B is the set of all books we may define $h: B \rightarrow \mathbb{Z}$ by $h(b) =$ the number of pages in b.

EXERCISE 40   Give an example of a function $f: \mathbb{Z} \rightarrow \mathbb{Z}$ of the form

$$f(x) = a_3 x^3 + a_2 x^2 + a_1 x + a_0$$

which is such that $f(1) = f(2) = f(3) = 1$. (*Hint:* Is $f(x) = (x - 1)(x - 2)(x - 3)$ any good?)

Here are some slightly more interesting examples.

Two common functions

One function which occurs fairly frequently is $f: \mathbb{R} \rightarrow \mathbb{R}$ given by

$$f(x) = |x| = \begin{cases} x & \text{if } x \geq 0 \\ -x & \text{if } x < 0 \end{cases}$$

Part of its graph is shown in Figure 4.10 (Note that, alternatively, $f(x) = +\sqrt{(x^2)}$.) The function f is called the **modulus** function, $|x|$ being the *modulus* of x. As examples, note that $f(3.142) = |3.142| = 3.142; f(-\frac{22}{7}) = |-\frac{22}{7}| = \frac{22}{7}$. (Informally, each number is 'made positive'.)

EXERCISE 41   Let $a \in \mathbb{R}^+$. Prove that

$$(\forall x \in \mathbb{R})[|x| < a \longleftrightarrow -a < x < a]$$

Another commonly occurring function is $g: \mathbb{R} \rightarrow \mathbb{R}$ given by $g(n) = [n]$ where $[n]$ is the greatest integer not exceeding n.

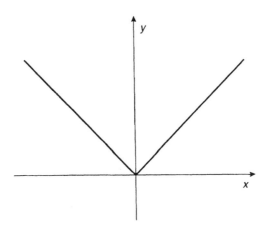

Figure 4.10

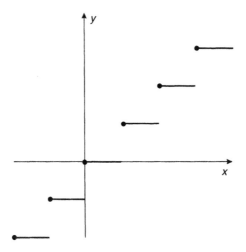

Figure 4.11

Thus $g(1.234) = [1.234] = 1$, whilst $g(-1.234) = [-1.234] = -2$. Its graph is shown in Figure 4.11.

EXERCISE 42   Show that the number of zeros at the end of the integer $n!$ (n factorial) is

$$\left[\frac{n}{5}\right] + \left[\frac{n}{25}\right] + \left[\frac{n}{625}\right] + \left[\frac{n}{3125}\right] + \cdots$$

Onto and one-to-one functions

Suppose $f: A \to B$ is given. We say that f is **onto** B iff each $b \in B$ is 'visited', i.e. to each $b \in B$ there is *at least one* $a \in A$ such that $b = f(a)$. In symbols:

$$(\forall b \in B)(\exists a \in A)[b = f(a)]$$

We say that f is **one-to-one (1–1)** iff distinct elements of A are sent to distinct elements of B, i.e. iff $a_1 \neq a_2$ implies $f(a_1) \neq f(a_2)$. In symbols:

$$(\forall a_1 \in A)(\forall a_2 \in A)[\neg\{a_1 = a_2\} \to \neg\{f(a_1) = f(a_2)\}]$$

We usually *check* that f is 1–1 by showing the contrapositive, i.e. that

$$(\forall a_1 \in A)(\forall a_2 \in A)[\{f(a_1) = f(a_2)\} \to \{a_1 = a_2\}]$$

These ideas may be conveyed pictorially as in Figures 4.12 and 4.13.

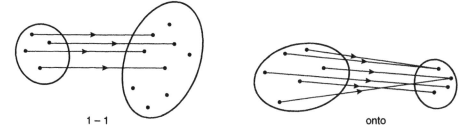

1 – 1   onto

Figure 4.12   **Figure 4.13**

Examples 10(a) and 10(c) are 1–1, but only 10(c) is onto. Here are some more examples, including rough sketches of parts of (a) to (e).

Example 11

	A	B	defn of function	onto?	1–1?
(a)	$\mathbb{R}$	$\mathbb{R}$	$f(a) = a^2$	✗	✗
(b)	$\mathbb{R}$	$\mathbb{R}$	$g(a) = a^3$	✓	✓
(c)	$\mathbb{R}$	$\mathbb{R}^+ \cup \{0\}$	$h(a) = a^2$	✓	✗
(d)	$\mathbb{R}$	$\mathbb{R}$	$j(a) = e^a$	✗	✓
(e)	$\mathbb{R}$	$\mathbb{R}^+$	$k(a) = e^a$	✓	✓
(f)	$\mathbb{C}$	$\mathbb{C}$	$l(a) = a^2$	✓	✗

EXERCISE 43 Which of the following functions are (i) onto; (ii) 1–1? For (a), (b), (c) make your deductions from rough sketches. For (d) *prove* your claim.

(a) $f\colon \mathbb{R} \to \mathbb{R};\quad f(a) = a + |a|$

(b) $g\colon \mathbb{R} \to \mathbb{R};\quad g(a) = \begin{cases} -a^2 - 1 & \text{if } a < 0 \\ a^2 & \text{if } a \geqslant 0 \end{cases}$

(c) $h\colon \mathbb{R} \to \mathbb{R};\quad h(a) = \begin{cases} -a^2 + 1 & \text{if } a \geqslant 0 \\ a^2 & \text{if } a < 0 \end{cases}$

(d) $k\colon \{la, so\} \to \{365, 375\};\quad k(la) = 375, \; k(so) = 365.$[12]

EXERCISE 44 Let $f\colon \mathbb{Z}^+ \to \mathbb{Z}$ be defined by $f(2n) = n - 1$ and $f(2n - 1) = -n$. Let $g\colon \mathbb{Z}^+ \to \mathbb{Z}^+$ be defined by $g(n) = n^2$. Is f 1–1? Is it onto? What about g?

Equality of functions/domain/codomain/range (image)

Examples 11 were chosen to make some points quite forcefully. Functions $f\colon A \to B$, $g\colon C \to D$ are said to be **equal** iff $A = C$ *and* $B = D$ *and*, for all $a \in A \; (= C)$ we have $f(a) = g(a)$. Since A is called the **domain** of f and B is called its **codomain**, we see that $f = g$ iff f and g have the same domains, the same codomains *and* are defined

---

[12] Cricket enthusiasts shouldn't need to have this function explained to them!

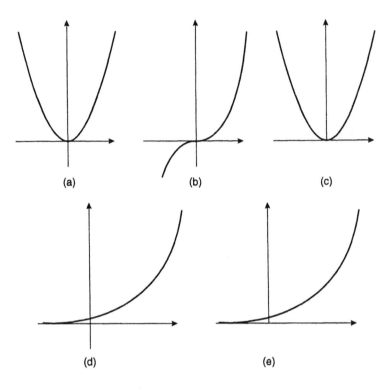

Figure 4.14

the same way on their (identical) domains. In particular, the functions f and h of Example 11 are definitely *not* equal. What *are* the same for f and h are their ranges. The **range** (or **image**) of f is the set $\{f(a): a \in A\}$ of *all* values of the $f(a)$ which we (naturally) denote by $f(A)$. This is depicted in Figure 4.15. Clearly $f(A) = h(A) = \mathbb{R}^+ \cup \{0\}$.

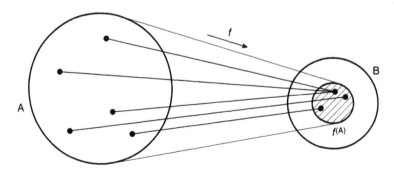

Figure 4.15

EXERCISE 45 Let f, g be the functions from $A = \{1, 2, 3\}$ to $\{2, 3, 4\}$ defined by $f(x) = -x + 5$ and $g(x) = -x^3 + 6x^2 - 12x + 11$. Show that $f = g$.

EXERCISE 46 Let $A = \{a, b, c, d\}$ and $B = \{1, 2, 3, 4, 5, 6, 7\}$.

(a) How many distinct functions are there (i) from A to B and (ii) from B to A.
(b) How many functions from A to B are 1–1?
(c) (Difficult.) How many functions from B to A are onto?

Functions as subsets of Cartesian products

We now show you how you can eliminate all the problems which arose with Definition 2 or, at least, lay them at set theory's door. We begin with a definition.

Definition 3

Given sets A and B we define $A \times B$, the **Cartesian product** of A and B, to be the set $\{(a, b): a \in A \wedge b \in B\}$ of **ordered pairs**. (The word 'ordered' tells us that pairs (a, b) and (c, d) are to be regarded as identical *when and only when $a = c$ and $b = d$*. This caution is only really needed when $A = B$, as in Example 12(b) below.)

Example 12

(a) If $A = \{1, \sqrt{2}, \pi\}$ and $B = \{1.4, \frac{22}{7}\}$ then $A \times B$ is the set of six elements $\{(1, 1.4),$
   $(1, \frac{22}{7}), (\sqrt{2}, 1.4), (\sqrt{2}, \frac{22}{7}), (\pi, 1.4), (\pi, \frac{22}{7})\}$.
(b) If $A = B = \mathbb{R}$ we get $\mathbb{R} \times \mathbb{R} = \{(x, y): x \in \mathbb{R} \wedge y \in \mathbb{R}\}$. Because the pairs are ordered, we have $(r, s) = (u, v)$ iff $r = u$ and $s = v$. (This reminds us of coordinates in the plane where the points (α, β) and (γ, δ) are one and the same iff $\alpha = \gamma$ and $\beta = \delta$.)

EXERCISE 47 Are the following statements true for all sets A, B, C and D?

(a) $A \times (B \cup C) = (A \times B) \cup (A \times C)$
(b) $(A \times B) \cap (C \times D) = (A \cap C) \times (B \cap D)$
(c) $(A \times B) \cup (C \times D) = (A \cup C) \times (B \cup D)$

Each function $f: A \to B$ can be thought of as the subset $\{(a, f(a)): a \in A\}$ of $A \times B$. This observation encourages us to eliminate the suspect word 'rule' in Definition 2 by *defining* a **function from A to B** as *a subset S of $A \times B$ such that*

1. $(\forall a \in A)(\exists b \in B)[(a, b) \in S]$, *and*
2. $(\forall a \in A)(\forall b_1 \in B)(\forall b_2 \in B)[\{(a, b_1) = (a, b_2)\} \to \{b_1 = b_2\}]$

That's it! Needless to say, it is much easier in practice to use the informal definition, but the mathematician is comforted by the knowledge that, should any unexpected anomalies appear, he can resort to the precise definition just given.

Counting

Inclusion–exclusion principle

Exercise 46(c) shows that not all counting problems are straightforward. Here is a result, suggested by studying Venn diagrams, which is fun and has both interesting, unexpected and even potentially profitable consequences.

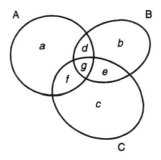

Figure 4.16

The problem we answer is: given three finite sets A, B and C, how many elements are there in $A \cup B \cup C$?

Letting $|X|$ denote the number of elements in the (finite) set X,[13] we see that, generally, $|A \cup B \cup C| < |A| + |B| + |C|$ since elements in $A \cap B$, in $B \cap C$ and in $C \cap A$ will have been counted twice. But how many are there *exactly*?

In Figure 4.16 each small letter denotes the *number* of elements in the section indicated. Thus c is the number of elements in C which are not in $A \cup B$, and d is the number of elements in $A \cap B$ which are not in C.

Now, $|A \cup B \cup C| = a + b + c + d + e + f + g$, $|A| = a + d + f + g$, $|B| = b + e + d + g$, $|C| = c + f + e + g$, $|A \cap B| = d + g$, $|B \cap C| = e + g$, $|C \cap A| = f + g$, $|A \cap B \cap C| = g$. One then easily checks that the following theorem holds.

Theorem 2
(Inclusion–exclusion principle) If A, B and C are finite sets then $|A \cup B \cup C| = |A| + |B| + |C| - |A \cap B| - |B \cap C| - |C \cap A| + |A \cap B \cap C|$.

Theorem 2 can be used to detect cheating! Here is an example.

Example 13
A researcher for Eatem's crisps found that, of 215 people interviewed, 98 people liked the arsenic-flavoured (A) crisps, 86 the botulism-flavoured (B) and 125 the cholera-flavoured (C); 30 liked both A and B flavours, 39 liked B and C, 42 liked C and A, and 18 liked all three flavours. Why did the researcher's employers believe that the researcher had made up his figures whilst sitting at home watching *Neighbours*?

SOLUTION   We have $|A \cup B \cup C| = 215$ whereas

$$|A| + |B| + |C| - |A \cap B| - |B \cap C| - |C \cap A| + |A \cap B \cap C|$$

$$= 98 + 86 + 125 - 30 - 39 - 42 + 18 = 216$$

Thus the alleged figures just don't 'add up'.

---

[13] This is standard notation. Unfortunately it is identical to the modulus sign introduced earlier. Note that each, in its own way, indicates that 'size' is being measured. If we take care then there will be little chance of confusion.

EXERCISE 48   Asked to determine the preferences of the intake of 214 new students, the departmental secretary found that of the 115 who wanted to study French, 42 wished to do French and German and 34 French and Italian (including, each case, 11 students who wished to do all three). Of 102 wishing to do German, 28 wished also to do Italian (including the 11 doing all three). In all, 85 wished to do Italian and possibly one or more of French and German). 'How many have failed to nominate at least one language, Miss Staik?' asked Dr Zdrazvutyer. 'They must do Russian.' 'I can't tell, sir', said Miss Staik, 'and I've thrown away the forms.' Can Miss Staik find that number?

COMMENT   The obvious generalization of Theorem 2 to more than three sets has several applications. Using it, we find that we do not need to know which integers are prime in order to count them. See Riesel (1985, p. 11). In 1830, Legendre attempted to find the number of primes less than $1\,000\,000$ by using such a generalization. (Unfortunately his answer was marginally further away from the truth than that given by the (incorrect) prime tables then available.)

A second, easily deduced, consequence of the inclusion–exclusion principle is this. Take two packs of cards each numbered from 1 to, say, 20. Turning pairs of cards over, one from each pack, simultaneously, what do you think are the chances of at least one pair of 'simultaneously turned' cards being identical? One chance in ten? One chance in five? No! Indeed with about eight or more cards the chances scarcely change and are not far short of two chances in three! Of course, you must not take my word for it – have a look at Slomson (1991, pp. 25–7).

Summary

A **set** is merely a collection of entities which are called its **elements** or **members**. Some sets which are used frequently are given special names and symbols; in particular, the set with *no* element is the **empty set**, Ø. Some (small) sets can be fully described by listing all their elements, others by identifying the properties characterizing their elements. The concepts of **equality, subset, intersection** and **union** of sets were introduced.

Sets, their intersections and their unions can often be represented conveniently by drawing a **Venn diagram**.

To prove that sets X and Y are **equal** it is often easiest to prove that both $X \subseteq Y$ and that $Y \subseteq X$. This is achieved by showing that a 'typical' element in X (in Y) must, necessarily, lie in Y (in X).

To aid clarity, it is useful to introduce symbols $\forall$ and $\exists$ for the **quantifiers** 'for all' and 'there exists', especially if seeking the negation of a multi-quantified statement. Results similar to those we have previously deduced from truth tables are valid here. For example, the assertion $(\forall x)\,[(A(x)\to B(x)]$ is equivalent to $(\forall x)\,[\neg A(x) \vee B(x)]$. Because of the rules for passing negations through quantifiers, we see that the negation of $(\forall x)\,[A(x)\to B(x)]$ may be written as $(\exists x)\,[A(x) \wedge \neg B(x)]$.

The **function** concept is just as basic in mathematics as is that of set. An informal definition can be replaced by a definition in terms of **ordered pairs**. **Equality** of functions implies, in particular, that their **domains** and their **codomains** are identical.

Functions which are **one-to-one (1–1)** and **onto** will be important in Chapter 9 in identifying a distinction between set $\mathbb{Q}$ of rational numbers and the set $\mathbb{R}$ of real numbers.

One rather interesting result, suggested by the drawing of Venn diagrams, is the **inclusion–exclusion principle** which has serious mathematical consequences besides the somewhat more frivolous ones given here.

The above discussion of functions barely gets us off the ground. Given functions f: $A \rightarrow B$ and g: $B \rightarrow C$, we can define a new function (denoted by $g \circ f$: $A \rightarrow C$, called the **composition** of f and g and defined by 'for all $a \in A$, $[g \circ f](a) = g\{f(a)\}$'. This then enables us to define the important concept of **inverse function**. When you study such ideas there will be many important results which will need *proving* (for example, if f and g are each 1–1 then so is $g \circ f$). These would seem good candidates for a book of this sort but, as we have limited space, and the proofs are possibly not the most exciting in the mathematical world, we shall cut our losses here and allow someone else the pleasure of presenting them to you.

CHAPTER 5

Proofs ... for All

Since my appeal says I did strive to prove ... SHAKESPEARE

We now consider the EACH statements, 7–11. In our Discussions we meet some familiar methods of procedure but also some new problems. For example, in looking for a particular type of example, how far do we search before changing our viewpoint by trying to prove that the sought-for example *doesn't exist* (see Statement 7)? We also come across 'proofs' which are no more than restatements of what is to be proved (Statement 9(a)) and others which rely on results equivalent to those to be proved (Comment 4 of Statement 11). We experience two promising ideas which fail (Statement 10) and offer an ungainly proof (Statement 8) because, as yet, we can see no way to do any better. Finally we emphasize a point of style. A theorem, whose proof occurs to us totally 'backwards', is better presented to the reader in the opposite direction – and once it is, a colleague finds a new proof 'out of the blue'.

Introduction

Although in this chapter, we shall concentrate only on the EACH statements (7–11), Statements 7 and 7.1 will immediately remind us of two observations made in Chapter 4 and Chapter 2, namely (i) that there is an intimate connection between the words *each* and *exists*, especially when statements involving these words are negated, and (ii) that the position of the word 'not' in a 'for all' statement is crucial, not least because of the method of proof one has to adopt to prove the statement.

We admit that, apart from Statements 7 and 10, none of the statements in this chapter is so exciting or remarkable that it will cause us sleepless nights. This is perhaps to the good in that we will be able to concentrate more closely on what is required by way of proof; in particular on why certain 'proofs' might be deemed unacceptable or inadequate.

The proofs

STATEMENT 7 Not every positive integer of the form $n^2 - n + 41$, where $n \in \mathbb{Z}^+$, is prime.

Compare the position of 'not' and 'every' in Statement 7 with the position of 'every' and 'not' in the following statement.

STATEMENT 7.1 For every positive integer n (other than $n = 1$) $n^4 + 4$ is not prime.

DISCUSSION OF STATEMENT 7

We don't need to review Chapter 4 to see that Statement 7 is equivalent to 'there exists some positive integer n for which $n^2 - n + 41$ is *not* prime'. Therefore, to prove Statement 7 correct, we have only to find such an integer n. One might wonder who but a fool would have expected $n^2 - n + 41$ always to be prime – a few trial values of n should soon sort this out. Putting $n = 1, 2, ...$ we find $n^2 - n + 41$ generates the values 41, 43, 47, 53, 61, 71, 83, 97, 113, 131, 151, 173, 197,... all primes! This looks pretty remarkable. Perhaps the 'fool' is right. Do we now change our stance and try to prove that $n^2 - n + 41$ *is* always prime, or do we battle on for a few more n and see if a non-prime turns up?

Actually we have here an opportunity to use computing facilities sensibly by investigating values of $n^2 - n + 41$ much more quickly than by hand. Indeed, we could let the computer search for non-primes whilst we try to prove that $n^2 - n + 41$ *is* prime for every n. In this case the computer wins easily: $n^2 - n + 41$ is prime for every n from 1 to 40 inclusive but is *not* prime when $n = 41$. Indeed, if $n = 41$, then $n^2 - n + 41 = 41^2 - 41 + 41 = 41^2$, which is clearly not prime.

Having discovered this you are permitted to thump your knees hard and say 'Why didn't I see that immediately?' Sometimes one doesn't – it's as simple as that. Anyway, at least Statement 7 is true – and its proof is (now) easy.

Proof of Statement 7 The statement is true as the case $n = 41$ shows.

COMMENTS

1. The single line of proof is all that is needed. Once the integer 41 has been identified it is expected that the reader can fill in the details for himself. However, 41 is not the only integer for which $n^2 - n + 41$ is not prime.
2. It was the great Swiss mathematician Leonhard Euler who discovered this amazing formula $n^2 - n + 41$. I have no idea how he found it.
3. You surely want to know if there is a polynomial formula of the form $An^k + Bn^{k-1} + ... + Ln^2 + Mn + N$, where $A, B, ..., M, N$ are integers, which produces only primes. We leave you in suspense until Chapter 14.

LEONHARD EULER
(15 April 1707-18 September 1783)

Leonhard Euler (1707–1783), the greatest mathematician of the eighteenth century, was born in Basel. He entered the University of Basel in 1720 and produced his first research paper when aged 18. Described, one year later, by Johann Bernoulli, his teacher in Basel, as that 'most gifted and learned man of science', Euler accepted an invitation to join the Academy of Sciences in St Petersburg. Euler's range, output and energy were phenomenal. In the *Dictionary of Scientific Biography* Euler's contributions to mathematics are listed under eleven separate headings.

Beginning in 1738, Euler won, on twelve occasions, the biennial prize awarded by the Academie des Sciences.

In 1741, Frederick the Great persuaded Euler, now blind in one eye, to move to Berlin. Somehow he found time to maintain contact with St Petersburg whilst producing huge amounts of research work and undertaking duties (in Berlin) related to the observatory, the botanical gardens, the publication of maps and improving canal navigation. He also produced a text *Letters to a German Princess* detailing the lessons on natural philosophy he had given to Frederick's niece.

In 1766 Euler returned to St Petersburg. When, in 1771, a fire burnt down Euler's house, he was totally blind. Although his house was lost, Euler was carried to safety along with his researches. Nor did his blindness slow his volume of publications. In 1776, with the help of Amenuenses, he published one thousand pages of research.

Euler was a man of moderation. He loved his home and his thirteen children. At teatime on 18 September 1783, whilst playing with grandchildren (and, presumably, simultaneously contemplating mathematics) he became ill and said 'I am dying'. He died six hours later.

EXERCISE 1 (For your computer) What is the least positive n for which $|36n^2 - 810n + 2753|$ is *not* prime? (Due to Russell Ruby: see Guy (1994, p. 37).)

DISCUSSION OF STATEMENT 7.1

For $n = 2, 3, 4, 5, 6, 7, 8, \ldots, n^4 + 4$ takes the values $20, 85, 260, 629, 1300, 2405, 4100, \ldots,$ so that it seems that $n^4 + 4$ might be a multiple of 5 – except when n itself is a multiple of 5. Indeed, if $n = 5k + r$ where $r = 1, 2, 3$ or 4 we easily check that $5 | n^4 + 4$. However, this still leaves all n which are multiples of 5 to be looked at. For $n = 5, 15, 25, 35$ we find $n^4 + 4 = 629 = 17 \times 37$, $50\,629 = 197 \times 257$, $390\,629 = 577 \times 677$, $1\,500\,629 = 13 \times 89 \times 1297$. (Why do all seem to end in 629 and why don't I need to look at $n = 10, 20, 30, \ldots$?) How can we prove that $n^4 + 4$ is composite for all $n > 1$? After all, we have infinitely many still left to check and no obvious pattern to follow. Here is a thought. Maybe a polynomial whose values are composite for each n takes these composite values *because it factorizes as a product of two polynomials of lesser degree (with integer coefficients)*? No! Exercise 2 shows this isn't true. And $n^4 + 4$ doesn't *look* as if it factorizes. Furthermore, if it did, the case of $n = 1$ would have to be very special since 5 does not factorize. I have no feeling as to whether $n^2 + 4$ will factorize or not. Nevertheless, let us see if it does. We should have to have

$$n^4 + 4 = (n + a)(n^3 + bn^2 + cn + d)$$

or

$$n^4 + 4 = (n^2 + an + b)(n^2 + cn + d)$$

where a, b, c, d are integers. Setting $n = -a$, the first factorization would imply $a^4 + 4 = 0$ and so is impossible. The second can be checked by testing a finite number of cases. For example, the pair (b, d) must be one of the pairs $(1, 4), (-1, -4), (4, 1), (-4, -1), (2, 2)$ or $(-2, -2)$. Further effort on our part yields a factorization into two quadratics

$$n^4 + 4 = (n^2 - 2n + 2)(n^2 + 2n + 2)$$

We therefore have our proof of Statement 7.1.

Proof of Statement 7.1 Since $n^4 + 4 = (n^2 - 2n + 2)(n^2 + 2n + 2)$ we see that $n^4 + 4$ is composite for all positive integers n except $n = 1$ (for which the smaller factor $n^2 - 2n + 2$ takes the value 1).

COMMENT Our old friend 'proof by cases' gives us what we wanted for $n = 5k + r$ with $r = 1, 2, 3, 4$ but not for $r = 0$. A not unreasonable (though generally incorrect) thought about polynomial factorization led to a proof not only in the remaining cases but for *all* n, so proof by cases was not needed after all. Nevertheless, getting rid of the 'easy' cases is often a good ploy when attempting a proof. You can then concentrate closely on the more stubborn cases which remain.

EXERCISE 2 Show that, for each positive integer n, $n^2 + n + 2$ is composite and yet $n^2 + n + 2$ does not factorize into a product of two polynomials (of degree 1) with integer coefficients. (Not even if we allow *real* coefficients.)

EXERCISE 3   Check that $(n^2 - 2n + 2)(n^2 + 2n + 2)$ is indeed $n^4 + 4$.

EXERCISE 4   Does the *polynomial* $n^4 + 9$ factorize nicely? Is the *integer* $n^4 + 9$ composite for each $n \in \mathbb{Z}^+$?

WHAT WE HAVE LEARNED

- If, in attempting a proof of a theorem or, as in Statement 7, in seeking an example of a particular kind, we believe the evidence is beginning to point in the opposite direction, it is perfectly proper to change one's stance and to try to find a counterexample to the alleged theorem or to give a proof that the sought-for example does not exist. Of course, failure to find the counterexample or proof in a reasonable period of time is of little significance but it may persuade us to return to our original plan.
- Sometimes, as in Statement 7.1, it can be helpful to get the 'easy' cases out of the way in order to identify those where the greatest difficulties lie. The solution in these extreme cases may come 'out of the blue'. In our case it also made the solution we previously obtained for the easy cases redundant.
- When looking for a specific example (as in Statement 7) a search by computer may be appropriate. Of course, failure to find an example may signal that there is none to be found or merely that the computer has not found it *yet* (cf. Final Exercise 3(g) in Chapter 1).

STATEMENT 8   If x is a positive integer, then $x^3 - x$ is a multiple of 3 and $x^5 - x$ is a multiple of 5.

DISCUSSION

Despite the fact that, compared with the version of Statement 8 in Chapter 2, the words 'for each x' have been replaced by 'if x is a positive integer', our task is still to investigate the truth of Statement 8 as a statement about *all* positive integers. We therefore should expect to apply the by now familiar method of arguing about a typical 'x' of which we assume nothing other than that it is a positive integer.

How would we set out to prove just the first part, namely whether for each positive integer x, $x^3 - x$ is a multiple of 3? Probably the most obvious first move is to factorize $x^3 - x$ as $x(x-1)(x+1)$. Since this is the product of three consecutive integers and since, of each set of three consecutive integers, exactly one is a multiple of 3,[1] we see that $x^3 - x$ is a multiple of 3 no matter what integer x might designate.

Can we use the same technique on $x^5 - x \{= x(x-1)(x+1)(x^2+1)\}$? If x or $x - 1$ or $x + 1$ is a multiple of 5, that is, if x is of the form $5k$ or $5k + 1$ or $5k + 4$ (equivalently $5k - 1$), we clearly have no problem. This leaves undone the cases where x is of one of the forms $5k + 2$, $5k + 3$ (equivalently $5k - 2$). Now it is not

---

[1] I think we may now regard this as sufficiently well known not to require any proof.

difficult to check that in these two cases $x^2 + 1$ (and therefore $x^5 - x$) is, again, a multiple of 5. Since, for each x we have proved that $3|x^3 - x$ and, for each x, we have proved that $x|x^5 - x$ it appears that we have proved Statement 8.

Proof of Statement 8   Since $x^3 - x = x(x - 1)(x + 1)$ is a product of three consecutive integers it is certainly a multiple of 3 for each x. Similarly, taking x to be, in turn, of the form $5k$, $5k + 1$, $5k - 1$, $5k + 2$, $5k - 2$ we readily check that $x^5 - x = x(x - 1)(x + 1)(x^2 + 1)$ is a multiple of 5 in each case.

COMMENTS

1. If $A(x)$ and $B(x)$ denote the predicates $3|(x^3 - x)$ and $5|(x^5 - x)$ respectively, we have proved Statement 8, $(\forall x \in \mathbb{Z}^+)[A(x) \wedge B(x)]$, by proving $(\forall x \in \mathbb{Z}^+)[A(x)] \wedge (\forall x \in \mathbb{Z}^+)[B(x)]$. This seems to be fair – but see Exercise 5.
2. I just don't feel at ease with the proof just given. It's too cumbersome – one of Hardy's despised ugly pieces of mathematics. It can't be the correct way to do it, especially as factorizing $x^n - x$ for large n would be fairly unpleasant.
3. The use of the five forms $5k$, $5k \pm 1$, $5k \pm 2$ (rather than $5k$, $5k + 1, ..., 5k + 4$) saves a little work here since one only has to square 0, ± 1 and ± 2 rather than $0, 1, 2, 3$ and 4. Such a saving of effort is very noticeable with Exercise 3(a) in Chapter 3.

EXERCISE 5

(a) Find predicates $A(x)$ and $B(x)$ such that $(\forall x \in \mathbb{Z}^+)[A(x) \vee B(x)]$ is *not* equivalent to $(\forall x \in \mathbb{Z}^+)[A(x)] \vee (\forall x \in \mathbb{Z}^+)[B(x)]$. (*Hint*: x odd, x even.)
(b) Do we have $\{(\forall x \in \mathbb{Z}^+)[A(x) \to B(x)]\} \longleftrightarrow \{(\forall x \in \mathbb{Z}^+)[A(x)] \to (\forall x \in \mathbb{Z}^+)[B(x)]\}$?

EXERCISE 6   Is it true, for every positive integer x, that $x^4 - x$ is divisible by 4. Investigate the corresponding question when '4' is replaced by (a) 2; (b) 6; (c) 7; (d) 8; (e) 9. Is any general result emerging?

WHAT WE HAVE LEARNED

● A proof of $(\forall x \in \mathbb{Z}^+)[A(x) \wedge B(x)]$ may be accomplished by proving $(\forall x \in \mathbb{Z}^+)[A(x)] \wedge (\forall x \in \mathbb{Z}^+)[B(x)]$ but a corresponding attempt to prove $(\forall x \in \mathbb{Z}^+)[A(x) \vee B(x)]$ would be invalid.
● If a proof feels 'ugly' then there may be a more natural one available if you are prepared to look for it.

Next a result which will strike you as obvious – but it has a role to play precisely because of that.

STATEMENT 9(a)   Let x, y be real numbers. If $x^2 = y^2$ then $x = y$ or $x = -y$.

DISCUSSION

The first sentence indicates that what follows is claimed to be true for *all* pairs x, y of real numbers. So our proof should begin 'Let x and y be real numbers'. We are given that $x^2 = y^2$: how are we to argue? Perhaps we could take the square root of each side obtaining x (or $-x$) $= y$ (or $-y$). The desired conclusion then (seemingly) follows from the four possibilities $x = y$ or $-x = y$ or $x = -y$ or $-x = -y$. However, if you think about it, this 'proof' seems to be little more than rewriting the statement in a different form. (Surely Statement 9(a) says 'If a and b are real numbers then the square root of a is plus or minus the square root of b'? So, by taking square roots, we are not proving that Statement 9(a) is correct, merely restating it.) If you feel that we must not simply accept Statement 9(a) as 'obvious', then we must offer a deeper understanding as to why it is true. Let us think again.

We could rewrite $x^2 = y^2$ as $x^2 - y^2 = 0$ and then factorize the left-hand side to get $(x - y)(x + y) = 0$. Is this any use? We now have two real numbers $x - y$ and $x + y$ whose product is 0. Hence $x - y = 0$ or $x + y = 0$. That is, $x = y$ or $x = -y$. (Of course, $x - y = 0$ and $x + y = 0$ *may* hold simultaneously, namely when x and y are both zero.) To get our desired conclusion we are assuming as obvious the assertion that if a product of two real numbers is 0 then at least one of the two must be 0. If you are not prepared to accept *that* then you must be prepared to give a proof of it based on even more fundamental properties of the real numbers.

Proof of Statement 9(a)   Let x and y be real numbers such that $x^2 = y^2$. It then follows that

$$(x - y)(x + y) = x^2 - y^2 = 0$$

Hence, either $x - y = 0$ or $x + y = 0$. That is, either $x = y$ or $x = -y$.

COMMENT   Your scepticism about not accepting Statement 9(a) as obvious may be matched by your scepticism about using, unproved, the result concerning real numbers whose product is 0. Nevertheless, one feels that, in calling upon this latter (unproved) result one is getting nearer to the *real* reason as to why Statement 9(a) is true. At least we have seen why Statement 9(a) ought not to be regarded as 'totally, unprovably, obvious'.

EXERCISE 7   Rewrite Statement 9(a) in contrapositive form. Is it any easier to prove in that form, rather than directly?

EXERCISE 8   Show that for each pair of real numbers x, y, if $x^3 = y^3$ then $x = y$. (*Hint:* You may need to know that if x, y are not *both* zero then $x^2 + xy + y^2 \neq 0$. If you use this you should *prove* it. Hint: $(x + y/2)^2 + \frac{3}{4}y^2 = \ldots?$)

STATEMENT 9(b)   Let x be a real number. If $x^2 > 9$ then $x > 3$ or $x < -3$. (*Note:* This 'or' is exclusive since one clearly cannot have both $x > 3$ and $x < -3$ simultaneously.)

DISCUSSION

This statement is so similar to that of Statement 9(a) that we shall try to follow the same line of argument. We try a direct proof. From $x^2 > 9$ we first deduce that

$x^2 - 9 > 0$, that is, $(x - 3)(x + 3) > 0$. Now if a product of two real numbers is positive then either each is positive or each is negative. We are therefore forced to deduce that *either* (i) $x - 3$ AND $x + 3$ are both > 0 (in which case we must conclude that $x > 3$ AND $x > -3$, a conjunction which clearly simplifies to the condition $x > 3$), *or* (ii) $x - 3$ AND $x + 3$ are both < 0 (in which case we must conclude that $x < 3$ AND $x < -3$, a conjunction which simplifies to $x < -3$).

Proof of Statement 9(b)   Let $x \in \mathbb{R}$. From $x^2 > 9$ we obtain

$$(x - 3)(x + 3) = x^2 - 9 > 0$$

Consequently either (i) $x - 3 > 0$ AND $x + 3 > 0$, which implies that $x > 3$, or (ii) $x - 3 < 0$ and $x + 3 < 0$, which implies that $x < -3$.

COMMENTS

1. As in Statement 9(a), our (once again direct) proof depends on an assumption concerning products of real numbers, namely that if a product of two real numbers a and b is positive then a and b are either both positive or both negative. Are you prepared to accept this? If not, you will have to give a proof of it (based on somewhat more fundamental properties of the real numbers).
2. The above argument would clearly work for any real number, not just 3 – right? No, wrong! For example, from $x^2 + 5 > 0$ you may *not* infer that either $x > \sqrt{-5}$ or $x < -\sqrt{-5}$. If you cannot see why, you will have to await the section on complex numbers and the rules of arithmetic in Chapter 10.

EXERCISE 9   What can you say about real numbers x for which $x^2 < 4$? Write down a statement as near in form to Statement 9(b) as you can, and then prove it.

EXERCISE 10   Give direct proofs of the following: (a) For each integer x, if x is even then x^2 is even. (b) For each integer y, if y is odd then y^2 is odd. (*Hint:* If x is even then $x = 2m$ for some suitable integer m. Then $x^2 = \dots$. If y is odd then $y = 2n + 1$ for some suitable integer n, hence $y^2 = \dots$.)

EXERCISE 11   Criticize the following 'proof' that, if $x = y$ then $x^2 = y^2$. *Proof:* Square each side. (Cf. the Discussion of Statement 9(a).)

WHAT WE HAVE LEARNED

- We must ensure that what we present as a proof is not merely a restatement of the result in question.
- Rewriting $X = Y$ and $X > Y$ as $X - Y = 0$ and $X - Y > 0$ can be helpful.

In Chapter 2 we called the statement $B \rightarrow A$ the *converse* of the statement $A \rightarrow B$. Statement 10 is the converse of the (false) assertion in Final Exercise 2(a) of Chapter 1.

STATEMENT 10   Let n be a positive integer. If $2^n - 1$ is prime then n is prime.

DISCUSSION

It is easy to check, for $n \leqslant 11$, that the statement holds. As we have no reason to suppose Statement 10 is false let us, initially, try to prove it true. Unfortunately, I just *cannot* see how I might use the assumed primeness of $2^n - 1$ to give a direct proof of the required primeness of n. After a period of mental inactivity it occurs to me that perhaps a contrapositive attack on Statement 10 may offer hope. That is, we should assume that n is *not* prime and try to prove that $2^n - 1$ is *not* prime.

Accordingly, we suppose that n is a product of two integers r and s each greater than 1. Since we need to show that $2^n - 1 \; (= 2^{rs} - 1)$ is *not* prime we must exhibit a factorization for it too. How might $M \; (= 2^n - 1)$ factorize? Trying $n = 4, 6, 8, 9$, we find that $M = 3 \times 5, 3 \times 3 \times 7, 3 \times 5 \times 17, 7 \times 73, \ldots$, so no obvious pattern here. Nevertheless, after a bit of thinking, it might occur to you that $2^n - 1$ 'looks like' $x^n - 1$ (with 2 replacing x) and, just as $x^2 - 1$ factorizes as $(x - 1)(x + 1)$, you may come to the conclusion (or you may know) that $x^n - 1$ factorizes as $(x - 1)(x^{n-1} + x^{n-2} + \ldots + x^2 + x + 1)$. Likewise $2^n - 1$ will factorize as $(2 - 1)(2^{n-1} + 2^{n-2} + \ldots + 2^2 + 2 + 1)$. Oh dear! This is not a proper factorization of $2^n - 1$ – since the 'factor' $(2 - 1)$ is just 1. However, you may then recall that *we have not yet used our assumption that n is composite* (i.e. > 1 and not prime). You may also realize that, since $y^s - 1 = (y - 1)(y^{s-1} + y^{s-2} + \ldots + y^2 + y + 1)$ we may put $y = 2^r$, to obtain

$$\{2^r\}^s - 1 = (\{2^r\} - 1)(\{2^r\}^{(s-1)} + \{2^r\}^{(s-2)} + \ldots + \{2^r\}^2 + \{2^r\} + 1)$$

(We have used the brackets $\{\}$ just for clarity.) Now, since $r > 1$, we see that $2^{rs} - 1$ has two factors, each different from 1. Thus $2^n - 1$ is *not* prime, and we may now give our formal proof.

Proof of Statement 10   Suppose that n is *not* prime. Then $n = rs$ for suitable integers r, s each greater than 1. It follows that

$$2^n - 1 = (2^r)^s - 1 = (\{2^r\} - 1)(\{2^r\}^{(s-1)} + \{2^r\}^{(s-2)} + \ldots + \{2^r\}^2 + \{2^r\} + 1)$$

Since $r > 1$, we see that $2^r - 1 > 1$, as is the other factor. Hence the statement is proved.

COMMENT   If you *didn't* know the above factorization of $x^n - 1$ you may rightly complain that the above proof wasn't within your grasp. All the more reason to file the factorization away in your memory in case it proves useful (as I am confident it will) in several other areas of mathematics.

WHAT WE HAVE LEARNED

● If a first attempt to solve a problem fails it may take some time for a different approach to occur to you. Accordingly don't give up until you are fairly convinced you are going nowhere. Ask yourself if you have used *all* the hypotheses of the problem.

We now come to another result (already used in the proof of Statement 1) which is included mainly because it is interesting and instructive to see on what the proof of such an 'obvious' result depends.

STATEMENT 11   Let x be an integer. Then x^2 is odd if and only if x is odd.

DISCUSSION

Here there are two things to prove about each integer x: (i) if x is odd then x^2 is odd; (ii) if x^2 is odd then x is odd. Part (i) is part of Exercise 10 so we shall assume that we already have a proof of that. So now for (ii). Very informally one would probably argue: 'x must surely be odd because, if it weren't, it would have to be even – and then x^2 would have to be even (by Exercise 10) which it isn't (because we are *told* it isn't). So x *can't* be even. Hence it must be odd.' This contrapositive point of view is almost forced on us here as being the obvious way to argue. Hence our formal proof may be given as follows.

Proof of Statement 11   For the 'if' part: if x is odd then x^2 is easily seen to be odd (see Exercise 10). Likewise, for the 'only if' part: if x is even then x^2 is even. Consequently, if x^2 is *odd* then so too must x be.

COMMENTS

1. In Statement 11 the proof of $(\forall x \in \mathbb{Z})[A(x) \longleftrightarrow B(x)]$, that is, the proof of (a) $(\forall x \in \mathbb{Z})[A(x) \rightarrow B(x)]$ *and* (b) $(\forall x \in \mathbb{Z})[B(x) \rightarrow A(x)]$, is most naturally accomplished by proving (b) and then $(\forall x \in \mathbb{Z})[\neg B(x) \rightarrow \neg A(x)]$ in place of (a).
   In the proof of Statement 11.1 below, it would be more natural to prove $(\forall x \in \mathbb{Z})[A(x) \longleftrightarrow B(x)]$ by proving both (a) and (b) directly.
2. 'If x is odd then ...' is just another way of saying 'Let x be an odd integer. Then ...'. This is just a matter of style.
3. The reader might feel slightly uneasy if he thinks that, in proving the implication $(x^2 \text{ odd}) \rightarrow (x \text{ odd})$, we are making use of the no-more-obvious implication $(x \text{ even}) \rightarrow (x^2 \text{ even})$. In fact, I would contend that the first of these implications is not *quite* as obvious as the second since the implication $(x \text{ even}) \rightarrow (x^2 \text{ even})$ is most easily proved by direct proof whereas a direct proof of $(x^2 \text{ odd}) \rightarrow (x \text{ odd})$ is perhaps a bit harder to find. (See Exercise 12.)
4. Here are two *failed* attempts at a *direct* proof of 'if x^2 is odd then x is odd'.
   (a) Assuming x^2 is odd then $x^2 = 2m + 1$ for some integer m. But then $x = \sqrt{(2m + 1)}$ [?] which is not even [?]. This attempt is clearly getting nowhere.
   (b) Assume x^2 is odd. Then $(x - 1)(x + 1) = x^2 - 1$ is even. Thus at least one (and, in fact, then *both*) of $x - 1$ and $x + 1$ is even. Hence (from either of these) x is odd.

   This second 'proof' surely *is* a bit of a cheat since we appear to be assuming that if a product of two integers (here the integers $x - 1$ and $x + 1$) is even then at least one of the integers is even. Since we are trying to prove that if a product, x^2, of two (*equal*) integers, namely x and x, is odd then x is odd, we appear to be *assuming* a result whose validity is at least as doubtful as that of the result we are attempting to prove.

EXERCISE 12   Construct a direct proof of the 'only if' part of Statement 11 using the following outline: (a) Let $y = x(x - 1)$; (b) show that y is even; (c) hence show that $x(=x^2 - y)$ is odd.

EXERCISE 13   Criticize the following proof that if x^2 is odd then so is x.
   Proof: Since x^2 is odd we have $x^2 = 4m^2 + 4m + 1$ for some integer m. Thus $x^2 = (2m + 1)^2$ and so $x = \pm(2m + 1)$, which is odd.

EXERCISE 14

(a) Prove, by the contrapositive ($\neg B \rightarrow \neg A$) method, that if the product ab of two integers a and b is even then at least one of a and b is even. In symbols

$$(\forall a \in \mathbb{Z})(\forall b \in \mathbb{Z})[ab \text{ is even} \longleftrightarrow \{a \text{ is even} \vee b \text{ is even}\}]$$

(b) Deduce from (a) that, if x is an integer such that x^2 is even then x is even.
(c) Deduce the result stated in (b) directly from Statement 11.

EXERCISE 15   If it is obvious that 'if x^2 is even then x is even', that is, 'if $2|x^2$ then $2|x$', is it just as obvious that 'if $6571|x^2$ then $6571|x$'?

STATEMENT 11.1   For each real number ϑ: $\sin \vartheta + \cos \vartheta = 0$ if and only if $\vartheta = -(\pi/4) + n\pi$, n being any integer.

DISCUSSION
Here we have to prove: (i) that if $\sin \vartheta + \cos \vartheta = 0$ then $\vartheta = -(\pi/4) + n\pi$ and, *conversely*, (ii) that if $\vartheta = -(\pi/4) + n\pi$ then $\cos \vartheta + \sin \vartheta = 0$. Now the converse part is so straightforward that we can surely 'leave it to the reader'. For part (i) we proceed naively, rewriting the equation $\sin \vartheta + \cos \vartheta = 0$ as $\sin \vartheta = -\cos \vartheta$ and deducing that $\tan \vartheta = -1$. Then $\vartheta = -(\pi/4) + n\pi$, immediately.
   Before writing out the above formally we ask the reader if the above discussion overlooks anything.

Proof of Statement 11.1   We leave the 'if' part (i.e. (ii) above) to the reader to prove. For the 'only if' part, given that $\sin \vartheta + \cos \vartheta = 0$ we see that $\cos \vartheta \neq 0$ (for if $\cos \vartheta = 0$ then $\vartheta = (\pi/2) + k\pi$ (k an integer), but then $\sin \vartheta = 1$ or -1 and so $\sin \vartheta + \cos \vartheta$ would not be 0). Also $\sin \vartheta + \cos \vartheta = 0$ implies that $\sin \vartheta = -\cos \vartheta$ and hence $\tan \vartheta = -1$ (*since* $\cos \vartheta \neq 0$). It then follows immediately that

$$\vartheta = -(\pi/4) + n\pi \cdot$$

So, what *did* we overlook in the Discussion?

WHAT WE HAVE LEARNED

- We saw in the Discussion of Statement 9(a) that it could be useful to rewrite $X = Y$ as $X - Y = 0$. Here we find the reverse strategy, that is, rewriting $X + Y = 0$ as $X = -Y$, equally useful.
- There is often no harm in proceeding naively.

A backwards proof

As a finale we discuss both a point of style and exhibit another case of 'how did they think of that?'.

We claimed earlier that proofs are rarely discovered in the order in which they are finally presented. Sometimes proofs are discovered in an entirely backwards fashion in the form: 'I could prove S *if only* I could prove R. I could prove R *if only* I could prove Q. I could ...'

I should think there are quite a few readers who, by experimenting, will have noticed that if you take two integers, for example 2 and 6, then their *arithmetic mean* (or average) $(2 + 6)/2$ is greater than their *geometric mean* $\sqrt{(2 \cdot 6)}$ and will have wondered, quite naturally, if this is true for all pairs of numbers – maybe real numbers, not just integers. If so, how would you prove it? There follows an attempt at a proof, together with an immediate 'tidying up' and then an inspired 'out of the blue' proof.

Theorem 1

For every two positive real numbers a and b, $(a + b)/2 \geqslant \sqrt{(ab)}$.

> DISCUSSION
>
> Our first attempt at a proof might go as follows. Let us get rid of the nasty $\sqrt{\ }$. We will surely have proved that $(a + b)/2 \geqslant \sqrt{(ab)}$ provided we can show that $\{(a + b)/2\}^2 \geqslant ab$ (since we can then take positive square roots). *This* will be true provided that $(a + b)^2 \geqslant 4ab$. But *this* will be true provided $a^2 + 2ab + b^2 \geqslant 4ab$ which will be true provided $a^2 - 2ab + b^2 \geqslant 0$. But this *is* true since $a^2 - 2ab + b^2 = (a - b)^2$ which, being a square, is certainly non-negative. Hence $(a + b)/2 \geqslant \sqrt{(ab)}$, as required.
>
> Isn't this cumbersome? We certainly can't offer it as a proof without tidying it up. Notice the increased readability – as well as the total *lack of description* concerning how the proof was discovered – of the following proof which is, in essence, the first version written out 'backwards'.

Second proof For all real numbers a and b we have

$$a^2 - 2ab + b^2 = (a - b)^2 \geqslant 0$$

Hence, adding $4ab$ to each side we get

$$(a + b)^2 = a^2 + 2ab + b^2 \geqslant 4ab$$

Taking square roots (which is allowable since a and $b \geqslant 0$) we get

$$a + b \geqslant +2\sqrt{(ab)}$$

Dividing each side by 2 gives

$$(a + b)/2 \geqslant \sqrt{(ab)}$$

COMMENT In oversimplified symbolic form the above proof is *discovered* as: $A \leftarrow$ (i.e. 'provided') B, B$\leftarrow$C, C$\leftarrow$D... It is *presented* as D$\rightarrow$C, C$\rightarrow$B, B$\rightarrow$A. Each is a proof of D$\rightarrow$A. But do take care! If you have a proof of the form X$\rightarrow$Y, Y$\rightarrow$Z, Z$\rightarrow$T you do have a proof of X$\rightarrow$T. However, just as A$\rightarrow$B and B$\rightarrow$A are not, in general equivalent, the steps in this proof may not be reversible. Consequently, you don't have a proof of T$\rightarrow$X. Indeed T$\rightarrow$X may be false. Moral: *Watch, carefully, the direction in which the arrows are pointing!*

Once a theorem has been established, mathematicians like to find even snappier versions of its proof. Here is a very short proof of the theorem from which all hint of the means of its original discovery has vanished.

Third proof Let a and b be positive real numbers. Then

$$0 \leqslant (\sqrt{a} - \sqrt{b})^2 = a - 2\sqrt{(ab)} + b$$

This does it!

COMMENT This proof is very easy to follow and it is easy to agree that it is correct. Its main defect from the learner's viewpoint is summarized by 'what can I learn from this proof?' Little more than 'squares of real numbers are non-negative'.

Summary

The Discussion of Statement 7 raises the problem of how long you should spend searching for a particular example with desirable properties. Here I can offer but little advice. If, after a long search (what is 'long'?), the desired example has failed to show, maybe one should now try to *prove* that the sought-for example *doesn't exist*. If that fails, one may think of making a further search. Thus one's strategy may vary from day to day.

The Discussion of Statement 7.1 also shows various changes of attack which can occur as one tackles a problem. This is quite natural. However, these changes *will not readily occur if you don't write something down*. So, if you are stuck, try any idea, even if it doesn't look too promising. Some ideas really do strike you right out of the blue – but they are often sparked off by an earlier period of mulling over. Perhaps the mulling over gets rid of the dead wood and shows you the proper path *between* the trees.

The proof of Statement 8 is very ugly. It can't be the best way to proceed. (We find a much better way in Chapter 8.)

In the Discussion of Statements 9(a) and (b) the trivial changes of viewpoint (from

$x^2 = y^2$ to $x^2 - y^2 = 0$ and from $x^2 > 9$ to $x^2 - 9 > 0$) prove useful at suggesting 'the next step'. Notice the reverse move: from $\sin \vartheta + \cos \vartheta = 0$ to $\sin \vartheta = -\cos \vartheta$ in the Discussion of Statement 11.1. Another alternative when given or looking to prove $a = b$ is to work with, or towards, $-a/b = 1$. (Once again, see Statement 11.1.)

Statements 10 and 11 show again the importance of *indirect proof*. Since the statements $A \rightarrow B$ and $\neg B \rightarrow \neg A$ are equivalent it is often useful to adopt the different viewpoint of beginning from an assumption of $\neg B$ than from A itself. Indeed, as the Discussion of both statements shows, the contrapositive method can in some cases be the more natural approach.

The eventual presentation of a proof may be little more than a repetition of the path of its discovery, written down 'backwards'. Take care, though. The truth of $X \rightarrow Y$ does not imply that $Y \rightarrow X$ is also true.

CHAPTER 6

There Exist ... Proofs

When you have eliminated the impossible, whatever remains,
however improbable, must be the truth.
SHERLOCK HOLMES

This chapter tackles problems of existence and uniqueness. Although it is frequently possible to discover more than one proof for a particular theorem, several here seem to rely, naturally, on *reductio ad absurdum*. The proof of the fairly straightforward Statement 13 appears to require a mathematical sledgehammer to crack it, whilst the proof of Theorem 15 generates a moment's uncertainty – because the proof we are confident is the right one does not seem to allow for the exceptional cases which we know exist.

Introduction

We now consider Statements 12–15, all of which concern existence. We begin with the famous result which allegedly induced the Pythagoreans to murder![1]

---

[1] The Pythagoreans' principle that 'Everything is [positive whole] number', implies that each pair of lengths is commensurable (i.e. both lengths are a whole number of times the length of one common unit of length). But one Pythagorean pointed out that the diagonal and side of a square are *not* commensurable – since the ratio of their lengths is $\sqrt{2}$ – a fact which follows from *Pythagoras' theorem*! Fearing this truth would get out, the discoverer was, it is said, taken out to sea and drowned.

The proofs

STATEMENT 12   There is no rational number whose square is 2.

DISCUSSION

Many readers will already know that Statement 12 is true. Symbolically it can be rewritten

$$\neg(\exists x \in \mathbb{Q})[x^2 = 2]$$

The idea of trying to prove the equivalent version, $(\forall x \in \mathbb{Q})[x^2 \neq 2]$, appears, on reflection, not too inviting and we are led, quite naturally, to attempting the contrapositive approach by supposing its negation $(\exists x \in \mathbb{Q})[x^2 = 2]$. Accordingly, we begin, 'Just suppose there *were* such an x. Then ...'. Once again, this approach at least gives us something to get our teeth into.

What happens if we follow this approach? We suppose there *is* some rational number a/b such that $(a/b)^2 = 2$. What can we say? There seems little alternative to rewriting this as $a^2 = 2b^2$ and then to noticing that a^2, and hence a itself, is even (by Exercise 14(b) of Chapter 5). So we may as well write $a = 2m$ for some integer m. (It feels as if we *are* making definite progress.) Pressing on, we write $(2m)^2 = 2b^2$, that is $4m^2 = 2b^2$, in other words $2m^2 = b^2$. This means that b, too, is even. Wait a minute! Both a and b have been proved to be even and yet, surely, at the start, we could have assumed that a/b was expressed *in lowest terms*? It looks as if there is a contradiction here. Now if our argument is logically sound (is it?), it will follow that a contradiction could only arise from our having made a false assumption. But we've only made one.... A formal proof might read as follows.

Proof of Statement 12   Suppose that a/b is a rational number, *written in its lowest terms*, such that $(a/b)^2 = 2$. It follows that $a^2 = 2b^2$. Hence a^2 and, consequently, a is even. Writing $a = 2m$, for suitable integer m, we obtain $(2m)^2 = 2b^2$, that is $2m^2 = b^2$. But this shows that b, too, must be even. Therefore the supposition that $\sqrt{2}$ is a rational number must be wrong since it leads to the contradiction that the integers a and b, whilst having no common divisor greater than 1, are both divisible by 2.

COMMENTS

1. The method of proof here is worth noting. It is one of **contradiction**. Indeed, to prove 'there is *not* a rational a/b such that $(a/b)^2 = 2$' we showed that its negation, 'there *is* a rational number a/b such that $(a/b)^2 = 2$' is *false* by showing that the assumption of this negation leads to a contradiction concerning the possible values of a and b. If we let $S(x, y)$ and $X(x, y)$ denote the predicates $S(x, y)$: $(x/y)^2 = 2$ and $X(x, y)$: x and y have greatest common divisor 1, then the contradiction we have obtained takes the form

$$(\forall x \in \mathbb{Z}^+)(\forall y \in \mathbb{Z}^+)[\{S(x, y) \wedge X(x, y)\} \rightarrow \{\neg X(x, y) \wedge X(x, y)\}]$$

the conclusion $\{\neg X(x, y) \wedge X(x, y)\}$ being impossible. This form of proof by contradiction may be compared with that used in the proof of Statement 1.

2. Statement 12 is an example of a **non-existence theorem**: 'Rational numbers with square equal to 2 do not exist'. Such theorems are important in mathematics as they can save time (and, in some computing circumstances, money) by advising you not to engage in what would be a futile search. One famous non-existence theorem was obtained by two mathematicians whose combined lives totalled less than 49 years.[2] They showed, amongst other things that, unlike the general quadratic equation $ax^2 + bx + c = 0$, whose roots are given by the well-known formula

$$x = \frac{-b \pm \sqrt{(b^2 - 4ac)}}{2a}$$

and the general cubic and quartic equations, where similar but more complicated formulae are available (see Chapter 1), there is no similar formula for the roots of a general quintic equation.

A non-existence theorem which it would profit many a young child to know (although it would spoil a lot of fun!) is that *there is no solution to the 'utilities problem'* (Figure 6.1) where three houses have to be connected to gas, water and electricity supplies in such a way that no two of the supply lines cross each other. (For a proof of this non-existence see Liu (1968, p. 210).) Yet another non-existence result, due to Euler, features in Chapter 11.

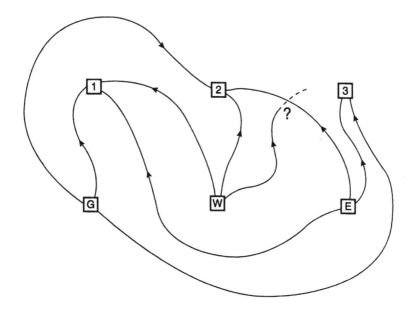

Figure 6.1

[2] Niels Henrik Abel (5 August 1802–6 April 1829) and Evariste Galois (25 October 1811–31 May 1832).

WHAT WE HAVE LEARNED

- Non-existence theorems can be important. To show some specific type of object cannot exist, a good method is to assume that it *does* exist and aim for a contradiction – which may come in the form $X \wedge \neg X$.

EXERCISE 1   Prove that neither $x = \sqrt{2}/2$ nor $x = 1 + \sqrt{2}$ is rational. (*Hint:* Assume $x = a/b$ is rational and obtain a contradiction.)

EXERCISE 2   Prove that $x = \sqrt{3} - \sqrt{2}$ is not rational by writing $x + \sqrt{2} = \sqrt{3}$ and squaring each side.

EXERCISE 3   Show that if m is a rational number such that $m^2 < 2$, then $(m + 1/n)^2 < 2$ provided that $n > ?$. (Of course, $m + 1/n$ is nearer to 2 than is m). (*Hint:* $(m + 1/n)^2 < m^2 + 2m/n + 1/n$ and *this* is less than 2 provided $n > \ldots$.)

EXERCISE 4   (a) Copy the proof of Statement 12 word for word as far as you can to prove that $\sqrt{3}$ is not a rational number. (b) Do the same for $\sqrt{4}$, identifying where the 'proof' that $\sqrt{4}$ is not rational first breaks down. (*Hint:* For (a) you will need to know that if $a^2 = 3b^2$ then $3|a$. To use this you must first *prove* it.)

EXERCISE 5   Let $a, b, c, d \in \mathbb{Q}$ and suppose that $a + b\sqrt{2} = c + d\sqrt{2}$. Prove that $a = c$ and $b = d$. (*Hint:* Put $a - c = (d - b)\sqrt{2}$.)

STATEMENT 13   Let $f(x) = 2x^3 - 9x^2 + 6x + 3$. Then there is some real number $a > 1$ for which $f(a) = 0$.

COMMENT   Statement 13 is an **existence theorem**. It says that, if you are seeking positive roots of $f(x) = 2x^3 - 9x^2 + 6x + 3$ then you are not looking in vain. There is at least one to the right of $x = 1$. It is surely useful to know, before you spend lots of computer time trying to find, numerically, a (perhaps approximate) solution to an equation, that there is one there to be found. Such approximations can be found by well-established methods.

DISCUSSION
Here is an interesting problem. We are claiming there is some root of the given equation and yet we cannot name it. On what grounds are we confident of its existence? Perhaps a picture might help? Because x^3 is, for numerically large values of x, the dominant term in the given expression, it is clear that for large values of x the graph of f will lie in the first (north-east) quadrant of the plane, whilst for numerically large negative values of x, the graph of f will lie in the third (south-west) quadrant of the plane. Accordingly, in order that these bits should join up somewhere, it seems obvious that the graph of f must look something like one of those shown in Figure 6.2.

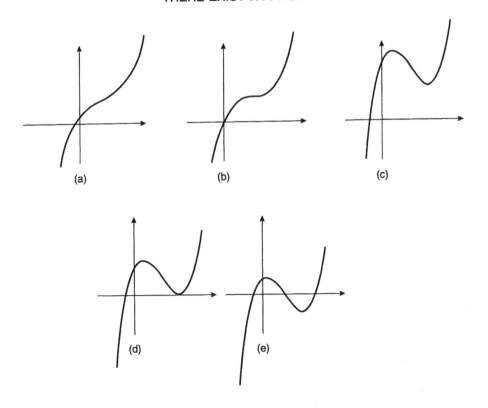

Figure 6.2

The graph will meet the x-axis one or three times. (A tangential contact, as in (d), is regarded as giving a repeated root of the curve and x-axis.)

It then seems clear that, if a and b are real numbers such that either (i) $f(a) < 0$ and $f(b) > 0$ or (ii) $f(a) > 0$ and $f(b) < 0$, then such a crossing point must lie between the points $x = a$ and $x = b$. Thus, if we can find pairs of points (a, b) satisfying (i) or (ii) we shall know that there is a root of $f(x)$ somewhere in between. A formal proof might read as follows.

Proof of Statement 13 Since f is a continuous function[3] and since $f(1) = 2 \cdot 1^3 - 9 \cdot 1^2 + 6 \cdot 1 + 3 = 2$ and $f(2) = 2 \cdot 2^3 - 9 \cdot 2^2 + 6 \cdot 2 + 3 = -5$ we see that there must be some real number a, between 1 and 2, for which $f(a) = 0$.

COMMENT

1. We have proved that there is a root *without being able to identify it exactly.* (Finding an approximation to the root will probably have to do.) Such a proof

---

[3] 'Continuous function' is a technical term which informally means that the graph of the function can be drawn without lifting the pen off the paper. In Chapter 4 the first of the 'two common functions' is continuous, the second is not.

is called a **non-constructive (existence) proof** because no exact recipe is given for determining the number *a* which is claimed to exist. (Perhaps more surprising is the following non-constructive existence story. Recall F.N. Cole's silent lecture in which he factorized $2^{67} - 1$. Many years earlier, Lucas had devised a test which *proved* that $2^{67} - 1$ factorized without being able to name either factor. So Lucas's proof that $2^{67} - 1$ has proper factors is *non-constructive*. Cole's proof of factorization is, of course, *constructive*.)

2. If I mention the most serious objection to the above proof, you may think I have taken leave of my senses. It is this. Although the diagrams in Figure 6.2 and arguments are pretty convincing, how do I know *for certain* that there will be a point which lies both on the graph of *f* and on the *x*-axis? In fact, there is a theorem called the **intermediate value theorem** (IVT) which you will eventually meet in a course of Analysis. It states that *if f is a function which is continuous throughout the range* $a \leqslant x \leqslant b$ *and if t is any real number lying between f(a) and f(b) then there is a real number* x_0 *such that* $a \leqslant x_0 \leqslant b$ *and* $f(x_0) = t$. *(That is, for each real number t intermediate between f(a) and f(b) there is at least one value, say* x_0, *of x lying between a and b such that* $f(x_0) = t$. This theorem needs more than a proof by pictures (because pictures can often lie, and in any case, look at Exercise 10) and the proof is quite difficult.

WHAT WE HAVE LEARNED

● Existence proofs, even non-constructive ones, are useful sorts of proof. When you can't find an exact theoretical answer to a problem, it is often helpful to look for an approximation to it. It is then useful to know that what you are attempting to approximate does actually exist.

EXERCISE 6 Show that the polynomial $p(x) = 2x^3 - 9x^2 + 6x + 3$ has three real roots.

EXERCISE 7 Apart from $p = 5$, there exists a prime p such that $(p - 1)! + 1$ is a multiple of p^2. Give a *constructive* existence proof of this claim. That is, *exhibit this p and show that it works.*

EXERCISE 8 Show that $\cos x + \sin x = \frac{1}{2}$ has a solution between $\pi/2$ and π? Can you state this solution *exactly*? Has it a solution in the range 0 to $\pi/2$?

EXERCISE 9 Make it clear, by means of a diagram, that the intermediate value theorem may not be applicable if the given function is not continuous.

EXERCISE 10 Consider the function *f* given by $f(x) = x^2 - 2$. Show that

(a) $f(x)$ is rational whenever *x* is rational
(b) $f(1) < 0$
(c) $f(2) > 0$

(d) there is no *rational* number a for which $f(a) = 0$. (Since Pythagoras thought that all numbers were rational numbers, he would have had to have admitted that the graph of $f(x)$ did not meet the x-axis. This shows that, to prove the intermediate value theorem you need to call upon properties that the real numbers (specifically the *completeness property* – see Chapter 9) possess that which the rational numbers do not.)

STATEMENT 14   There are infinitely many prime numbers.

DISCUSSION
Of course there is no way that mere experiment can tell you if the primes will ever 'run out'. Even after a succession of 10 000 non-prime numbers (see Exercise 12) there may be another prime waiting just around the corner.

Let us suppose that we are going to try to prove that Statement 14 is true. If there seems, initially, no obvious *direct* proof, where we try to produce an endless list of integers which are primes (though Exercise 12 looks promising), perhaps we could once again try the old 'trick' of a proof by contradiction? Thus we begin by assuming that there are only *finitely many primes* $p_1 (=2)$, $p_2 (=3), ..., p_t$, say, and seeing what, if anything, this implies. It certainly implies that *all* integers larger than p_t are *not* primes. Now we have it at the backs of our minds that *every integer can be expressed as a product of primes* (for example, $3\,185\,306\,544 = 2\cdot2\cdot2\cdot2\cdot3\cdot7\cdot7\cdot7\cdot31\cdot79\cdot79$) – indeed as a product involving some or all of the $p_1, p_2, ..., p_t$ (recall, there are, supposedly, *no other primes*). However, *some* integers greater than p_t don't appear to be divisible by *any* of these p_i. For example, the integer $N = (2\cdot3\cdot5\cdot...\cdot p_t) + 1$ certainly isn't since $2\cdot3\cdot5\cdot...\cdot p_t$ clearly *is* divisible by each p_i whilst 1 (equally clearly) is not. It looks, then, as if we have reached some kind of contradiction. A formal proof may take the following form.

Proof of Statement 14   Suppose, to the contrary, that there are only a finite number of primes. Let these primes be $p_1 = 2$, $p_2 = 3, ..., p_{25} = 97, ..., p_t$. Now let N denote the integer $p_1 \cdot p_2 \cdot ... \cdot p_t + 1$. Certainly N is not prime, since it is clearly bigger than the supposed largest prime p_t. Hence N is a product of primes. Let q be a prime dividing N. Then q must be one of the p_j (since $p_1, p_2, ..., p_t$ are the only primes – by assumption), $q = p_r$, say. But then $q | N$ whilst $p_r | (N - 1)$. Consequently q divides both N and $N - 1$ and, hence, their difference $N - (N - 1)$, i.e. $q | 1$. However, this is impossible since q is prime. It follows that our supposition (that the number of primes is finite) must be wrong. Hence the number of primes is infinite.

COMMENTS

1. Yet another proof by *reductio*, the contradiction coming about via the (false) conclusion that there is a prime which divides 1.
2. The most obvious defect in the above proof is its reliance on the (equally contentious?) assertion that every integer is expressible as a product of primes. Is this true? How do you know? Can we prove it? Fortunately we can – see Theorem 1 of Chapter 8.)

WHAT WE HAVE LEARNED

● We have been reminded yet again of the power of proof by contradiction. To prove that there are infinitely many objects of a certain type, it may be easier to show that there cannot be only finitely many.

EXERCISE 11   Are $2 + 1$, $2 \cdot 3 + 1$, $2 \cdot 3 \cdot 5 + 1$, $2 \cdot 3 \cdot 5 \cdot 7 + 1 \dots$ all primes? That is, does the formula $p_1 \cdot p_2 \cdot \dots \cdot p_t + 1$ generate *only* primes? (You may use your calculator.)

EXERCISE 12

(a) How many primes are there in the following sequence of numbers?

$$100! + 2, \ 100! + 3, \dots, 100! + 98, \ 100! + 99, \ 100! + 100$$

Make a conjecture! (Recall that $n!$ denotes $1 \cdot 2 \cdot \dots \cdot n$. Notice that $n!$ is divisible by 2, by 3, by 4, $\dots$, by $n - 1$ and by n.)

(b) Now exhibit two integers a and $a + 10^6$ such that none of the integers in between is a prime.

EXERCISE 13   Rewrite the proof of Statement 14 to turn it into a direct proof, that is, prove that to each prime p_k there is always a larger one.

Statements 12, 13 and 14 deal with the existence of none, of 'at least one' and of infinitely many objects of the type under consideration. Statement 15 deals in uniqueness.

STATEMENT 15   There is exactly one integer t (say) which is such that t, $t + 2$ and $t + 4$ are all primes.

DISCUSSION

Recall that 1 is not regarded as a prime. I'm intrigued by this allegedly unique t. So once again I'm inclined to try a few early examples to see if I can find such a t, and to see if any general pattern emerges which might indicate why this t is unique. Trying just values of t up to, say, 10, we soon see that t, $t + 2$ and $t + 4$ are certainly not all primes if t is even, not all primes if $t = 1$ (since 1 is not prime) and *are* all primes if $t = 3$. So the statement claims that for each odd $t \geqslant 5$ the triple t, $t + 2$, $t + 4$ contains at least one composite number. Looking for such composites in the triples $5, 7, 9$; $7, 9, 11$; $9, 11, 13$; $11, 13, 15$; $13, 15, 17$, etc. we find $9, 9, 9, 15, 15$, and so on. Perhaps each triple contains a number which is not prime *because it is divisible by 3*. You then ask 'Does *every* triple (for t odd, $t > 3$) contain a multiple of 3?' If we can prove this then we're in business!

Using what is, by now, a well-worn idea, we note that each integer is of one of the forms $3k$ or $3k + 1$ or $3k + 2$. Now, if t is of the form $3k$ then t itself is the required multiple of 3; if $t = 3k + 1$ then $t + 2$ ($= 3k + 3$) is a multiple of 3; if $t = 3k + 2$ then $t + 4$ ($= 3k + 6$) is a multiple of 3. Therefore (using the method of 'proof by cases') we have obtained the desired result. Or have we? This argument seems to apply to *every* odd t; so if each triple, with t odd, contains a multiple of 3 then surely *every* triple contains a non-prime? Yet this doesn't hold for the triples $1, 3, 5$ and $3, 5, 7$. We'd better sort this out! Oh, yes! I see why: $1, 3, 5$ and $3, 5, 7$ are special cases to which our argument doesn't apply.... A brief formal proof leaving a bit of work for the reader is given here.

Outline proof of Statement 15   For each odd t, one of $t, t + 2, t + 4$ is always a multiple of 3. Only if[4] $t = 1$ or $t = 3$ is that multiple actually *equal* to 3. For $t = 3$ we get a triple of primes; for $t = 1$ the corresponding triple contains the non-prime 1.

COMMENT   Here we reverted to the direct proof method ('if $t = 3$ all three integers are primes; if $t = 1$ or $t \geqslant 5$ they aren't'). Is there an indirect one beginning 'suppose there is *not* a unique t such that'. You would have to tackle two cases: (i) that there is *no* solution t and then (ii) that there is more than one solution t, and then rule out each of these as impossible. This looks a bit messy! (We shall come across a similar 'none/unique/more than one' argument in Theorem 6 in Chapter 12.)

EXERCISE 14   Prove there is a unique prime p such that $p^2 + 2$ is also prime. (*Hint:* All odd primes (except 3) are of the form $6k + 1$ or $6k - 1$ (why?).)

EXERCISE 15   Prove that there are *exactly* three positive integers n such that $n!$ has exactly n digits.

EXERCISE 16   Find three integers $t, t + 2, t + 4$ such that each is the product of exactly *two* distinct primes. Is it the only such triple?

Summary

This chapter exhibited two rather strange ideas. First, Statement 12 showed that **non-existence** can be proved. (Imagine how difficult it would be to show that a certain haystack *doesn't* contain a needle.) Then Statement 13 proved that a sought-for object did exist *but couldn't say exactly what it was*. (Statement 14 is similar. It tells us that there are infinitely many primes greater than $2^{1\,257\,787} - 1$ but, if asked, couldn't name a single one.) Statements 12 and 14 made use of the important method of **proof by contradiction**, the contradiction in the first case being of the form $X \wedge \neg X$ and in the second case a simple absurdity, namely that some prime number divided by 1.

The strategy of determining what would *have* to happen if a certain assumption is made (even if you don't know whether or not the assumption is valid) once again proved useful. (It was first used by the Greeks in geometrical arguments.)

---

[4] Do I mean 'only if' or do I mean 'if', or even 'iff'?

CHAPTER 7

Principle of Mathematical Induction

The case was ... so strong I determined to act as if it were actually proved.

SHERLOCK HOLMES

This chapter introduces a method of proof which is particularly appropriate for establishing many (but by no means all) claims which assert that some particular property holds for all positive integers. This method, (badly) named 'mathematical induction', is not to be confused with the much older concept of induction (see below). We present several examples of mathematical induction in use, as well as some cautionary examples illustrating its limitations.

Introduction

We now investigate the truth of Statements 16, 17 and 18. When presented with a statement which expresses some assertion about *all* positive integers, there is one method by which one may attempt a proof which stands out as being particularly appropriate, namely the method of **mathematical induction**.[1] What is this method?

Suppose that, for each positive integer n, $S(n)$ is a statement involving the integer n. For example, $S(n)$ might be the statement that $1 + \frac{1}{2} + \frac{1}{3} + ... + 1/n < 10$. We want to know, is the statement $S(n)$ true for *all* positive integers n? In order to

---

[1] Generally credited to Fermat and Pascal (19 June 1623–19 August 1662), but occasionally used earlier.

answer this question the method of mathematical induction requires you to proceed as follows:

1. Prove that S(1) is true.
2. Prove that, for each positive integer k, IF S(k) is true THEN S($k + 1$) is true. In symbols, this is $(\forall k \in \mathbb{Z}^+)[S(k) \rightarrow S(k + 1)]$.

If both (1) and (2) can be achieved, then the principle of mathematical induction allows us to conclude that S(n) will be true for *every* positive integer n.

The reason that we can draw this conclusion is that (1) shows that S(n) is true if $n = 1$. Then, using (2) and *modus ponens*, the truth of S(1) implies that of S(2). Using (2) again, the truth of S(2) implies that of S(3), that of S(3) implies that of S(4) and so on.

A nice analogy is the toppling an infinite row of dominoes.[2] Suppose that dominoes are arranged so that, as each falls, it can knock the 'next one' down. Then clearly, if the first domino can be made to fall, then every domino will necessarily be knocked down.

The principle of mathematical induction is badly named because of the existence of the much older method of **induction** which may be described as the inference of a general law from the observation of sufficiently many special cases. In mathematics, the essential difference between induction and mathematical induction is that, at best, the former may strongly suggest a conjecture whereas, of the two, only mathematical induction is capable of proving it.

Application of the method

Let us try the method of mathematical induction on Statement 16.

STATEMENT 16   For each positive integer n,

$$1 + 2 + \ldots + n = \frac{n(n + 1)}{2}$$

COMMENT   The dots between the 2 and the n indicate that you should continue adding 1, 2, 3, ... (and so on) until you finally add n. For $n = 1$ and 2, the left-hand side of the above formula (meaning 'the sum of all positive integers from 1 to n inclusive') must be read with a little care. In those cases this sum is supposed to read 1 and $1 + 2$. A better notation for that sum is $\sum_{t=1}^{n} t$, which is also read as 'the sum of all (integers) t from 1 to n (inclusive)'.

DISCUSSION

It is always essential to be clear precisely what S(n) is. Here S(n) is the statement

$$1 + 2 + \ldots + n = \frac{n(n + 1)}{2}$$

---

[2] Domino-toppling has become less popular of late. In 1988 a push on a single domino knocked over a further 1 382 100. This latter number is scarcely 'equal' to infinity, but it must have semed pretty close to it when the dominoes had finally been set up!

For example, S(4) states that the sum $1 + 2 + 3 + 4$ of the first four integers is equal to

$$\frac{4 \cdot 5}{2} = 10$$

As a consequence, the statement S(4) is true. We take steps 1 and 2 in turn.

Now S(1) asserts that $1 = (1 \cdot 2)/2$, which is correct, so that induction step 1 has been checked. To check induction step 2 we *assume*, for the unspecified[3] (positive) integer k, the truth of S(k):

$$1 + 2 + ... + k = \frac{k(k + 1)}{2} \tag{I}$$

and try to deduce the truth of S($k + 1$):

$$1 + 2 + ... + k + (k + 1) = (k + 1)\{(k + 1) + 1\}/2$$

This deduction is most naively achieved by adding $\{k + 1\}$ to each side of the *assumed* equality (I) and then checking that

$$\frac{k(k + 1)}{2} + \{k + 1\} = \frac{\{k + 1\}(\{k + 1\} + 1)}{2}$$

However, the right-hand side of this equality is just S(n) with n replaced by $k + 1$ throughout. Hence S($k + 1$) holds. A formal proof of Statement 16 might read as follows.

Proof of Statement 16 For each positive integer n let S(n) denote the statement

$$1 + 2 + ... + n = \frac{n(n + 1)}{2}$$

Since $1 = (1 \cdot 2)/2$ we see that S(1) holds.

Now suppose that S(k) holds for the integer k. We consider the sum $1 + 2 + ... + k + \{k + 1\}$.[4] By the induction hypothesis

$$1 + 2 + ... + k = \frac{k(k + 1)}{2}$$

Hence

$$1 + 2 + ... + k + \{k + 1\} = \frac{k(k + 1)}{2} + \{k + 1\} = \{k + 1\}\left(\frac{k}{2} + 1\right) = \frac{(k + 1)(k + 2)}{2}$$

The first and last terms of this equality tell us that S($k + 1$) holds.

Since parts 1 and 2 of the induction requirements hold, we deduce that S(n) holds for all positive integers n.

COMMENTS

1. A *very* common error made by students in writing out proofs by mathematical induction is to confuse the statement S(n) with a number. For example, the

---

[3] So that the proof of the implication will apply to each $k \in \mathbb{Z}^+$.
[4] It seems helpful to highlight the 'extra' term by the use of brackets { }.

sentence beginning 'Hence' in the above proof is often replaced (wrongly, since it is meaningless) by 'Hence $S(k) + (k + 1) = ... = S(k + 1)$.' Remember, $S(k)$ is a *statement. It is not (here) the number $k(k + 1)/2$.*

2. Another common error is to set out the above proof of part 2, from '$S(1)$ holds' onwards, as follows.

'Proof'... Now $S(k)$ says that

$$1 + 2 + ... + k = \frac{k(k + 1)}{2}$$

and $S(k + 1)$ says that

$$1 + 2 + ... + k + \{k + 1\} = \frac{(k + 1)(k + 2)}{2}$$

Subtracting the first from the second gives

$$\{k + 1\} = \frac{(k + 1)(k + 2)}{2} - \frac{k(k + 1)}{2} = \frac{(k + 1)}{2}(\{k + 2\} - k) = k + 1$$

which is true. Therefore $S(n)$ is true for all n. (Is this proof valid? If not, why not? See Exercise 2.)

3. There is a rather nice proof of Statement 16 which does not use the induction method *and which is the idea of an 8-year-old boy.* The idea goes as follows.

Suppose that $1 + 2 + ... + n = N$, say. Then

$$
\begin{array}{l}
N = 1 \quad\quad + \quad\quad 2 + ... + (n - 1) + n \\
N = n \quad\quad + (n - 1) + ... + 2 \quad\quad + 1 \\
\hline
2N = (n + 1) + (n + 1) + ... + (n + 1) + (n + 1)
\end{array}
$$

Adding top to bottom

$$= n(n + 1)$$

Hence $N = n(n + 1)/2$.

Who was the boy? Carl Friedrich Gauss. How did Gauss get the wonderful idea? I don't know – he probably just thought it was the obvious thing to do.

EXERCISE 1

(a) For each positive integer n, set $A(n) = 1 + 3 + ... + (2n - 1)$. Calculate $A(1)$, $A(2)$, $A(3)$, $A(4)$. Make a conjecture. Try to prove your conjecture correct by mathematical induction.

(b) Let $a, d \in \mathbb{Z}$ and $n \in \mathbb{Z}^+$. Find the sum $B(n)$ of the terms of the *finite arithmetic progression* $a, a + d, a + 2d, ..., a + nd$.

EXERCISE 2   What is wrong with the following 'proof' that, for all $n \in \mathbb{Z}^+$

$$\frac{1}{n} \leqslant \frac{1}{2^{n-1}}?$$

(The claimed equality is false for all $n \geqslant 3$.)

'Proof': Let S(n) be

$$\frac{1}{n} \leqslant \frac{1}{2^{n-1}}?$$

Clearly S(1) is true. Now S(k) is

$$\frac{1}{k} \leqslant \frac{1}{2^{k-1}}$$

and S(k + 1) is

$$\frac{1}{k+1} \leqslant \frac{1}{2^k}$$

Adding the first of these inequalities to twice the second we obtain

$$\frac{1}{k} + \frac{2}{k+1} \leqslant \frac{1}{2^{k-1}} + \frac{2}{2^k} = \frac{1}{2^{k-1}} + \frac{1}{2^{k-1}} = \frac{2}{2^{k-1}}$$

But, as is easily checked, if $k \geqslant 1$ then

$$\frac{2}{k} \leqslant \frac{1}{k} + \frac{2}{k+1}$$

We have therefore shown that, from $\{S(k) \text{ AND } S(k + 1)\}$ there follows

$$\frac{2}{k} \leqslant \frac{2}{2^{k-1}}$$

which is true. Since S(k) is true by assumption, and since $2/k \leqslant 2/2^{k-1}$ is true we deduce that S(k + 1) is true. Hence $\{S(n): 1/n \leqslant 1/2^{n-1}\}$ is true for all $n \in \mathbb{Z}^+$, as claimed.

An assertion which can be proved by mathematical induction but not by Gauss's method is Statement 16.1.

STATEMENT 16.1 For each positive integer n, $1^3 + 2^3 + ... + n^3 = (1 + 2 + ... + n)^2$.

DISCUSSION
We try mathematical induction on $\{S(n): 1^3 + 2^3 + ... + n^3 = (1 + 2 + ... + n)^2\}$.

Proof of Statement 16.1 Let S(n) be the statement

$$1^3 + 2^3 + ... + n^3 = (1 + 2 + ... + n)^2$$

S(1) is true since $1^3 = (1)^2$. Now suppose that S(k) holds for some integer k. We consider the sum $1^3 + 2^3 + ... + k^3 + \{k + 1\}^3$. By the induction hypothesis $1^3 + 2^3 + ... + k^3 = (1 + 2 + ... + k)^3$ and, by Statement 16,

$$(1 + 2 + ... + k)^2 = \left(\frac{k(k+1)}{2}\right)^2$$

Hence

$$1^3 + 2^3 + \dots + k^3 + \{k + 1\}^3 = \left(\frac{k(k + 1)}{2}\right)^2 + \{k + 1\}^3$$

$$= \{k + 1\}^2\left(\frac{k^2}{4} + \{k + 1\}\right)$$

$$= \frac{\{k + 1\}^2}{4}(k^2 + 4k + 4)$$

$$= \frac{\{k + 1\}^2}{4}(k + 2)^2$$

The first and last terms in this equality show that $S(k + 1)$ holds, and hence that $S(n)$ holds for all positive integers n.

COMMENT *Proving* Statement 16.1 is all very well, but how do you ever begin to *suspect*, never mind *conjecture*, that the equality in Statement 16.1 holds? There are ways, other than guessing, in which the correct formula for $1^t + 2^t + \dots + n^t$ can be found for each positive integer t but, sometimes you just have to guess, often following (ordinary) induction, what you think the correct result is. (See, especially, Exercise 4 below.) Certainly the method of mathematical induction cannot *find* the formula for you. Of course, if your guess is incorrect, neither mathematical induction nor anything else can *prove* it correct (see Statement 16.3 below).

EXERCISE 3

(a) Show that for each positive integer n

$$1^2 + 2^2 + \dots + n^2 = \frac{n(n + 1)(2n + 1)}{6}$$

(b) Find, and prove correct, a formula for the sum $1^3 + 3^3 + \dots + (2n - 1)^3$. (*Hint*: $1^3 + 3^3 + \dots + (2n - 1)^3 = 1^3 + 2^3 + 3^3 + 4^3 + \dots + (2n - 1)^3 + (2n)^3 - \{2^3 + 4^3 + \dots + (2n)^3\}$.)

EXERCISE 4 Prove that, for each positive integer n, the following equality holds

$$1(1!) + 2(2!) + \dots + n(n!) = (n + 1)! - 1$$

EXERCISE 5 Evaluate

$$A(n) = \frac{1}{2!} + \frac{2}{3!} + \frac{3}{4!} + \dots + \frac{n}{(n + 1)!}$$

for several values of n. Guess a formula for $A(n)$. Now use a mathematical induction to try to prove your guess correct.

Here is yet another example to consider.

STATEMENT 16.2 For each positive integer n, $3^{3n-2} + 2 \cdot 8^n$ is divisible by 19.

DISCUSSION

Induction step 2 requires us to prove that if $3^{3k-2} + 2 \cdot 8^k$ ($=$ A, say) is an integer multiple of 19 then so is $3^{3(k+1)-2} + 2 \cdot 8^{k+1}$ ($=$ B, say). Perhaps we can achieve this most easily by writing B in the form B $=$ (some multiple of A) $+$ (whatever is left over) and then showing that the left over bit is also divisible by 19. Indeed, if we write B $= 3^3$A $+$ (what is left over), (namely B $-$ 27A) then the term 3^3A gives rise to the exact power of 3 in B.

Proof of Statement 16.2 Let S(n) be the statement $19|(3^{3n-2} + 2 \cdot 8^n)$. S(1) is true because $3^{3-2} + 2 \cdot 8^1 = 3 + 16 = 19$. Now suppose that S($k$) holds for the integer k. We consider the quantity B $= 3^{3(k+1)-2} + 2 \cdot 8^{(k+1)}$ which appears in S($k + 1$).

$$B = 3^3(3^{3k-2} + 2 \cdot 8^k) + (-3^3 \cdot 2 \cdot 8^k + 2 \cdot 8^{k+1})$$

Now A $= 3^{3k-2} + 2 \cdot 8^k$ is a multiple of 19, by the induction assumption that S(k) holds, whilst the 'bit left over' is

$$(-3^3 \cdot 2 \cdot 8^k + 2 \cdot 8^{k+1}) = (-27 + 8) \cdot 2 \cdot 8^k$$

Since $-27 + 8$ is a multiple of 19, then so is B. Hence the assumption of S(k) leads to the deduction of S($k + 1$) and Statement 16.2 is proved, by mathematical induction.

EXERCISE 6   Prove that $3|2^{67} + 1$ and that $5|2^{224466} + 1$. (*Hint*: Show, for every positive integer n, that $3|2^{2n-1} + 1$ and that $5|2^{4n-2} + 1$. Then deduce the results claimed as special cases.)

GENERAL COMMENTS

1. In an earlier comment we made the obvious remark that no incorrect 'formula', even one which *is* correct for many small values of n, can be shown to be valid for *all* n by use of mathematical induction. But where *exactly* would such a 'proof' break down? Statement 16.3 gives an example.
2. Induction step 1 does *not have to* involve $n = 1$. Indeed, sometimes a result may only be claimed to be true for all $n > 11$, say. Then induction step 1 would be to prove S(12) true. Statement 16.4 gives an example.

With regard to General Comment 1, let us consider the following assertion.

STATEMENT 16.3   For each positive integer n,

$$1^2 + 2^2 + \ldots + n^2 = \frac{n(n^2 + 6n - 1)}{6}$$

'***Proof***'   Certainly, equality holds for $n = 1$. Now assume that the equality holds for $n = k$. We try to deduce that it also holds if $n = k + 1$. Thus we assume that

$$1^2 + 2^2 + \ldots + k^2 = \frac{k(k^2 + 6k - 1)}{6}$$

so that

$$1^2 + 2^2 + \ldots + k^2 + \{k + 1\}^2 = \frac{k(k^2 + 6k - 1)}{6} + \{k + 1\}^2$$

But the right-hand side of this equality is *not* equal to $\{k + 1\}(\{k + 1\}^2 + 6\{k + 1\} - 1)/6$ (except for $k = 0$ and $k = 1$). Hence step 2 of the induction procedure fails and so mathematical induction has failed to prove the claimed formula correct.

With regard to General Comment 2 consider the following assertion.

STATEMENT 16.4   For all integers $n > 12$, $5^n < n!$ (n factorial).

'**Proof**'   Let $S(n)$ be $5^n < n!$. Then, we easily check that $5^{13} < 13!$. Hence $S(13)$ is true. Now suppose that $5^k < k!$ (where k is some integer greater than 12). Then $5^{k+1} = 5 \cdot 5^k < 5 \cdot k!$ (by the induction hypothesis) $< (k + 1)!$ (since $5 < k + 1$). Consequently induction steps 1 and 2 hold if $n > 12$. Hence, by mathematical induction, $5^n < n!$ for all $n > 12$.

COMMENT   I have not claimed that 13 is the least n for which $5^n < n!$ Is it?

EXERCISE 7   For which positive integers n is $n! > n^{10}$? (And your proof, please!)

EXERCISE 8   For each positive integer n, let

$$C(n) = \frac{n(n - 1)(n^2 - 5n + 18)}{24} + 1$$

Prove by mathematical induction that, for all integers $n \geqslant 6$, $C(n) < 2^{n-1}$.

Fibonacci's sequence

In order to introduce an example of a 'mysterious formula proved by induction', consider the **Fibonacci sequence** 1, 1, 2, 3, 5, 8, 13, 21,..., in which each term after the second is the sum of the two preceding it. Letting $f(n)$ denote the nth term of this sequence, we have:[5] $f(1) = f(2) = 1$ and, for each integer $n \geqslant 3$, $f(n) = f(n - 1) + f(n - 2)$.

STATEMENT 16.5   For each integer $n \geqslant 1, f(n) = \{(1 + \sqrt{5})^n - (1 - \sqrt{5})^n\}/2^n\sqrt{5}$.

---

[5] This way of defining f is called *definition by induction*. This method is also used when defining $n!$ ($1! = 1$ and, for each $n \geqslant 1$, $n! = n \cdot (n - 1)!$) and similarly, the positive powers x^n of the number x ($x^0 = 1$ and $x^n = x \cdot x^{n-1}$).

LEONARDO FIBONACCI
(c. 1170–1240)

Leonardo Fibonacci was born in Pisa in around 1170 and is sometimes called Leonardo of Pisa or Leonardo Pisano. Travelling with his merchant father in Africa and around the Mediterranean, Fibonacci became acquainted with the advantage that the Hindu Arabic number system had over the Roman numerals.

On returning to Pisa around 1200, Fibonacci wrote his famous work, *Liber Abaci* (Book of Counting), in which he promoted the benefits of calculating with the 'new' number system. The *Liber Abaci*, reprinted in 1228, also described the method of false position (see Chapter 11) and many problems of a more theoretical nature, including the well-known 'rabbits problem' which gives rise to the *Fibonacci sequence*. Fibonacci was also aware of (and indeed gave a proof of) the formula (see Exercise 3(a) in this chapter) for the sum of the first n integer squares and dealt with problems of volumes taking π as $3\frac{1}{7}$, a value he later refined, in *Practica Geometricae*, to 3.141818 using Archimedes' polygonal method (see Chapter 3).

In *Liber Quadratorum* (1225) Fibonacci shows his prowess as a number theorist, solving the problem of finding numbers h such that both $x^2 + h$ and $x^2 - h$ are squares for some x. In *Flos* he tackles the problem of finding roots of equations such as $x^3 + 2x^2 + 10x = 20$, showing that such a root could be neither rational nor of any one of several other types of form. He then finds an approximate solution correct to nine decimal places. Although he did not always accept negative roots of quadratic equations, in *Flos* Fibonacci feels able to recognize negative quantities in financial problems as meaningful (that is, indicating a loss).

In his time, and for some period after, Fibonacci was the leading mathematical light in Europe. In number theory few, if any, surpassed him until Pierre de Fermat arrived on the scene.

DISCUSSION
Let S(n) be the statement

$$f(n) = \{(1 + \sqrt{5})^n - (1 - \sqrt{5})^n\}/2^n\sqrt{5}$$

S(1) is easily checked to be true. Assume S(k). We want to deduce S(k + 1). Now S(k + 1) involves $f(k + 1)$, but we cannot write $f(k + 1)$ in terms of $f(k)$ alone – we need to introduce $f(k - 1)$ too. Thus we have to assume that S(k) and S(k − 1) are *both* valid. We then check (see below) that the claimed formula holds for $f(k + 1)$ too. Hence we can infer that S(n) holds for all *n*. *No we can't!* Do you say, 'But the truth of S(k) implies the truth of S(k − 1) so, by assuming S(k) we are also assuming S(k − 1)'? Are we? Show me where! To make the k + 1 step work we have to be sure that each of the previous *two* steps works. Therefore, to proceed by induction, we have to be sure that S(k + 1) allows us to call upon both S(k) *and* S(k − 1), as follows.

Proof of Statement 16.5   For $n \geqslant 2$, let S(n) be the statement

$$f(n) = \{(1 + \sqrt{5})^n - (1 - \sqrt{5})^n\}/2^n\sqrt{5} \text{ and}$$
$$f(n - 1) = \{(1 + \sqrt{5})^{n-1} - (1 - \sqrt{5})^{n-1}\}/2^{n-1}\sqrt{5}$$

Then S(2), which requires us to check $f(2)$ *and* $f(1)$, is true (we leave this to you). Now suppose that S(k) is true. We must deduce that S(k + 1) is true. That is, we must check the above formula for $f(k + 1)$ *and* $f(k)$ are correct. Now that for $f(k)$ holds, by our assumption that S(k) is true. Therefore we have only to check that the formula for $f(k + 1)$ is correct. Now

$$f(k + 1) = f(k) + f(k - 1)$$

so that, using the induction hypothesis S(k) on $f(k)$ and $f(k - 1)$, we may write:

$$f(k + 1) = \{(1 + \sqrt{5})^k - (1 - \sqrt{5})^k\}/2^k\sqrt{5} + \{(1 + \sqrt{5})^{k-1}$$
$$- (1 - \sqrt{5})^{k-1}\}/2^{k-1}\sqrt{5}$$

$$= \{(1 + \sqrt{5})^{k-1}/2^k\sqrt{5}\}\{(1 + \sqrt{5}) + 2\}$$
$$- \{(1 - \sqrt{5})^{k-1}\}/2^k\sqrt{5}\}\{(1 - \sqrt{5}) + 2\} \tag{II}$$

But $\{(1 + \sqrt{5}) + 2\} = \frac{1}{2}(1 + \sqrt{5})^2$ and $\{(1 - \sqrt{5}) + 2\} = \frac{1}{2}(1 - \sqrt{5})^2$. Putting this into (II) and tidying up shows that

$$f(k + 1) = \{(1 + \sqrt{5})^{k+1} - (1 - \sqrt{5})^{k+1}\}/2^{k+1}\sqrt{5}$$

as required.

COMMENT   The formula in Statement 16.5 was not obtained as the result of a lucky guess but rather by a method outlined in Chapter 11 (see Examples 5 and 6).

EXERCISE 9   Prove (by mathematical induction) that, for each positive integer *n*,

$f(5n)$ is a multiple of 5. (*Hint:* make use of the relation

$$f(n) = f(n-1) + f(n-2)$$

to express $f(5k + 5)$ in terms of $f(5k)$ together with some multiple of 5.)

EXERCISE 10   Prove that for each positive integer n, no integer greater than 1 divides both $f(n)$ and $f(n + 1)$.

EXERCISE 11   Prove that for all positive integers n

$$f(1) + f(3) + \dots + f(2n-1) = f(2n)$$

and that

$$f(2) + f(4) + \dots + f(2n) = f(2n+1) - 1$$

EXERCISE 12   Prove that $\sum_{t=1}^{n} f(t)^2 = f(n)f(n+1)$.

The binomial theorem

Another result which similarly involves 'mysterious' numbers and which invites an attack by induction is the formula for the **binomial theorem**.

STATEMENT 16.6   Let n be a positive integer. Then

$$(x+y)^n = \binom{n}{0}x^n + \binom{n}{1}x^{n-1}y^1 + \binom{n}{2}x^{n-2}y^2 + \dots + \binom{n}{n-1}x^1y^{n-1} + \binom{n}{n}y^n$$

Here $\binom{n}{r}$ is the **binomial coefficient**

$$\frac{n!}{r!(n-r)!}$$

where, for convenience, we *define* 0! to be 1 so that both $\binom{n}{0}$ and $\binom{n}{n}$ turn out to be 1. In fact $\binom{n}{r}$ is the number of different ways of choosing r items from a collection of size n and is, consequently, an integer (see Slomson (1991, pp. 6–9)). Further, for each integer r such that $1 \leqslant r \leqslant n$ we can prove that

$$\binom{n}{r} = \binom{n-1}{r-1} + \binom{n-1}{r}$$

Indeed, for each n, the $\binom{n}{r}$ $(0 \leqslant r \leqslant n)$ are the numbers on the nth row of Pascal's Triangle, the first seven rows of which are

```
            1
          1   1
        1   2   1
      1   3   3   1
    1   4   6   4   1
  1   5  10  10   5   1
1   6  15  20  15   6   1
```

We shall ask you to prove Statement 16.6 by mathematical induction in Chapter 14. We show you now, by a *different route*, *why* Statement 16.6 is true.
    Writing

$$(x + y)^n = (x + y)(x + y)\ldots(x + y) \tag{III}$$

there being n factors in the product, each term $x^r y^{n-r}$ in the expansion of $(x + y)^n$ is obtained by making a choice of x from r of the n factors (and, therefore, a forced choice of y from each of the other $n - r$ factors of the product). The number of such choices (of r brackets from n brackets) is $\binom{n}{r}$. This, therefore, is the appropriate coefficient $x^r y^{n-r}$ in the binomial expansion.

EXERCISE 13   Prove, by mathematical induction, that, for each positive integer x, we have $5 \mid x^5 - x$.

EXERCISE 14   Let h be a fixed real number such that $h > -1$. Prove, for each positive integer n, that

$$(1 + h)^n \geqslant 1 + nh$$

Two inequalities

Statement 17, if true, looks like a likely candidate for proof by induction.

STATEMENT 17   For each positive integer n

$$\frac{1}{1^2} + \frac{1}{2^2} + \frac{1}{3^2} + \ldots + \frac{1}{(n-1)^2} + \frac{1}{n^2} < 2$$

DISCUSSION
We naturally let S(n) be

$$\frac{1}{1^2} + \frac{1}{2^2} + \frac{1}{3^2} + \ldots + \frac{1}{(n-1)^2} + \frac{1}{n^2} < 2$$

Then S(1) is $1/1^2 < 2$. Hence S(1) is true. Now we suppose S(k) is true, that is

$$\frac{1}{1^2} + \frac{1}{2^2} + \frac{1}{3^2} + \ldots + \frac{1}{(k-1)^2} + \frac{1}{k^2} < 2$$

To check S($k + 1$) we must look at

$$\frac{1}{1^2} + \frac{1}{2^2} + \frac{1}{3^2} + \ldots + \frac{1}{(k-1)^2} + \frac{1}{k^2} + \left\{\frac{1}{(k+1)^2}\right\}$$

which is *not* obviously less than 2. So our induction attempt fails. What can we do? It may seem crazy but we can try to prove a *stronger* result. (There is no reason why this idea should have occurred to you.) Sometimes, but perhaps not very often, attempting to prove a result stronger than that sought will work, so it is worthwhile seeing an example. Here we replace S(n) by the stronger

$$T(n): \frac{1}{1^2} + \frac{1}{2^2} + \frac{1}{3^2} + \ldots + \frac{1}{(n-1)^2} + \frac{1}{n^2} \leqslant 2 - \frac{1}{n}$$

If we can prove T(n) for all n then certainly S(n) will be true for all n. A formal proof may proceed as follows.

Proof of Statement 17   Let T(n) be

$$\frac{1}{1^2} + \frac{1}{2^2} + \frac{1}{3^2} + \ldots + \frac{1}{(n-1)^2} + \frac{1}{n^2} \leqslant 2 - \frac{1}{n}$$

Clearly T(1) is true. Now

$$\frac{1}{1^2} + \frac{1}{2^2} + \frac{1}{3^2} + \ldots + \frac{1}{(k-1)^2} + \frac{1}{k^2} + \left\{\frac{1}{(k+1)^2}\right\} \leqslant 2 - \frac{1}{k} + \left\{\frac{1}{(k+1)^2}\right\}$$

$$\leqslant 2 - \frac{1}{k+1}$$

as is easily confirmed. We have therefore shown that, if T(k) holds, so does T(k + 1). Hence the stronger result is true by mathematical induction and the desired result follows immediately.

COMMENT   What helps this 'strengthened' version to work where the 'weaker' version doesn't? It is the fact that if you strengthen the conclusion you are able to call upon a correspondingly strengthened hypothesis.

EXERCISE 15   Let S(n) be

$$\frac{1}{2} + \frac{1}{2^2} + \frac{1}{2^3} + \frac{1}{2^4} + \ldots + \frac{1}{2^n} < 1$$

Prove that S(n) is valid for all $n \in \mathbb{Z}^+$: (i) by replacing S(n) by

$$T(n): \frac{1}{2} + \frac{1}{2^2} + \frac{1}{2^3} + \frac{1}{2^4} + \ldots + \frac{1}{2^n} = 1 - \frac{1}{2^n}$$

(ii) by supposing

$$\frac{1}{2} + \frac{1}{2^2} + \frac{1}{2^3} + \frac{1}{2^4} + \ldots + \frac{1}{2^k} < 1$$

deducing that

$$\frac{1}{2^2} + \frac{1}{2^3} + \frac{1}{2^4} + \ldots + \frac{1}{2^{k+1}} < \frac{1}{2}$$

and, finally, adding $\frac{1}{2}$ to each side.

EXERCISE 16   Show, for all positive integers n that

$$\frac{1}{1!} + \frac{1}{2!} + \ldots + \frac{1}{n!} < 2$$

(*Hint:* Prove $n! \geqslant 2^{n-1}$ and then use Exercise 15.)

EXERCISE 17   Define the function $h: \mathbb{Z}^+ \to \mathbb{R}$ by

$$h(1) = 1$$

$$h(n) = \frac{1}{2}h(n-1) + \frac{1}{h(n-1)}$$

Show, for $n \geqslant 2$, that $h(n) < 2$. (*Hint:* Prove the stronger result, $1 \leqslant h(n) < 2$.)

Statements 16 and 17 said 'For each integer n' and then something about n. Statement 18 is slightly different. It says 'For each integer n there exists an integer t such that' and then something about t and n.

STATEMENT 18   For each positive integer n there exists a positive integer t such that

$$1 + \frac{1}{2} + \frac{1}{3} + \dots + \frac{1}{t} > n$$

DISCUSSION

Despite the form, this looks to be another candidate for an induction proof – provided it is a true statement. We let $S(n)$ be 'There exists an integer $t(n)$ which depends on n and which is such that

$$1 + \frac{1}{2} + \frac{1}{3} + \dots + \frac{1}{t(n)} > n'$$

Then $S(1)$ is clearly true (just take $t(1) = 2$). Now we suppose that there exists a $t(k)$ such that $S(k)$ is true. How can we show that there exists a $t(k+1)$ which will ensure that $S(k+1)$ is true? All we want is an integer $t(k+1)$ such that

$$\frac{1}{t(k) + 1} + \frac{1}{t(k) + 2} + \dots + \frac{1}{t(k+1)} > 1$$

But this doesn't seem to be any easier than the problem that we started with. What can we do? We could look at Problem 25 in Chapter 14, but on this occasion we follow the proof given by Nicole Oresme.[6]

Proof of Statement 18   For each positive integer s we see easily by comparing corresponding terms that:

$$\frac{1}{1} + \frac{1}{2} + \frac{1}{3} + \frac{1}{4} + \frac{1}{5} + \frac{1}{6} + \frac{1}{7} + \frac{1}{8} + \frac{1}{9} + \frac{1}{10} + \frac{1}{11} + \frac{1}{12} + \frac{1}{13} + \frac{1}{14} + \frac{1}{15} + \frac{1}{16} + \frac{1}{17} + \frac{1}{18} + \dots + \frac{1}{2^s} \tag{IV}$$

$$> \frac{1}{1} + \frac{1}{2} + \frac{1}{4} + \frac{1}{4} + \frac{1}{8} + \frac{1}{8} + \frac{1}{8} + \frac{1}{8} + \frac{1}{16} + \frac{1}{16} + \frac{1}{16} + \frac{1}{16} + \frac{1}{16} + \frac{1}{16} + \frac{1}{16} + \frac{1}{16} + \frac{1}{32} + \frac{1}{32} + \dots + \frac{1}{2^s} + \frac{1}{2^s} + \dots + \frac{1}{2^s}$$

$$= \frac{1}{1} + \frac{1}{2} \quad + \frac{1}{2} \qquad + \frac{1}{2} \qquad\qquad\qquad\qquad + \frac{1}{2} \qquad\qquad + \dots + \qquad + \frac{1}{2} = \frac{s+2}{2}$$

[6] Bishop Nicole Oresme (c. 1350).

Hence, to be sure that

$$1 + \frac{1}{2} + \frac{1}{3} + \ldots + \frac{1}{t(n)} > n$$

we have only to take $t \geqslant 2^s$ where $(s + 2)/2 \geqslant n$, for example $s = 2n - 2$ will do. In other words, given n, we may take $t(n) = 2^{2n-2}$.

EXERCISE 18 (For your computer.) What is the least value of t for which

$$1 + \frac{1}{2} + \frac{1}{3} + \ldots + \frac{1}{t} \geqslant 10$$

Because of possible rounding errors, can you rely on your computer's answer?

Tricky inductions

EXERCISE 19 The following results are all true. (See Allenby (1989, pp. 170, 67, 164).) Trying to prove them by mathematical induction might be rather tricky. Where do *you* get stuck?

(a) Every positive integer n is a sum of four integer squares.
(b) For each integer $n > 1$ there is a prime between n and $2n$.
(c) If p is a prime of the form $4k + 1$ then $p = a^2 + b^2$ where $a, b \in \mathbb{Z}$.

An invalid argument

What is wrong with the following argument by mathematical induction that everyone in the world is the same height?

'**Proof**' Let R be a (very big) room. I shall prove

{H(n): Whenever n people are in room R then they all have the same height}

(i) H(1) is true. For, whenever just 1 person is in room R then all people in room R at that time have the same height.
(ii) Now suppose H(k) is true. That is, assume that, whenever k people are assembled in room R then all people present have the same height. We deduce that H($k + 1$) is true, as follows. (Does Figure 7.1 help?)

Take any $k + 1$ people into room R. Now send out person 1. This leaves the k people $2, 3, \ldots, k + 1$ in room R and so, by the induction hypothesis, they are all of the same height. Now ask person 1 to return but invite person $k + 1$ to leave. Again, the induction hypothesis tells us that people $1, 2, \ldots, k$ are all the same height. But person 1 has the same height as persons $2, 3, \ldots, k$, and person $k + 1$ has the same height as persons $2, \ldots, k$. Thus *all $k + 1$ people* in room R have the same height.

Since induction steps 1 and 2 hold, we see that, for each positive integer n, if there are n people in the room, then they all have the same height.

The fallacy is revealed in Chapter 15.

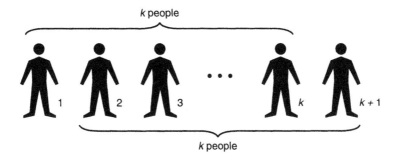

Figure 7.1

Summary

1. This chapter has concentrated on one method of proving certain results 'for all positive integers n' (or from some integer onwards) – the method of **mathematical induction**. The method requires that you show that both the two induction steps are true. However, the method is not guaranteed to work even when the result you are trying to prove is true. (In the proof of Statement 17 it is difficult to prove directly that step 2 is true. However, attempting to prove an even stronger statement may work.) Step 1 need not concern the case $n = 1$. We may start an induction proof by proving S(n) for some $n > 1$ (Statement 16.4).
2. If S(n) concerns a formula giving a numerical value do *not* confuse the statement S(n) with that value occurring in that statement (see comment 1 after Statement 16).
3. Always set out clearly the condition S(n). Sometimes the exact formulation of S(n) requires more careful attention than is obvious at first glance (see the Discussion of Statement 16.5).
4. Avoid presentations as given in Comment 2 following Statement 16. The logical structure of the argument ('From {S(k) AND S($k + 1$)} deduce something true. Hence, since S(k) is true, so is S($k + 1$) true') is unsound.
5. Mathematical induction cannot find the correct formula for you, nor can it prove correct a guess which is wrong.

CHAPTER 8

The Integers and the Rational Numbers

God made the integers, the rest is the work of man.

KRONECKER

This chapter is concerned mainly with division properties of the integers the highlight being the proof of the Fundamental Theorem of Arithmetic. To obtain this result, several lesser results, interesting in their own right, are first proved. These proofs depend heavily on the well-ordering principle for $\mathbb{Z}$ with which we begin.

Introduction

In previous chapters, we have used the integers (and rational and real numbers) as vehicles to introduce various methods of proof. In the next three chapters we reverse these roles, using our now greater facility with proof construction to demonstrate further properties of these numbers. In this chapter these properties relate mainly to the integers. We begin with what appears to be a triflingly obvious statement.

Well-ordering principle

The **well-ordering principle** of the integers states that

Each non-empty set of positive integers contains a least member.

Example 1

(a) The set of all odd primes has least member 3.
(b) The set $\{n \in \mathbb{Z}: n > 1$ and $1 + 2 + \ldots + n$ is a perfect square$\}$ has least member 8.

Some least members are more difficult to identify.

Example 2

The set P of all positive integers in the set $\{117x^3 - 55y^3 : x, y \in \mathbb{Z}^+\}$ clearly contains 62 $(x = y = 1)$, 613 $(x = 4, y = 5)$. It is not (to me) immediately obvious what the smallest member of P is. We should note immediately that the well-ordering principle certainly distinguishes $\mathbb{Z}^+$ from the set of positive rational numbers, $\mathbb{Q}^+$, which has no smallest member. (For if α were the supposed 'least positive rational' then $\alpha/2$ would be an even smaller (positive) one.)

EXERCISE 1   What is the least positive integer in the set $\{217x - 84y : x \in \mathbb{Z}^+, y \in \mathbb{Z}^+\}$?

Which principle – well-ordering or mathematical induction – do you find the easier to accept? In fact, in $\mathbb{Z}$, the well-ordering principle is equivalent to the principle of mathematical induction in that each of these principles can be proved on assumption of the other (see Hirst (1995, pp. 21, 22)). Statements concerning 'all n' can sometimes be proved more easily by using the well-ordering principle (w.o.) rather than by using mathematical induction (m.i.).

   To show these two principles in action, we offer two proofs of one and the same result, first using the principle of m.i. and then using w.o.

Products of primes

Earlier we defined the concept of prime number. Primes are important because they are 'building blocks' (with respect to multiplication) from which all integers are constructed. This is shown in Theorem 1.

Theorem 1
Every integer greater than 1 is expressible as a product of primes.

   DISCUSSION
   To make this theorem work we have to allow that each prime p is expressible in the 'product' form 'p'. That is, we extend the usual meaning of the word 'product' to the case where there is only one term in the product. If we didn't do this then our theorem would have to read 'Every integer greater than 1 is either a prime or a product of primes.' There would be no harm in this, it's just that the earlier way of stating the theorem is preferred.

   How to prove the theorem? The words 'every integer' suggest mathematical induction. So let us try to prove, for each integer $n > 1$, the truth of the statement

$$S(n): \text{The integer } n \ (>1) \text{ is a product of primes}$$

Step 1 here requires that S(2) be true. It is (since 2 is prime). We now suppose that S(k) is true (so that k is, in our extended sense, a product of primes) and we try to prove S(k + 1).

If k + 1 is prime then S(k + 1) is certainly true. What happens if k + 1 is *not* prime? Then k + 1 must be a product, k + 1 = mn, say, where 1 < m, n < k + 1. Since m and n are products of primes... (why are they? *All you know from assuming S(k) – the induction hypothesis – is that k is a product of primes. S(k) tells you nothing about m and n.*)

After some thought we see that, to prove that S(k + 1) is true we need to know that *all* integers, from 2 to k inclusive, are products of primes (since we don't know into which integers k + 1 will factorize). Eventually the idea occurs of replacing S(n) by the stronger hypothesis T(n),

T(n): Every integer from 2 to n inclusive is a product of primes

Now we are ready to present our proof.

First proof of Theorem 1   Let S(n) be 'every integer from 2 to n inclusive is a product of primes'. Then S(2) is (trivially) true. Now suppose that S(k) is true and consider the integer k + 1. Now either k + 1 is prime, in which case there is nothing left to prove, or there exist integers m, n, each less than k + 1 such that k + 1 = mn. Because we are assuming S(k) to be true we can deduce that both m and n can be factorized into products of primes. Multiplying these factorizations together provides the required product for k + 1.

COMMENT   Is that it? Nearly, but not quite. Recall, S(k + 1) says that 'all integers from 2 to k + 1 inclusive'. We have just shown that k + 1 has the required factorization, but our statement form S(n) requires us to deduce (from S(k)) that *all* integers from 2 to k + 1 inclusive have the required factorization. Do we know that similar factorizations hold for all the other integers from 2 to k inclusive? In fact we do, because it is the induction hypothesis S(k). Consequently the problem is resolved easily, but we ought at least to comment on it, if for no other reason than that it will encourage us to check that *all* the induction hypotheses assumed under S(k) are still valid for S(k + 1). (You may recall that a similar situation arose with the Fibonacci sequence in Chapter 7.)

Now how about a proof of Theorem 1 using w.o.? Here it is – with no preliminary Discussion.

Second proof of Theorem 1   Suppose that Theorem 1 is *false*. Then there is at least one integer (> 1) which is *not* expressible as a product of primes. The set of all such integers – let us denote it by C and call it the set of all *criminals* (!) – is, therefore, not empty and so, by w.o., C contains a smallest member (affectionately known as the *least criminal*) c, say.

Now c itself cannot be a prime since, being a criminal, it has no factorization as a product of primes. Hence c is expressible as a product, c = mn, say, where 1 < m, n < c. But each of m and n is expressible as a product of primes (since each is smaller than the least criminal c) and a prime factorization for c follows by multiplying those for m and n together.

!Thus, the assumption that C is non-empty (and, therefore, has a smallest member) leads to a contradiction. Hence C *is* empty, that is, *there are no criminals.*

COMMENT This second proof has turned out a bit longer than the first but one has the feeling that it tackles the problem more directly and presents the argument more clearly than does the first proof. Indeed, whilst proofs by mathematical induction are usually convincing with regard to correctness, they often obliterate any possible hope of seeing *why* the result being proved is true. As a consequence some mathematicians feel that, if there is a method of proof other than by m.i., one should use it.

EXERCISE 2 Theorem 1 tells us that, in $\mathbb{Z}$, one cannot keep on factorizing an integer > 1 into longer and longer products of smaller and smaller integers. Show that this conclusion is false in $\mathbb{Q}$ by finding rational numbers a_1, a_2; b_1, b_2, b_3; c_1, c_2, c_3, c_4; etc., all greater than 1, for which

$$\frac{3}{2} = a_1 \cdot a_2 = b_1 \cdot b_2 \cdot b_3 = c_1 \cdot c_2 \cdot c_3 \cdot c_4 \ldots$$

EXERCISE 3 From w.o. deduce the following trivial but useful extension: 'every non-empty set of *non-negative* integers contains a least member'.

The division theorem

One nice application of w.o. is in the proof of a result anticipated in Chapter 3 (see Comment 1 following Statement 1.1).

Theorem 2
(The Division Theorem) Let a and b be given integers with $b \neq 0$. Then there exist integers m and r such that $a = mb + r$ where $0 \leqslant r < |b|$.[1]

Example 3

(a) Given $a = 47$ and $b = 14$ we have $47 = 3.14 + 5$.
(b) Given $a = 4\,000\,001$ and $b = -6317$ we have

$$4\,000\,001 = (-633) \cdot (-6317) + 1340$$

Note that $0 \leqslant 1340 < |-6337|$.
(c) Given $a = 1\,111\,111$ and $b = 4649$ we have $1\,111\,111 = 239 \cdot 4649 + 0$.

COMMENT In the statement of Theorem 2 we use the sensible letters m and r to remind us that they stand for 'multiplier' and 'remainder' respectively. Well-chosen notation can be very effective in reminding the reader what objects the symbols stand for.

---

[1] Recall from Chapter 4 that $|b|$ is equal to b if $b \geqslant 0$ and is equal to $-b$ if $b < 0$. Thus $|b| \geqslant 0$ for every b.

Do you believe Theorem 2? Of course you do! I still have some schoolwork exercises on division with remainder which I did when I was nine years old. I expect that you too have 'known' Theorem 2 from an equally early age. Surely one can't prove it – it's too basic to have a proof? Nonsense! Here is one, based on w.o.

Proof of Theorem 2 Let T be the set $\{a - m|b|: m \in \mathbb{Z}\}$ and S the subset of non-negative integers which belong to T. It is easy to see (is it? why?) that S is non-empty. Hence, by Exercise 3, S possesses a least non-negative member $c = a - m_0|b|$, say. Now we must have $c < |b|$ (for, if $c \geqslant |b|$ then $c - |b|$ ($= a - \{m_0 + 1\}|b|) \geqslant 0$ would, because it is non-negative and of the correct form, be an element of S which is smaller than S's supposed smallest member c). Rewriting $c = a - m_0|b|$ in the form $a = m_0|b| + c$, we see that, if $b > 0$, we have $a = m_0 b + c$ whereas, if $b < 0$, $a = (-m_0)b + c$. Thus, in all cases (that is, $b > 0$ and $b < 0$) we can find suitable integers m and r such that $a = mb + r$ where $0 \leqslant r < |b|$, as claimed.

EXERCISE 4 Find suitable m, r as in Theorem 2 if: (a) $a = 198\,128$, $b = 991$; (b) $a = -111\,211$, $b = 311$.

EXERCISE 5 Show that if, following Theorem 2, I write $a = mb + r$ and you write $a = m_1 b + r_1$, where $0 \leqslant r < |b|$ and $0 \leqslant r_1 < |b|$, then $r = r_1$ and $m = m_1$ (that is, a and b determine m and r uniquely).

(Greatest) common divisors

Theorem 1 is not by any means the end of the story. In fact, the above representation of each integer as a product of primes is, in a very strong sense, unique. To prove this we need some preliminary results which are fascinating in their own right and depend heavily on Theorem 2. We begin with a definition.

Definition 1
Let a and b be given integers, not both zero. Then the positive integer d is called the **greatest common divisor** (**gcd**)[2] of a and b if and only if: (i) $d|a$ and $d|b$ *and* (ii) *if* c is a positive integer such that $c|a$ *and* $c|b$ then $c \leqslant d$. This gcd we denote by (a, b).

Any number c as in (i) is called a **common divisor** of a and b. The gcd of a and b is, therefore, that common divisor of a and b which is numerically largest.

Definition 2
If $(a, b) = 1$ we say that a and b are **coprime** (sometimes **relatively prime**).

Example 4

(a) $(154, 649) = (-649, 154) = 11$ – since 11 is the greatest integer dividing both 154 and 649

(b) $(585, 847) = 1$. Thus 585 and 847 are coprime.

[2] Also sometimes called, in schools, *highest common factor* (*hcf*).

EXERCISE 6   Show that the pair of integers 0 and 0 has all integers as common divisors – and, hence, *no* gcd.

EXERCISE 7   Show that, for integers a, b not both zero,

$$(a, b) = (-a, b) = (a, -b) = (-a, -b) = (b, a) = \ldots$$

i.e. show that $(a, b) = (|a|, |b|)$.

A pretty little function, which is of great importance in public key cryptography, is given in Definition 3.

Definition 3
For each positive integer n, we define $\phi(n)$ to be the number of integers k such that $1 \leqslant k \leqslant n$ such that $(k, n) = 1$. This function $\phi: \mathbb{Z}^+ \to \mathbb{Z}^+$ is called **Euler's** ϕ **(phi) function**.

Example 5
$\phi(1) = 1$, $\phi(5) = 4$, $\phi(12) = 4$, $\phi(36) = 12$, $\phi(7519) = 7344$. (How do I know this last one? Try the last part of the next exercise with $p = 73$.)

EXERCISE 8   Show that, if p and q are unequal primes, then (a) $\phi(p) = p - 1$; (b) $\phi(p^2) = p(p - 1)$; (c) $\phi(pq) = (p - 1)(q - 1)$.

It may seem that to find (a, b) we only need to factorize a and b into products of primes – as we know we may do, by Theorem 1 – and pick out common primes. For example, $1\,987\,788 = 2 \cdot 2 \cdot 3 \cdot 11 \cdot 11 \cdot 37 \cdot 37$ and $91\,464 = 2 \cdot 2 \cdot 2 \cdot 3 \cdot 37 \cdot 103$. Hence $(1\,987\,788, 91\,464) = 2 \cdot 2 \cdot 3 \cdot 37$. What if I tell you that $1\,987\,788$ can be expressed in a different way as a product of primes? Do I hear you say 'But, *it can't be*'? Why can't it? We shan't know this for *certain* until we have proved Theorem 8 later in this chapter. (See the surprising Exercise 26.) Nevertheless, there is a way to find (a, b) – and you *don't need to factorize a and b to get it!*

The Euclidean Algorithm[3]

Given integers a and b (with $b \neq 0$) suppose we repeatedly apply the division theorem as follows (it is helpful to change the notation very slightly):

> Step 1 finds integers m_1 and r_1 such that $a = m_1 b + r_1$ where $0 \leqslant r_1 < |b|$
> Step 2 finds integers m_2 and r_2 such that $b = m_2 r_1 + r_2$ where $0 \leqslant r_2 < r_1$
> Step 3 finds integers m_3 and r_3 such that $r_1 = m_3 r_2 + r_3$ where $0 \leqslant r_3 < r_2$

and so on.

Continuing in this way we obtain a decreasing sequence $|b| > r_1 > r_2 > r_3 > \ldots$ of non-negative integers. However, by the well-ordering principle, each such sequence, continued as far as possible, must stop when, for some integer k, we arrive at $r_k = 0$. Now, in our repeated division theorem we finally arrive at step k when

---

[3] An algorithm is a procedure described by a sequence of steps which, if followed unquestioningly, will lead to the desired result.

Step k finds integers m_k and r_k such that $r_{k-2} = m_k r_{k-1} + r_k$ where $0 = r_k$

that is $r_{k-2} = m_k r_{k-1}$. This equality tells us that $r_{k-1} | r_{k-2}$. In other words, $(r_{k-2}, r_{k-1}) = r_{k-1}$. Theorem 4 below will show that

$$(a, b) = (b, r_1) = (r_1, r_2) = \ldots = (r_{k-2}, r_{k-1}) = r_{k-1}$$

We will then have proved Theorem 3.

Theorem 3
Let a and b ($\neq 0$) be integers. Repeatedly applying the division theorem, the last non-zero term in the sequence of remainders is (a, b).

An example should make everything clear (even if it *proves* very little).

Example 6

$$649 = 4 \cdot 154 + 33$$
$$154 = 4 \cdot \ 33 + 22$$
$$33 = 1 \cdot \ 22 + 11 \leftarrow \text{last non-zero remainder}$$
$$22 = 2 \cdot \ 11 + \ 0$$

This confirms what we already knew (cf. Example 4) – that $(154, 649) = 11$. The procedure of finding gcds in this way is known as the **Euclidean Algorithm**.

EXERCISE 9   Find $(312\,497, 453\,919)$ without factorizing either integer.

We still have to complete the proof of Theorem 3. We show Theorem 4.

Theorem 4
Let a and b ($\neq 0$) be integers. Writing $a = mb + r$ as in the division theorem, we have $(a, b) = (b, r)$.

DISCUSSION
How can one prove that two numbers x and y are equal? One might prove that $x - y = 0$. Alternatively, one might show that $x \leqslant y$ *and* $y \leqslant x$. Here we use the latter approach. Specifically we prove (i) $(a, b) \leqslant (b, r)$ and then (ii) $(b, r) \leqslant (a, b)$. This will ensure that $(a, b) = (b, r)$, as required.

Proof of Theorem 4   For convenience only, let us write (a, b) as u and (b, r) as v. Then, $u | a$ and $u | b$. Since $r = a - mb$ we deduce that $u | r$. But since $u | b$ we see that u divides both b and r, and, consequently, $u \leqslant v$. In an identical manner we prove $v \leqslant u$.[4] This does it.

EXERCISE 10   Let a and b be integers such that $(a, b) = 1$. Is it necessarily true that

(a) $(2a, 3b) = 1$
(b) $(a + b, a^2 + b^2) = 1$

[4] 'Let symmetry work *for* you.' (DTC, schoolmaster.)

Going backwards

If we reverse the steps of the Euclidean Algorithm we obtain several interesting consequences. As applied to Example 6, we use the equalities in the reverse order successively to write (i) 11 in terms of 33 and 22; (ii) 22 in terms of 154 and 33, and (iii) 33 in terms of 649 and 154. We obtain

$$
\begin{aligned}
11 = \quad & 33 - 1 \cdot 22 = 33 & - 1 \cdot (154 - 4 \cdot 33) \\
= 5 \cdot \ & 33 - 1 \cdot 154 = 5 \cdot (649 - 4 \cdot 154) - 1 \cdot 154 \\
= 5 \cdot & 649 - 21 \cdot 154
\end{aligned}
\tag{II}
$$

We have therefore written the gcd $(649, 154)$ as a **linear combination** of 649 and 154. (Note that the multipliers 5 and -21 are by no means unique. Indeed, adding and subtracting $154 \cdot 649 \cdot k$, we see that, for *every* integer k, we have $11 = (5 - 154k) \cdot 649 + (649k - 21) \cdot 154$.

The proof that this procedure always works is a bit messy to write out (there are too many suffices flying around). The first couple of steps are

$$
\begin{aligned}
(a, b) = r_{k-1} = r_{k-3} \qquad & -m_{k-1} r_{k-2} = r_{k-3} - m_{k-1} (r_{k-4} - m_{k-2} r_{k-3}) \\
= (m_{k-1} m_{k-2} + 1) r_{k-3} \ & -m_{k-1} r_{k-4} \cdots
\end{aligned}
$$

We could try a proof by m.i., but it's a bit dry so we'll give it a miss! In any case, a much sweeter proof using w.o. is available. (See Chapter 12, Theorem 4.)

The following result, whose validity we have indicated rather than proved, then seems clear.

Theorem 5

Let a and b be integers, not both 0. Then there exist integers s and t, say, such that (a, b) may be expressed in the form $sa + tb$.

EXERCISE 11

(a) Find integers x and y such that $649x + 154y = 22$.
(b) Show that there are no integers z and t such that $649z + 154t = 23$.

EXERCISE 12   Find $(39\,001, 38\,141)$ and express it in the form $39\,001s + 38\,141t$
(a) for suitable integers s and t; (b) using integers s, t such that $s > t$.

EXERCISE 13   Show that if $(a, b) = d$ then

$$
\left(\frac{a}{d}, \frac{b}{d} \right) = 1
$$

Must $(a, b/d)$ also equal 1?

EXERCISE 14   Show that there exist integers x and y such that $ax + by = c$ iff $(a, b) | c$.

EXERCISE 15   Show that

$$
d = (a, b) \text{ iff } d | a \text{ AND } d | b \text{ AND } (\forall c \in \mathbb{Z}) [\{c | a \wedge c | b\} \rightarrow c | d]
$$

(In words: show that (a, b) is that (positive) common divisor of a and b which is a *multiple* of every other common divisor.)

We can use Theorem 5 to obtain an important result concerning primes.

Theorem 6
Let a and b be integers and let p be a prime number. Then:

$$p|ab \rightarrow p|a \text{ or } p|b \quad \text{(or both)}$$

DISCUSSION
The statement form $\{A \rightarrow (B \text{ or } C)\}$ should remind you of Statement 6. Accordingly, the method to try is: assume that $p|ab$ AND $p \nmid a$ and then try to show $p|b$. Assuming $p \nmid a$, what can we say? First, $(a, p) = 1$. Hence, if it helps, there exist integers s and t such that $sa + tp = 1$. Does it help? Yes! Watch this, with pleasure!

Proof of Theorem 6   Assume that the prime number p divides the product ab and that $p \nmid a$. Then there exist $s, t \in \mathbb{Z}$ such that $sa + tp = 1$. It follows that $sab + tpb = b$. Now $p|ab$ (given) and $p|p$. Hence $p|(sab + tpb)$. That is, $p|b$.

COMMENT   If ever you find integers a, b with $(a, b) = 1$ you should regard it almost as a duty to write, immediately, 'Hence there exist integers s, t such that $sa + tb = 1$'. It is often a useful move to make.

EXERCISE 16   Let $n, p \in \mathbb{Z}^+$, p being a prime. Show that if $p|n^2$ then $p|n$ and $p^2|n^2$.

EXERCISE 17

(a) Give an example of three integers a, b and n (all $\geqslant 2$) such that $n|ab$ but $n \nmid a$ and $n \nmid b$.
(b) Show that, if $(n, a) = 1$ then $n|ab \rightarrow n|b$. (*Hint:* $\exists r, s \in \mathbb{Z}$ such that $rn + sa = 1$.)

EXERCISE 18   Let a, b, c be integers. Given that $(a, bc) = (b, ca) = 1$ is it necessarily true that $(c, ab) = 1$? (Proof or counterexample please.)

The following extension of Theorem 6 is used in Theorem 8 below.

Theorem 7
Let $a_1, a_2, ..., a_n$ be integers and let p be a *prime* integer. Then

$$p|a_1 a_2 ... a_n \rightarrow p|a_1 \text{ or } p|a_2 \text{ or } ... \text{ or } p|a_n$$

(The 'or's are inclusive so that, informally, the theorem states: 'if $p|a_1 a_2 ... a_n$ then p divides at least one of the a_i'.)

DISCUSSION
Again, we have a statement of the type 'for each positive integer n'. So, what is the method of proof? Try induction. Notice, though, that induction is on the *number* of a_is, not on the a_i themselves. Theorem 6 does the case $n = 2$, but we might as well begin at $n = 1$ which says $p|a_1 \rightarrow p|a_1$ (which is trivially true). Even though we

are getting used to induction proofs I think it best to begin by stating clearly our induction hypothesis.

Proof of Theorem 7   Let S(n) be 'if $p|a_1 a_2 \ldots a_n$ then ($p|a_1$ or $p|a_2$ or ... or $p|a_n$)'. The case $n = 1$ is trivial. Assume that S(k) holds and suppose that p divides $a_1 a_2 \ldots a_k a_{k+1}$, that is, $p|(a_1 a_2 \ldots a_k)a_{k+1}$. By Theorem 6 we may deduce that either $p|a_1 a_2 \ldots a_k$ or $p|a_{k+1}$. Hence, by the induction hypothesis, either ($p|a_1$ or $p|a_2$ or ... or $p|a_k$) or $p|a_{k+1}$. This proves S(k + 1) holds and completes the proof of the theorem.

The Fundamental Theorem of Arithmetic

We have already seen that every integer greater than 1 can be expressed as a product of primes. Theorem 7 helps us to confirm that (if we disregard the *order* in which the primes are written down) this representation is *unique*.

Theorem 8 (The Fundamental Theorem of Arithmetic)
Let a be an integer ≥ 2. Then a may be expressed as a product of primes. Further, if $p_1 p_2 \ldots p_m$ and $q_1 q_2 \ldots q_n$ are two such representations, then $m = n$ and the ps and the qs can be paired off in a one-to-one manner so that paired ones are *equal*. (*Note:* There is no assertion that the p_i are distinct. Similarly for the q_j.)

DISCUSSION
Part of the theorem has already been proved (as Theorem 1). We include this part in the present statement just for completeness. Here we have another theorem asserting something about *all integers* (≥ 2). So let us attempt a proof by mathematical induction using the induction statement

S(a): Each integer from 2 to a inclusive is expressible as a product of a uniquely determined collection of primes

Proof of Theorem 8   Let S(a) be the statement set out above. Taking $a = 2$ we see that induction step 1 is trivially satisfied. Now suppose that S(k) is true and consider the integer $k + 1$. We know from Theorem 1 that $k + 1$ has at least one factorization into a product of primes. Suppose that

$$k + 1 = \alpha_1 \alpha_2 \ldots \alpha_r = \beta_1 \beta_2 \ldots \beta_s$$

are two such factorizations of $k + 1$. Since $\alpha_1|\alpha_1 \alpha_2 \ldots \alpha_r$ we see that $\alpha_1|\beta_1 \beta_2 \ldots \beta_s$. Hence by Theorem 7 we may deduce that $\alpha_1|\beta_i$ for some i. Dividing through the equality $k + 1 = \alpha_1 \alpha_2 \ldots \alpha_r = \beta_1 \beta_2 \ldots \beta_s$ by $\alpha_1 (= \beta_i)$ we obtain

$$\frac{k + 1}{\alpha_1} = \alpha_2 \ldots \alpha_r = \beta_1 \beta_2 \ldots \hat{\beta_i} \ldots \beta_s$$

the ^ indicating that the term β_i is now missing. Now $(k + 1)/\alpha_1$ is less than $k + 1$ and so our induction hypothesis may be applied. That tells us that $\alpha_2, \ldots, \alpha_r$ and $\beta_1, \ldots, \beta_s$ (excluding β_i) pair off in a 1–1 fashion so that paired ones are *equal*. Finally, pairing α_1 with β_i the theorem is proved (given the proviso in the comment to Theorem 1.)

COMMENT   In writing an integer as a product of primes it is common to gather together all equal primes. For example, we should write $999\,702 = 2 \cdot 3^5 \cdot 11^2 \cdot 17$. In this format Theorem 8 says that if

$$n = p_1^{\alpha_1} p_2^{\alpha_2} \cdots p_r^{\alpha_r} = q_1^{\beta_1} q_2^{\beta_2} \cdots q_s^{\beta_s}$$

where $p_1 < p_2 < \cdots < p_r$ and $q_1 < q_2 < \cdots < q_s$, the ps and qs being primes, and all αs and βs being positive integers, then

$$r = s, \ p_1 = q_1, \ p_2 = q_2 \ldots, p_r = q_r \ (=q_s)$$

and

$$\alpha_1 = \beta_1, \ \alpha_2 = \beta_2, \ldots, \alpha_r = \beta_r \ (=\beta_s)$$

EXERCISE 19   To show the usefulness of the definition of the word 'prime' try to state Theorem 8 without using that word.

EXERCISE 20   The other day I was told that $n(n + 1)(n + 2)$ is a perfect square if $n = 1033$. How did I know *immediately* that this must be wrong?

EXERCISE 21   Let $a, b, p \in \mathbb{Z}^+$ with p being a prime. Use Theorem 8 to show that, if $a^2 = pb^2$ then $p|a$ and $p|b$. Deduce that $\sqrt{p} \notin \mathbb{Q}$. Likewise prove $\sqrt[3]{p} \notin \mathbb{Q}$.

EXERCISE 22   Use Theorem 8 to show that, if a, b, u are positive integers such that $ab = u^2$ and $(a, b) = 1$, then each of a and b is a perfect square.

EXERCISE 23

(a) Let $a, b \in \mathbb{Z}$. Use Theorem 8 to show that if p, q are distinct primes such that $p|a$ and $q|a$ then $pq|a$.
(b) By choosing specific values for p, q and a show that the conclusion in (a) may be false if (i) p and q are not primes or if (ii) $p = q$. (Thus we do need both hypotheses, of primeness and of distinctness.)

EXERCISE 24   *Claim*: 'Let a, b be integers. If $a + b|ab$ then there exists a prime p such that $p|a$ and $p|b$.' What is wrong with the following proof?

'***Proof***':   Take any p dividing $a + b$. Then $p|ab$. Hence $p|a$ or $p|b$. From $p|a$ (respectively $p|b$) and $p|a + b$ we deduce that $p|b$ (respectively $p|a$). Hence p divides both a and b.

Why do we not allow 1 as a prime?

Since, for example

$$6 = 2 \cdot 3 = 1 \cdot 2 \cdot 3 = 1^2 \cdot 2 \cdot 3 = 1^3 \cdot 2 \cdot 3 = \ldots$$

the uniqueness part of Theorem 8 would be spoilt if we admitted 1 as a prime.

A nice application

The following theorem is useful in its own right, but even more important is the method of procedure Example 7 then suggests.

Theorem 9 (The rational root test)
Let $p(x) = a_n x^n + a_{n-1} x^{n-1} + \ldots + a_1 x + a_0$ be a polynomial with integer coefficients and suppose that r/s is a rational root of $p(x)$ written in its lowest terms. Then $r|a_0$ and $s|a_n$.

Proof of Theorem 9 Because r/s is a root of $p(x)$ we have

$$p\left(\frac{r}{s}\right) = a_n \left(\frac{r}{s}\right)^n + a_{n-1}\left(\frac{r}{s}\right)^{n-1} + \ldots + a_1\left(\frac{r}{s}\right) + a_0 = 0$$

Tidying this up we obtain

$$a_n r^n + a_{n-1} r^{n-1} s + \ldots + a_1 r s^{n-1} + a_0 s^n = 0$$

which may be rewritten as

$$-a_n r^n = s(a_{n-1} r^{n-1} + \ldots + a_1 r s^{n-2} + a_0 s^{n-1})$$

This shows that $s|a_n r^n$. We are given that $(r, s) = 1$, consequently $(r, s) = 1$ and so, by Exercise 17(b), we deduce that $s|a_n$. In an identical manner we can show that $r|a_0$.

Example 7
(We keep it simple to avoid heavy arithmetic.) Find all the rational roots of

$$p(x) = \tfrac{1}{3}x^5 - \tfrac{1}{6}x^4 + \tfrac{5}{6}x^3 - x^2 - \tfrac{1}{2}x - \tfrac{3}{2}$$

These will be identical to those of

$$6p(x) = 2x^5 - x^4 + 5x^3 - 6x^2 - 3x - 9$$

By Theorem 9, all rational roots must be of the form r/s where $r|-9$ and $s|2$. All *potential* roots are in the list $\frac{1}{1}, -\frac{1}{1}, \frac{3}{1}, -\frac{3}{1}, \frac{9}{1}, -\frac{9}{1}, \frac{1}{2}, -\frac{1}{2}, \frac{3}{2}, -\frac{3}{2}, \frac{9}{2}, -\frac{9}{2}$. Of these potential roots only $\frac{3}{2}$ is a root. Incidentally, my computer program tells me that $\frac{3}{2}$ is *not* a root; rather that $6p(\frac{3}{2}) = 3.469\,447 \cdot 10^{-18}$.

COMMENT The important method of procedure? It is a variant on the method of contradiction. We assume $p(x)$ has a root r/s, *not with the aim of showing no such root can exist* but rather with the aim of determining any *restrictions* which might apply. We then have to check all *potential* roots to see if they are actual roots.

EXERCISE 25 Find all rational roots of $p(x) = x^4 + 2x^3 + 2x^2 + x - 2$.

A cautionary example

You must not assume that Theorem 8 is obviously true and does not, therefore, need proof. The following is due to David Hilbert.

EXERCISE 26 Let H be the set $\{1, 5, 9, 13, 17, 21, 25, 29, 33, \ldots\}$ of positive integers which, on division by 4, leave remainder 1. A member of H (other than 1) which is *not* a product of two smaller elements of H is called an *H-prime*. List all the H-primes less than 50. Show that there are three distinct H-primes, a, b and c such that $441 = ab = c^2$. Find distinct H-primes p, q, r, s such that $pq = rs$.

An exercise on Fermat's Little Theorem

Our next exercise concerns a theorem at the forefront of mathematical research on computer factorization of very large integers. We precede it with a theorem on the divisibility of certain binomial coefficients.

Theorem 10

Let p be a prime. Then, for each integer r which is such that $0 < r < p$, we have $p \mid \binom{p}{r}$.

Proof of Theorem 10 Let

$$u = \binom{p}{r} = \frac{p \cdot (p - 1) \cdot \ldots \cdot (p - r + 1)}{1 \cdot \quad 2 \quad \cdot \quad \cdot \quad r}$$

It follows that

$$1 \cdot 2 \cdot \ldots \cdot r \cdot u = p \cdot (p - 1) \cdot \ldots \cdot (p - r + 1)$$

By Theorem 7 we may deduce that p divides (at least) one of $1, 2, \ldots, r, u$. But $p > r$, hence we must have $p \mid u$.

EXERCISE 27 **Fermat's Little Theorem** states:

Let p be a prime and a any (positive) integer. Then $p \mid a^p - a$.

Arrange the following sentences (all of which are true) in an appropriate order so as to produce a proof. (Write the sentences in full, not just their numbers. Then read over your 'proof'.)

(a) But $p \mid \binom{p}{1}, \binom{p}{2}, \ldots, \binom{p}{p-1}$ and $p \mid k^p - k$.

(b) Now suppose that $S(k)$ is true; that is $p \mid k^p - k$.

(c) Hence, by m.i. $S(a)$ is true for all a.

(d) Hence $S(1)$ is true.

(e) We prove the result by induction on a.

(f) Let $S(a)$ be the assertion that $p \mid a^p - a$.

(g) Trivially, $p \mid 1^p - 1$.

(h) Hence $p \mid (k + 1)^p - (k + 1)$.

(i) Consider the statement $S(k + 1)$: $p \mid (k + 1)^p - (k + 1)$.

(j) We have $(k + 1)^p - (k + 1) = k^p + \binom{p}{1} k^{p-1} + \ldots + \binom{p}{p-1} k + 1 - (k + 1)$.

DAVID HILBERT
(23 January 1862–14 February 1943)

David Hilbert was born in 1862 in Konigsberg, now Kaliningrad, the son of a judge. After obtaining his doctorate in 1885 he became a professor at the University of Konigsberg in 1892, moving to a similar position in Göttingen (where Gauss had been) in 1895. Hilbert researched in many areas of mathematics and, later, in mathematical physics, usually contributing greatly to one subject for some years, then moving on. In particular, apart from one instance, described below, his interest in number theory lasted from 1892 to 1899. Just prior to this period Hilbert had completely rewritten the subject of invariant theory by proving his *Basis Theorem*. Towards the end of the period he reorganized Euclid's geometry – eliminating its deficiencies by using the fully axiomatic method he wished to apply to all of mathematics. Hilbert showed that Euclidean geometry is consistent if and only if ordinary arithmetic is. His great wish was to prove the consistency of arithmetic but this was thwarted by Godel's *Incompleteness Theorem* which shows that this hope cannot be realized.

In 1909 Hilbert proved Waring's conjecture that, to each positive integer k there corresponds a positive integer $g(k)$ such that *every* positive integer is expressible as a sum of no more than $g(k)$ kth powers. (Lagrange's four squares theorem in Chapter 1 shows that $g(2) = 4$. The point is that Hilbert's proof was non-constructive: no recipe for determining $g(k)$ for each k was given.)

At the International Conference of Mathematicians in 1900, Hilbert drew up a list of 23 important problems he thought mathematicians should try to solve. Some remain unsolved to this day.

Arithmetic modulo *n* – the arithmetic of remainders, congruences

Fermat's Little Theorem (and many other results) can be nicely expressed if we use the arithmetic of remainders as introduced by Gauss. We cover this section quite quickly. See what you can follow – it's not too hard.

Definition 4

Let a, b and n be integers such that $n > 0$. We say that **a is congruent to b modulo n**, and we write $a \equiv b \pmod{n}$, precisely when $n|(a - b)$. If $n \nmid (a - b)$ we say **a is not congruent** (or **is incongruent**) to **b modulo n**, and write $a \not\equiv b \pmod{n}$.

It is clear that

1. $a \equiv b \pmod{n}$ iff a and b leave the same remainder r (with $0 \leqslant r < n$) on division by n.
2. $a \equiv 0 \pmod{n}$ iff $n|a$.

Example 8

(a) $87 \equiv 12 \pmod{15}$;
(b) $101 \equiv -195 \pmod{37}$;
(c) $5 \not\equiv 83 \pmod{11}$
(d) $1591 \equiv 0 \pmod{43}$.

Example 9

166 days after next Saturday the day will be Thursday because $7, 14, 21, \ldots$ days after a Saturday is a Saturday, and on dividing 166 by 7 we get a remainder of 5, and 5 days after a Saturday it is always a Thursday. (Notice that it is only the remainder 5 which is important here; in the equality $166 = 23 \cdot 7 + 5$ the multiplier 23 is irrelevant.)

EXERCISE 28

(a) By trial and error find all integers x such that $0 \leqslant x < 7$ and for which $3x \equiv 1 \pmod{7}$.
(b) Describe, in set theoretic notation, *all* integers x for which $3x \equiv 1 \pmod{7}$.

COMMENT   A brute-force approach is not recommended if the integers involved in a congruence are large. See Example 10 below.

Further properties of the congruence symbol $\equiv$ are given in Theorem 11.

Theorem 11
Let $a, b, c, n \in \mathbb{Z}$ with $n > 0$. Then

(i) $a \equiv a \pmod{n}$
(ii) If $a \equiv b \pmod{n}$ *then* $b \equiv a \pmod{n}$
(iii) If $a \equiv b \pmod{n}$ and *if* $b \equiv c \pmod{n}$ *then* $a \equiv c \pmod{n}$

Proof of Theorem 11

(i) Clearly $n|(a - a)$
(ii) *If $n|(a - b)$ then $n|(b - a)$*
(iii) *If $n|(a - b)$ and if $n|(b - c)$ then $n|\{(a - b) + (b - c)\}$*

Regarding the arithmetic of $\equiv$ we have Theorem 12.

Theorem 12

Let $a, b, c, d, n, t \in \mathbb{Z}$ be such that $n > 0, t > 0, a \equiv b \pmod{n}$ and $c \equiv d \pmod{n}$. Then, dropping the '(mod n)', merely for convenience:

(i) $a + c \equiv b + d$
(ii) $a - c \equiv b - d$
(iii) $ac \equiv bd$
(iv) $a^t \equiv b^t$

(In particular, given $a \equiv b$ and $c \in \mathbb{Z}$, we have $a + c \equiv b + c$ and $ac \equiv bc$.)

Proof of Theorem 12

To say $u \equiv v \pmod{n}$ is to say that $u = v + rn$ for some integer r. Hence we are given $a = b + \alpha n$ and $c = d + \beta n$ $(\alpha, \beta \in \mathbb{Z})$. Then

(i) $a + c = b + d + (\alpha + \beta)n$
(ii) $a - c = b - d + (\alpha - \beta)n$
(iii) $ac = (b + \alpha n)(d + \beta n) = bd + (b\beta + d\alpha + \alpha\beta n)n$

These equalities show the truth of (i), (ii), (iii). For (iv) the case $t = 2$ is just (iii) with $c = a$ and $d = b$. Proof for *all* t would formally be carried out by m.i. As this is a bit boring we shall omit it!

Cautionary example

$189 \equiv 357 \pmod{14}$, that is, $7 \cdot 27 \equiv 7 \cdot 51 \pmod{14}$ *but* $27 \not\equiv 51 \pmod{14}$. So the message is: *division is not always possible in a congruence*. (Exercise 30 describes one occasion when it is safe to do it.)

COMMENT   Expressions of the form $a \equiv b \pmod{n}$ are called **congruences (mod n)**. Notice that, except for division, $\equiv$ behaves very much like simple equality. This is why Gauss chose the symbol $\equiv$ for congruence to resemble that for equality.

Example 10

Find x such that $17x \equiv 23 \pmod{44}$.

SOLUTION $17x \equiv 23 \pmod{44}$ means that $44|(17x - 23)$ or, that, for some integer y, $44y = 17x - 23$. That is $17x - 44y = 23$. (Does this format remind you of anything? It should!)

We easily find that $44 = 2 \cdot 17 + 10$, that $17 = 1 \cdot 10 + 7$, that $10 = 1 \cdot 7 + 3$ and that $7 = 2 \cdot 3 + 1$. By these means we find that $(44, 17) = 1$ and that $1 = 13 \cdot 17 - 5 \cdot 44$.

This may be rewritten as $1 \equiv 13 \cdot 17 \pmod{44}$ or, if you wish, as $17 \cdot 13 \equiv 1 \pmod{44}$. Then, by Theorem 12 we may multiply through by 23 to obtain $17 \cdot (13 \cdot 23) \equiv 1 \cdot 23$ (mod 44). Hence $x = 13 \cdot 23$ will solve the problem. Since $13 \cdot 23 \equiv 35 \pmod{44}$ we see that $x = 35$ is the only x with $0 \leqslant x < 44$ which solves the equation.

EXERCISE 29   Find, if there are any, all non-negative values of x less than the modulus for each of the following congruences.

(a) $5x \equiv 1 \pmod{11}$ (f) $7x \equiv 8 \pmod 9$
(b) $5x \equiv 3 \pmod{11}$ (g) $17x \equiv 1 \pmod{73}$
(c) $15x \equiv 9 \pmod{11}$ (h) $51x \equiv 3 \pmod{219}$
(d) $4x \equiv 9 \pmod{11}$ (i) $51x \equiv 4 \pmod{219}$
(e) $6x \equiv 7 \pmod 8$

EXERCISE 30   Writing $sa \equiv sb \pmod n$ in the form $n \mid s(a - b)$, use Exercise 17(b) to deduce that, if $(n, s) = 1$ then $n \mid (a - b)$. Hence, when can you certainly divide in a congruence? Answer: when the divisor s is coprime to the mo....

Does it divide by 9?

We now use the convenience of congruences to prove the general version of Statement 5(a), namely the test for an integer to be divisible by 9.
   We first remind ourselves that the integer 1066, for example, has *digits* 1, 0, 6 and 6, and that 1066 is shorthand for $1 \cdot 10^3 + 0 \cdot 10^2 + 6 \cdot 10^1 + 6 \cdot 10^0$.

Theorem 13
Let $N = a_n \cdot 10^n + a_{n-1} \cdot 10^{n-1} + \ldots + a_2 \cdot 10^2 + a_1 \cdot 10 + a_0$ where each a_i is an integer satisfying $0 \leqslant a_i \leqslant 9$. The N is exactly divisible by 9 IFF $a_n + a_{n-1} + \ldots + a_2 + a_1 + a_0$ is exactly divisible by 9.

Proof of Theorem 13   It is easy to see that, modulo 9, we have $10^0 \equiv 1$, $10^1 \equiv 1$, $10^2 \equiv 1, \ldots$ $10^n \equiv 1$. Therefore (still modulo 9) we have, by Theorem 12(iii), the congruences

$$a_0 \equiv a_0, \; a_1 \cdot 10 \equiv a_1, \; a_2 \cdot 10^2 \equiv a_2, \ldots, a_{n-1} \cdot 10^{n-1} \equiv a_{n-1} \text{ and } a_n \cdot 10^n \equiv a_n$$

Adding these, Theorem 12(i) shows

$$a_n \cdot 10^n + a_{n-1} \cdot 10^{n-1} + \ldots + a_2 \cdot 10^2 + a_1 \cdot 10 + a_0$$
$$\equiv a_n + a_{n-1} + \ldots + a_2 + a_1 + a_0 \pmod 9$$

Consequently N will be congruent to 0 modulo 9 iff $a_n + a_{n-1} + \ldots + a_2 + a_1 + a_0$ is congruent to 0 modulo 9. In other words,

$$9 \mid N \text{ iff } 9 \mid (a_n + a_{n-1} + \ldots + a_2 + a_1 + a_0)$$

as claimed.

Example 11

18 735 726 970 936 201 658 372 449 is *not* divisible by 9 since $1 + 8 + 7 + 3 + 5 + 7 + 2 + 6 + 9 + 7 + 0 + 9 + 3 + 6 + 2 + 0 + 1 + 6 + 5 + 8 + 3 + 7 + 2 + 4 + 4 + 9$ isn't.

EXERCISE 31   Can you think of a way to save work in Example 11? (*Hint:* $1 + 8 = ?, 7 + 2 = ?, 1 + 5 + 7 = 13 \equiv 4?...$ – hence the expression *casting out nines*.)

EXERCISE 32   In 1938, Sir Arthur Eddington announced that there are exactly

15 747 724 136 275 002 577 605 653 961 181 555 468 044 717 914 527 116 709
366 231 425 076 185 631 031 296 protons in the universe. Is this number divisible by 9?

EXERCISE 33   Show that $N = a_n \cdot 10^n + a_{n-1} \cdot 10^{n-1} + ... + a_2 \cdot 10^2 + a_1 \cdot 10 + a_0$ is divisible by 3 iff $a_n + a_{n-1} + ... + a_2 + a_1 + a_0$ is divisible by 3 and that N is divisible by 11 iff $(-1)^n a_n + (-1)^{n-1} a_{n-1} + ... + (-1)^2 a_2 + (-1)^1 a_1 + (-1)^0 a_0$ is divisible by 11 (i.e. iff $11 | a_0 - a_1 + a_2 - a_3 + a_4 - a_5 + a_6 - a_7...$).

EXERCISE 34   Show that N is divisible by 8 iff 8 divides the number $a_2 \cdot 10^2 + a_1 \cdot 10 + a_0$.

EXERCISE 35   4 643 957·67 649 047 is equal neither to (a) 314 159 265 358 973 nor to (b) 314 159 365 358 979. How do I know? (For (a) work mod 10; for (b) work mod 9.)

EXERCISE 36   The ten digits $0, 1, 2, ..., 9$ are used (once each) in the form $30 + 25 + 16 + 9 + 8 + 7 + 4$ to make the sum 99. Can you similarly use the digits to make the sum 100? (You must either give a specific example showing how it can be done or explain why it can't.)

Some history (for information only)

In congruence notation, Fermat's Little Theorem (FLT) becomes:

Let p be a prime and a be a positive integer. The $a^p \equiv a$ (mod p)

If $p \nmid a$ (so that $(a, p) = 1$), Exercise 30 shows that we may divide through the congruence, by a, to obtain the alternative formulation of FLT:

Let p be a prime and a be a positive integer such that $(a, p) = 1$. Then $a^{p-1} \equiv 1$ (mod p)

This theorem can be used to show that each prime divisor of an integer of the form $2^m - 1$, where m is prime, is of the form $2km + 1$ and that any (prime) divisor of an integer of the form $2^{2^n} + 1$ is of the form $2^{n+1}k + 1$. (This meant that Euler only had to check primes 193, 257, 449, 577 and 641, each being of the form $64k + 1$, before finding the divisor of $2^{32} + 1$ which eluded Fermat – the other *potential* divisors less than 641, namely 65, 129, 321, 385, 513, are all composite.)

To *confirm* that $641 | 2^{32} + 1$ (once 641 has been found!) we may simply work out $2^{32} + 1$ and attempt the division ... or note that $5 \cdot 2^7 = 640 \equiv -1 \pmod{641}$. Hence $5^4 \cdot 2^{28} \equiv (-1)^4 = 1 \pmod{641}$. But $5^4 = 625 \equiv -16 = -(2^4) \pmod{641}$. Therefore $5^4 \cdot 2^{28} \equiv -(2^4) \cdot 2^{28} \equiv -(2^{32}) \pmod{641}$ and so $-(2^{32}) \equiv 1 \pmod{641}$.

The FLT can also be used as a preliminary test for primeness, as in Example 12.

Example 12

Is 1333 prime? If so, then $2^{1332} \equiv 1 \pmod{1333}$. Now

$$1332 = 1024 + 256 + 32 + 16 + 4 \ (= 2^{10} + 2^8 + 2^5 + 2^4 + 2^2)$$

Working modulo 1333 we have

$$2^4 \equiv 16$$
$$2^{16} \equiv (2^4)^4 \equiv 16^4 \equiv 219$$
$$2^{32} \equiv (2^{16})^2 \equiv (219)^2 \equiv 1306 \equiv -27$$
$$2^{256} \equiv (2^{32})^8 \equiv (-27)^8 \equiv 188$$
$$2^{1024} \equiv (2^{256})^4 \equiv (188)^4 \equiv 47$$

Hence $2^{1332} \equiv 47 \cdot 188 \cdot (-27) \cdot 219 \cdot 16$. But

$$47 \cdot 188 \equiv 838 \equiv -495$$
$$-495 \cdot -27 \equiv 35$$
$$35 \cdot 219 \equiv 1000 \equiv -333$$
$$-333 \cdot 16 \equiv 4 \not\equiv 1$$

Hence 1333 cannot be prime.

EXERCISE 37

(a) Check, by hand, that $2^{37} - 1$ is not prime.
(b) Using the method of Example 12, check 323 and 561 for primeness.

Later Euler generalized Fermat's Little Theorem to incorporate a general modulus. He proved

Let a and m be integers with $m > 1$ and $(a, m) = 1$. Then $a^{\phi(m)} \equiv 1 \pmod{m}$,

ϕ being Euler's function (see Definition 3).

It is the case in which $m = pq$ is a product of two distinct primes which has been central (dare one say of *prime* importance?) in the development of **public key cryptography** (see, for example, Riesel (1985), Allenby and Redfern (1989), Salomaa (1990)).

EXERCISE 38 Show that $2^{10} \equiv 1 \pmod{11}$ and that $2^{10} \equiv 1 \pmod{31}$. Deduce (Theorem 12(iv)) that $2^{340} \equiv 1 \pmod{11}$ and $2^{340} \equiv 1 \pmod{31}$. Now deduce (Exercise 23) that $341 | 2^{340} - 1$. (341 is the least composite integer n for which $2^{n-1} \equiv 1 \pmod{n}$. If there were no nasty integers such as 341 then checking a given integer for primeness would be a lot easier.)

Summary

The **well-ordering principle**, which distinguishes $\mathbb{Z}^+$ from $\mathbb{Q}^+$, is equivalent to the principle of mathematical induction. Each method readily proves that every integer greater than 1 has a **prime decomposition**. Well-ordering also allows a proof of the **division theorem**. Repeated use of this leads to the **Euclidean Algorithm** which, in turn, permits determination of the **greatest common divisor (gcd)** of any two given integers (not both zero) without factorizing either.

Integers whose gcd is 1 are called **coprime**. The **uniqueness** of the factorization of integers into products of primes is *not obvious*, as Hilbert's example shows: it has to be proved. The proof of the **Fundamental Theorem of Arithmetic** uses various divisibility properties of prime numbers. These are obtainable from the expression of the gcd of a and b as a **linear combination** of a and b. A nice application of the Fundamental Theorem is the rational root test for polynomials.

In **arithmetic modulo n**, which is, in essence, the arithmetic of remainders (on division by n), **congruences** replace equalities – but share many of the same properties, with the notable exception of division. An amusing offshoot of this theory is a test for **division by 9**. A major result most conveniently stated in terms of congruences is **Fermat's Little Theorem**. The FLT and its generalization, which uses **Euler's ϕ function**, are key results in the problems of (computer) factorization of large integers and the sending of unbreakable secret messages.

CHAPTER 9

The Rational Numbers and the Real Numbers

We shall return ... from the lecture pleased that the irrational is rational.

WALLACE STEVENS

The concept of proof, vital in all of mathematics, is especially prominent in mathematical analysis. To study (real) analysis one needs to be familiar with certain important properties of the real numbers. As it is beyond the scope of this book to present an introduction to mathematical analysis, this chapter exhibits just two of these properties, completeness and uncountability which, amongst other things, highlight differences between the set of real numbers and the set of rational numbers. Initially we have some fairly informal fun with the decimal representations of real numbers which enables yet another distinction between rational and irrational numbers to be made.

Introduction

So far we have assumed that the reader has an intuitive feel for the real numbers. In fact, the set of real numbers can be defined in several ways, all of them (provably) equivalent. One way defines real numbers to be (infinite) sequences of rationals, another to be special subsets of the rationals, and a third to be an (essentially unique) system of entities satisfying a certain set of axioms. It is perhaps worth noting that it was only fairly late in the day that serious discussion of the nature of the set of real numbers was undertaken: Dedekind,[1] in 1858, remarking that 'up to now even the truth of the equality $\sqrt{2}\cdot\sqrt{3} = \sqrt{6}$ has not been satisfactorily established'.

Decimal and pictorial representation of real and rational numbers

All of the ways mentioned above lead to the conclusion that we may take the **real numbers** to be just those numbers which have a decimal representation, for example 383.617 759..., this being shorthand for

$$383 + \frac{6}{10} + \frac{1}{100} + \frac{7}{1000} + \frac{7}{10\,000} + \frac{5}{100\,000} + \frac{9}{1\,000\,000} + \dots$$

where the sequence of digits after the (decimal) point may stop (e.g. 18.1942) or may go on for ever. In this latter case this sequence may, eventually, fall into a repeating pattern (e.g. 73.295 641 $\overline{311\,722}$) where $\overline{41\,311\,722}$ indicates that, after the 2956, the eight digits 41 311 722 keep reappearing in that order for ever.

Given an infinite straight line and two points labelled 0 and 1, the totality of real numbers as described above can be paired off in a one-to-one fashion with points on the line as indicated in Figure 9.1, where the second and third lines are magnified sections of the previous one.

All the rational numbers are accommodated on this (infinite) line (indeed 1.6, 1.667 and 1.6673 are rational numbers), and it is not difficult to show that a real number r, say, is a rational number *if and only if* the decimal representation of r repeats as above (see Example 1 below). Thus 73.295 641 $\overline{311\,722}$ is a rational number, as is 1.667 300 000 00... (which may also be represented by 1.667 299 999 99.... This changing of a0000000... to $(a-1)$9999999... and vice versa can be shown to be the only possible 'double' representation of a real number.)

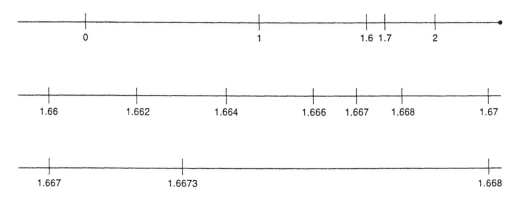

Figure 9.1

[1] Julius Wilhelm Richard Dedekind (6 October 1831–12 February 1916).

Those real numbers which are *not* rational are termed **irrational numbers**.

Example 1

(a) By long division $\frac{4}{13} = 0.307\,692\,307\,692\,30...$
(b) By setting $x = 0.3\overline{18} = 0.3181818...$, we see that $100x = 31.8181818....$ By subtraction

$$99x = 31.8181818... - 0.3181818... = 31.5$$

(We multiplied x by $100 = 10^2$ precisely because x's *repeating length*, that is, the number of digits under the bar, is 2.) Hence $x = 31.5/99$ which simplifies to 7/22.
(c) $0.101\,001\,000\,100\,001\,000\,001\,000\,000\,100...$ is irrational. If it were rational then it would from some point (say the mth digit) have to repeat in blocks of some length n, say. In the given decimal representation there are, beyond the mth digit, blocks of successive zeros of length $2n$ and longer. Thus the repeating length n block must contain nothing but zeros. However, this is impossible as, every so often, there appears the decimal digit 1. (Why do we look for blocks of zeros of length $2n$? Why not just length n?)

EXERCISE 1

(a) Find the (repeating) decimal expression for 1/17. (Can you think how to do this on your hand-calculator – which may not show sufficiently many decimal places?)
(b) Why is the repeating length of $1/n$ never more than $n - 1$?
(c) Express $73.295\,6\overline{41\,311\,722}$ as a rational number.

EXERCISE 2   Why is $0.123\,456\,789\,101\,112\,131\,415\,161\,718\,192\,021...$ an irrational number?

EXERCISE 3   Let a and b be real numbers. What can you say about the rationality or irrationality of $a + b$ if (a) a is rational and b is irrational; (b) if a and b are both irrational? (Proofs or examples are, of course, required.) (c) What kinds of proof did you use in answering (a)?

Ordering the real numbers

Because the real numbers may be thought of as strung out on a line, they obey certain **ordering axioms**.

1. For each pair of real numbers a and b *exactly one* of the following holds:
   (i) $a < b$; (ii) $a = b$; (iii) $a > b$. (**Law of trichotomy**)

For all real numbers a, b and c:

2. If $a < b$ and if $b < c$ then $a < c$. (**Law of transitivity**)
3. If $a < b$ then $a + c < b + c$. (**Law of addition**)
4. Suppose that $a < b$. If $c > 0$ then $ac < bc$. If $c < 0$ then $ac > bc$. (**Law of multiplication**)

We shall not give a complete list here of the algebraic axioms (such as $(\forall a \in \mathbb{R})(\forall b \in \mathbb{R})$ $(\forall c \in \mathbb{R})[\{a + b\} + c = a + \{b + c\}]$

$$(\forall a \in \mathbb{R})(\forall b \in \mathbb{R})[ab = ba])$$

which the real numbers obey.

COMMENT   All these axioms are also true for the set $\mathbb{Q}$ of rational numbers. So, you may ask 'What can distinguish the real numbers from the rational numbers?' We answer this question in two important ways – using the concepts of completeness and countability.

It is easy to show that between any two rational numbers you can always find another. Indeed, if a/b and c/d are rational numbers then their 'average' $\frac{1}{2}\left\{\dfrac{ad + bc}{bd}\right\}$ is also rational and lies midway between them. This generalizes to Theorem 1.

Theorem 1
Between any two rational numbers there are *infinitely many* more rational numbers.

DISCUSSION
By continually taking the average (i.e. mid-point) we obtain, given a and b, the sequence $c, d, e, f, g, \ldots$ (Figure 9.2). There is always a rational to the 'right' of the rational just constructed and to the 'left' of b. That is, there is always 'at least one more'.
      This proof can be written more formally using contradiction, as follows.

Proof of Theorem 1   Suppose that (strictly) between a and b there are only finitely many rationals. Then there will be a largest such – let's call it M (for 'max'?). However, then $\frac{1}{2}(M + b)$ is a rational number lying (strictly) between M and b, hence (strictly) between a and b – and yet greater than the supposed *greatest* such rational, M. This contradiction shows that our assumption (of only finitely many rationals lying between a and b) is untenable. Thus there must be *infinitely* many, as claimed.

EXERCISE 4   Find five rationals between $3\frac{10}{71}$ and $3\frac{1}{7}$.

EXERCISE 5   If $0 < a/b < c/d$ does it follow that $\dfrac{a}{b} < \dfrac{a + c}{b + d} < \dfrac{c}{d}$? (*Hint*: Be careful!)

Theorem 1 raises the obvious question: are there infinitely many rationals/irrationals between each pair of real numbers? Example 2 shows how to prove this, in principle. However, in order that important properties of the real numbers are not swamped by mere calculation, we later approach this question by a more sophisticated route.

a	c	d	e	f g b

Figure 9.2

Example 2

Between the irrational 1.672 710 305 070 901 101 301 501... and the rational $1.672\,710\,\overline{3}$
lie the rationals 1.672 710 306, 1.672 710 306 6, 1.672 710 306 66 and the irrationals
1.672 710 310 305 070 901 101 3... and 1.672 710 310 310 305 070 901 101 3....

EXERCISE 6   Find four rationals and five irrationals between the irrationals
1.234 567 891 011 121 314 15... and 1.234 567 891 010 121 314 15...

Upper and lower bounds

You may have come across the remark that the base of natural logarithms, e, is
given by the 'infinite sum'

$$e = 1 + \frac{1}{1!} + \frac{1}{2!} + \frac{1}{3!} + ... + \frac{1}{n!} + ...$$

What, if anything, might this mean? Informally it ought to mean that e is the number
you are heading towards as you consider the sequences:

$$s_0 = 1, \quad s_1 = 1 + \frac{1}{1!}, \quad s_2 = 1 + \frac{1}{1!} + \frac{1}{2!}, \quad s_3 = 1 + \frac{1}{1!} + \frac{1}{2!} + \frac{1}{3!}$$

etc. of **partial sums**. Certainly Exercise 16 in Chapter 7 tells us that, no matter how
far we go along this (increasing) sequence, we shall never pass the value 3. Therefore,
according to the following definition, 3 is an **upper bound** for the set $E = \{s_0, s_1, s_2, s_3, ...\}$
of partial sums.

Definition 1

Let A be a non-empty set of real numbers. The real number U is an **upper bound**
for A if and only if $U \geqslant a$ for every $a \in A$. We also say that A is **bounded above** by
U. (The concepts of **lower bound** of a set and of a set being **bounded below** are defined
analogously.)

COMMENT   If a set has an upper bound U, say, then it will clearly have infinitely
many upper bounds, namely all real numbers r for which $U \leqslant r$.

Example 3

(a) The set $\{1 - 1/n: n \in \mathbb{Z}^+\}$ has 1 (hence $2, 3, \pi, 871.22$, etc.) as upper bound(s); 0
    is an obvious lower bound.
(b) The set $T = \{t_1, t_2, t_3, ...,\}$ where $t_n = \frac{1}{1} + \frac{1}{2} + \frac{1}{3} + ... + 1/n$, is a set which has
    no upper bound since, as the proof of Statement 18 shows, there is no real
    number M, say, such that $t_n \leqslant M$ for all $n \in \mathbb{Z}^+$. Of course, 1 is a lower bound
    for T, as are $0, \frac{1}{2}, -77.391$ and all other real numbers $\leqslant 1$.

EXERCISE 7   Name five upper bounds and five lower bounds for each of the
following sets:

(a) $\left\{ \dfrac{1}{n}\left(1 - \dfrac{1}{n}\right): n \in \mathbb{Z}^+ \right\}$

(b) $\left\{ \cos x + 2 \sin x : 0 \leqslant x \leqslant \dfrac{\pi}{2} \right\}$

(c) $\left\{ \dfrac{1}{10^n} - \dfrac{1}{n!} : n \in \mathbb{Z}^+ \right\}$

EXERCISE 8 Give an example of a non-empty set which has neither an upper bound nor a lower bound.

One difference between $\mathbb{Q}$ and $\mathbb{R}$

Least upper and greatest lower bounds – the completeness of $\mathbb{R}$

To highlight the first of the two important differences between $\mathbb{Q}$ and $\mathbb{R}$, consider the sets $L = \{ x \in \mathbb{Q}^+ : x^2 < 2 \}$ and $R = \{ x \in \mathbb{Q}^+ : x^2 > 2 \}$ of positive rationals whose squares are, respectively, at most 2 and at least 2. Clearly, each member of R is an upper bound for the set L – and each member of L is a (guess what?) for the set R. However, just as to each member of L there exists a larger member of L (see Exercise 3 in Chapter 6) so, likewise, is there no smallest member in R. This means that *amongst the rational numbers* there is no smallest upper bound for the set L (of real numbers).

It is at precisely this point that the real numbers come to the rescue. To describe how, we introduce a definition.

Definition 2
The real number U is the **least upper bound (lub)** for the non-empty set A of real numbers if and only if (i) U is an upper bound for A *and* (ii) *if* U_1 is also an upper bound for A *then* $U \leqslant U_1$.

Likewise, the real number V is the **greatest lower bound (glb)** for the set A of real numbers if and only if (i) V is a lower bound for A *and* (ii) *if* V_1 is also a lower bound for A *then* $V_1 \leqslant V$.

Sometimes lub A and glb A are called the **supremum** and the **infimum** of A (in brief sup A and inf A).

COMMENT   Notice that Definition 1 talks of *an* upper bound of a set (because if there is one then there are infinitely many), but Definition 2 talks of *the* least upper bound – thereby indicating that it is unique. Why can't there be more than one least upper bound to a given set A? Exercise 12 gives a hint.

Example 4

(a) $\left\{ 1 - \dfrac{1}{n} : n \in \mathbb{Z}^+ \right\}$ has 0 for its glb and 1 for its lub.

(b) $\left\{ 1 - \dfrac{1}{n^2} : n \in \mathbb{Z}^+ \right\} \cup \left\{ -3 + \dfrac{1}{n} : n \in \mathbb{Z}^+ \right\}$ has -3 as its glb and 1 as its lub.

(c) The set $E = \{s_0, s_1, s_2, s_3, \ldots\}$, described above, where

$$s_n = 1 + \frac{1}{1!} + \frac{1}{2!} + \frac{1}{3!} + \ldots + \frac{1}{n!}$$

has glb equal to 1 and lub equal to e.

(d) As we observed in Chapter 4, the set $T = \{t_1, t_2, t_3, \ldots\}$ where

$$t_n = \frac{1}{1^2} + \frac{1}{2^2} + \ldots + \frac{1}{n^2}$$

has lub $\frac{\pi^2}{6}$.

Notice that, although all four sets above are sets of rational numbers, the lubs in parts (c) and (d) are not rational.[2]

EXERCISE 9   Show that the following sets do have lubs and glbs – by finding them:

(a) $\left\{\dfrac{1}{n} : n \in \mathbb{Z}^+\right\}$

(b) $\left\{(-1)^n\left(1 - \dfrac{1}{n}\right) : n \in \mathbb{Z}^+\right\}$

The various parts of Example 4 and Exercise 9 illustrate a major difference between the rational numbers and the real numbers. It is encapsulated in the **completeness property of the real numbers** which states that

Every non-empty subset of $\mathbb{R}$ which is bounded above has a least upper bound (in $\mathbb{R}$)

(We already know that replacing both occurrences of $\mathbb{R}$ by $\mathbb{Q}$ results in a false statement.)

QUESTION   Shall we also want a greatest lower bound version of the completeness property? Exercise 14 tells you how to get it 'for free'.

Example 5

Returning to the (non-empty) sets $L = \{x \in \mathbb{Q}^+ : x^2 < 2\}$ and $R = \{x \in \mathbb{Q}^+ : x^2 > 2\}$ introduced above, we see that L must have a least upper bound $U(L)$, say, and R must have a greatest lower bound $V(R)$, say. Surely $U(L) = V(R)$ since (surely?) $U(L)^2 = V(R)^2$ are both are equal to ... 2? Well, $U(L)^2$ cannot be *less* than 2, since, if it were, there would be (by Exercise 3 of Chapter 6) a member of L which is larger than $U(L)$. But then $U(L)$ wouldn't be an upper bound for L, never mind a *least* upper bound. Nor can $U(L)^2$ be *greater* than 2: if it were then, in a like manner, R would contain elements smaller than $U(L)$, so that $U(L)$ couldn't be a *least* upper bound for L. We therefore see that $U(L)^2 = 2$.

Likewise $V(R)^2 = 2$ and $U(L) = V(R)$ is *that unique real number greater than all members of L and less than all members of R and whose square is 2*. This is the number we call $\sqrt{2}$. (This was the approach adopted by Dedekind in his construction of the real numbers as 'cuts' in the set of the rational numbers.)

---

[2] The fact that neither e nor π^2 is a rational number is not obvious.

EXERCISE 10 Find lubs and glbs for the following sets. State whether or not these bounds belong to the set in question.

(a) $\left\{(-1)^n + \dfrac{1}{n} : n \in \mathbb{Z}^+\right\}$

(b) $\left\{2^n + \left(-\dfrac{1}{n}\right)^n : n \in \mathbb{Z}^+\right\}$

(c) $\{\cos \vartheta + \sin \vartheta : 0 < \vartheta < \pi\}$

(d) $\left\{1 - \dfrac{1}{n} + \dfrac{1}{n^2} : n \in \mathbb{Z} \text{ and } n > 1\right\}$

(e) $\left\{\cos \dfrac{n\pi}{4} + \dfrac{1}{n} : n \in \mathbb{Z}^+\right\}$

EXERCISE 11 Give an example of a set A with glb $A = \pi$ but $\pi \notin A$.

EXERCISE 12 Show that a non-empty set S of real numbers can have no more than one least upper bound. (*Hint:* Assuming X, Y are both lubs for S then $X \leqslant Y$ (why?) and $Y \leqslant X$ (why?)).

EXERCISE 13 Let A be a (non-empty) set of real numbers and let r be a real number. What element $r \in \mathbb{R}$ does the following describe?

$$(\forall a \in A)[a \leqslant r] \wedge (\forall h \in \mathbb{R}^+)(\exists b \in A)[r - h < b]$$

EXERCISE 14 Given that the non-empty set B of real numbers is bounded below, use the completeness property of $\mathbb{R}$ to show that B has a glb. (*Hint:* Consider the set $\{-b : b \in B\}$.)

In order to try to forestall a common misunderstanding, we take a few lines to introduce a definition, asking you to pay attention to condition (ii) in each case.

Definition 3
M is the **maximum** (value) of the non-empty set A if and only if (i) $a \leqslant M$ for all $a \in A$ and (ii) *M is an element of A.*
   Likewise m is the **minimum** (value) of the set A if and only if (i) $m \leqslant a$ for all $a \in A$ AND (ii) *m is an element of A.*

Example 6
In Exercise 10, sets (a) and (c) possess maximum elements but no minimum element; set (d) contains a minimum element but no maximum element.

From this example you should readily see that:

1. *Each set of real numbers which is bounded above has a lub in $\mathbb{R}$ but may not have a maximum element* – with similar remarks for glbs and minimum elements.

2. *If the non-empty set A of real numbers has a maximum element M then (i) M is the only maximum element and (ii) M = lub A* – with similar remarks for minimum elements and glbs.

Betweenness and the Archimedean property

Earlier we promised a more sophisticated proof that between any two real numbers there are infinitely many rational numbers and infinitely many irrational numbers. Here it is. However, we begin with an inferior result in order to emphasize a modification we have to make.

Theorem 2
Between each two rational numbers a and b there is one (and hence infinitely many) irrational numbers.

> DISCUSSION
> If this seems too hard, let's *specialize a* and b. For, if we can't prove our theorem when $a = 0$ and $b = 1$ (say) then we surely have no hope of establishing its truth in general. Is there any number between 0 and 1 which we *know* is irrational? No? Between 0 and 2? Yes, $\sqrt{2}$. So, between 0 and 1? What about $\sqrt{2}/2$? It surely lies between 0 and 1 – and Exercise 1 of Chapter 6 claimed it to be irrational. Is there any chance we can now pass back to 'general' a and b? From a to b there is the same gap as from 0 to $(b - a)$. Now $0 < \frac{\sqrt{2}}{2}(b - a) < (b - a)$ (why?). If $\frac{\sqrt{2}}{2}(b - a)$ were a rational number, r/s say, then we would have
>
> $$\sqrt{2} = \frac{2r}{s(b - a)}$$
>
> which is impossible, since $\sqrt{2}$ isn't rational. But then, from $0 < \frac{\sqrt{2}}{2}(b - a) < (b - a)$ we can deduce that $a < \frac{\sqrt{2}}{2}(b - a) + a < b$, and (surely?) $\frac{\sqrt{2}}{2}(b - a) + a$ is not rational (why not?).

Proof of Theorem 2   Since a and b are rational numbers, $\frac{\sqrt{2}}{2}(b - a)$ and $\frac{\sqrt{2}}{2}(b - a) + a$ are both irrational. But $0 < \dfrac{\sqrt{2}}{2} < 1$. Hence

$$a < \frac{\sqrt{2}}{2}(b - a) + a < b$$

EXERCISE 15   Can you now find *infinitely many* irrationals between a and b?

We can improve Theorem 2 as follows.

Theorem 3
Between any two *real* numbers $a < b$ there is (i) a rational number and (ii) an irrational number (and hence infinitely many of each).

DISCUSSION OF (i)

Here the methods of Theorem 2 seem to break down. For, even if we can find a rational number, r say, for which $0 < r < b - a$, so that $a < r + a < b$, we cannot be sure that $r + a$ is rational. (Indeed it will only be rational if a is!) *We need a new idea.* Once again such an idea may occur to you if you are lucky – or if you turn it over in your mind for long enough. The proof depends on the fact that: *given a positive real number* α *there exists a positive integer* n *such that* $0 < 1/n < \alpha$. This is called the **Archimedean property** of the real numbers. Thus, if n is an integer such that $0 < 1/n < b - a$, surely there must be a rational between a and b. Can you see why? (You can certainly write $a < a + 1/n < b$, but that's no good! Why not?) The correct approach follows.

Proof of Theorem 3(i)   Using the Archimedean property, choose an integer n such that $0 < 1/n < b - a$. Suppose that k is the largest integer such that

$$k \cdot \frac{1}{n} \leqslant a$$

Then

$$(k + 1)\frac{1}{n} > a$$

and yet, also

$$(k + 1) \cdot \frac{1}{n} < b$$

since $1/n < b - a$. Hence

$$a < \frac{k + 1}{n} < b$$

as required.

We leave the proof of (ii) to you in Exercise 18 below.

EXERCISE 16   Write down the Archimedean property in $\forall$–$\exists$ form.

EXERCISE 17   Show that the Archimedean property is equivalent to

(AP′)   *To each positive real number* α *there exists a positive integer* n *such that* $\alpha < n$.

That is, prove that the statement in italics can be deduced from the Archimedean property, and vice versa.

EXERCISE 18

(a) Use the method of Theorem 3(i) to show that between every two real numbers there is also an irrational number. (*Hint:* Replace the *rational* $1/n$ by the *irrational* …)
(b) Can you also prove this result directly from Theorems 2 and 3(i), that is, just using their statements?

The Archimedean property can be shown to be a logical consequence of the completeness property (CP). The intermediate value theorem (IVT) mentioned in the proof of Statement 13 in Chapter 6 also depends on CP.

EXERCISE 19   Show that $\mathbb{Q}$ satisfies AP but not CP.

Why do we need the real numbers?

One consequence of Theorem 3(i) is, given a real number r, we can find a rational number as near to it as we wish. Since, in real life, we can only take measurements approximately, why do we need to introduce the irrational numbers at all? There are several interrelated reasons. One is that a number system in which (simple) equations such as $x^2 - 2 = 0$ have *no solution* looks a bit feeble. (Put another way, the diagonal of a square of side 1 unit of length would have *no length* in this system!)

Another is that the only decimals we would be able to accept would be those which 'begin to repeat sometime' (thus $0.123\,456\,789\,101\,112\,13...$ would be meaningless).

Third, within the rational numbers we can find sequences which seem to 'converge' towards a *non-rational* answer, e.g. $0.1, 0.12, 0.123, 0.1234, 0.12345,...$, as above, or $1, 1.4, 1.41, 1.414, 1.4142,...$ converging to $\sqrt{2}$.

Last, even though it is hard enough to believe that between every two rationals there is 'room' for infinitely many more, removing irrational numbers from the real number line would leave, in a way to be made precise below, 'even more holes' in the line than points left on it!

Another difference between $\mathbb{Q}$ and $\mathbb{R}$

Sizes of infinity: countability and uncountability

We have just said that omitting the (infinitely many) irrational numbers from the real line would create 'even more holes' than there would be points remaining. How can this make sense when there are infinitely many rational and infinitely many *irrational* numbers? The following ideas, which are very important, first arose in connection with Cantor's investigations referred to at the beginning of Chapter 4.

Definition 4

An infinite set S of elements (in particular numbers) is said to be **countable** iff there is a one-to-one function mapping $\mathbb{Z}^+$ onto S.

An (infinite) set is **uncountable** iff it is not countable.

Example 7

(a) The set E of *even* integers is countable, as the function $f: \mathbb{Z}^+ \to E$ defined by $f(n) = 2n$ shows. We may picture this map as:

$$
\begin{array}{cccccccccccc}
1 & 2 & 3 & 4 & 5 & 6 & 7 & 8 & 9 & 10 & 11 & ... \\
2 & 4 & 6 & 8 & 10 & 12 & 14 & 16 & 18 & 20 & 22 & ...
\end{array}
$$

(b) The set Sq of all squares of integers is countable: define $f: \mathbb{Z}^+ \to$ Sq by $f(x) = x^2$. (This example worried Galileo. It seems to contradict Euclid's axiom that 'the whole is greater than the part'.) The picture is:

$$\begin{array}{ccccccccccc} 1 & 2 & 3 & 4 & 5 & 6 & 7 & 8 & 9 & 10 & 11 & \ldots \\ 1 & 4 & 9 & 16 & 25 & 36 & 49 & 64 & 81 & 100 & 121 & \ldots \end{array}$$

EXERCISE 20   Exhibit $1-1$ onto functions from $\mathbb{Z}^+$ which show that (a) the set of all odd integers (positive and negative) and (b) the set of all positive integers other than multiples of 3 are both countable.

EXERCISE 21   Let S and T be countable sets such that $S \cap T = \varnothing$. Show that $S \cup T$ is countable. (*Hint:* From $s_1, s_2, s_3, \ldots$ and $t_1, t_2, t_3, \ldots$ create $s_1, t_1, s_2, t_2, s_3, t_3, \ldots$. Furthermore, it is then surely no surprise that $S \cup T$ can be proved to be countable, if S and T are, even if $S \cap T \neq \varnothing$.)

We can now note another difference between $\mathbb{Q}$ and $\mathbb{R}$. It is quite a surprise!

Theorem 4
The set $\mathbb{Q}$ is countable, the set $\mathbb{R}$ is *not*.

Proof of Theorem 4

(a) Consider the scheme in Figure 9.3. The path indicated shows how to arrange all the (positive) rational numbers in such a way that a $1-1$ correspondence is set up between $\mathbb{Z}^+$ and $\mathbb{Q}$. (Of course the elements of $\mathbb{Q}$ are not in their natural order – we remind you that there is no *smallest positive rational number*.)

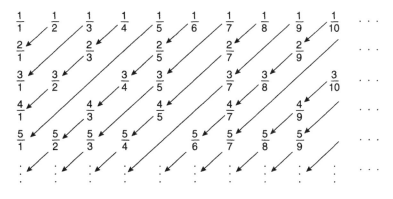

Figure 9.3

(b) To show that even the set of all reals x such that $0 < x < 1$ is *not* countable (i.e. it is uncountable) – *suppose it were* – and suppose that the assumed $1-1$ correspondence were as follows:

$$1 \longleftrightarrow A = 0.a_1 a_2 a_3 a_4 a_5 a_6 \ldots$$
$$2 \longleftrightarrow B = 0.b_1 b_2 b_3 b_4 b_5 b_6 \ldots$$

$$3 \longleftrightarrow C = 0.c_1 c_2 c_3 c_4 c_5 c_6 \ \ldots$$
$$4 \longleftrightarrow D = 0.d_1 d_2 d_3 d_4 d_5 d_6 \ldots$$
$$5 \longleftrightarrow E = 0.e_1 e_2 e_3 e_4 e_5 e_6 \ \ldots$$
$$\vdots \quad \vdots \quad \vdots \qquad \vdots \ \vdots \ \vdots \ \vdots \ \vdots \ \vdots \quad \ldots$$

where each $a, b, c, \ldots$ is an integer from 0 to 9.

We now construct the real number $x = 0.x_1 x_2 x_3 x_4 x_5 \ldots$ by taking x_1 to be any integer in the range from 0 to 9 *other than* a_1; x_2 to be any integer other than a_2; x_3 to be any integer other than a_3, etc. Then, clearly $x \neq A$, since x and A differ in the *first* decimal place, $x \neq B$, since x and B differ in the *second* decimal place, and so on. Consequently x is not equal to any of the members of the above list which was assumed to contain *all* real numbers between 0 and 1. This contradiction shows that that assumption must be false and that the set of real numbers lying strictly between 0 and 1 *isn't* countable.

(*Caution!* This argument is not *quite* good enough – can you see why not? *Hint:* $0.099\,999\,999\ldots = 0.100\,000\,000\ldots$)

EXERCISE 22 Let $S = \{s_1, s_2, s_3, \ldots\}$ and $T = \{t_1, t_2, t_3, \ldots\}$ be countable sets. Show that their Cartesian product $S \times T$ is countable.

(*Hint:*
$$
\begin{array}{llll}
(s_1, t_1) & (s_1, t_2) & (s_1, t_3) & \ldots \\
(s_2, t_1) & (s_2, t_2) & (s_2, t_3) & \ldots \\
(s_3, t_1) & (s_3, t_2) & (s_3, t_3) & \ldots \\
\vdots & \vdots & \vdots & \ldots
\end{array}
$$
)

To finish with, an amusing geometric problem.

EXERCISE 23

(a) Choose a positive real number h. Around each rational point on the real line draw a circle of radius h. Show that *every* real number is inside infinitely many circles.

(b) If the value of h is allowed to change from rational point to rational point show that the conclusion in (a) need no longer hold.

(c) Imagine a thin tree with a trunk of small, but non-zero, radius planted at each point (a, b) in the plane (other than the origin) with integer coordinates a and b. If you stand at the origin, is there a direction in which you can see past the trees, to 'infinity'?

Exercise 22 is the basis for showing that there are real numbers which are not roots of polynomial equations with rational coefficients. In essence the proof says: by Exercise 22 there are countably many polynomial equations of the form $rx + s$ where $r, s \in \mathbb{Q}$ and, hence, only countably many roots to such equations. Likewise for equations of degree 2, degree 3, degree 4, etc. Since it can be proved (cf. the proof of Theorem 4 and the addendum to Exercise 21) that the *set theoretic union of countably many countable sets is countable* then the totality of roots of all these rational polynomials is thus countable. Consequently there are uncountably many reals which are not roots of polynomials with rational coefficients.

CHALLENGE   Name just *one* (e, π?) and, of course, *prove* the named number is the root of no polynomial equation with rational coefficients. (Don't – it's too hard!)

Summary

We defined **real numbers** in terms of **decimal expansions**. As such, they can clearly be made to correspond to **points on a line** and, hence, can be **ordered**. The **rationals** are those amongst them which have a **repeating decimal** expansion. *Between* each pair of real numbers there are infinitely many rational numbers and infinitely many real numbers – indeed a **countable** infinity of the former and an **uncountable** infinity of the latter. In fact, the **Archimedean property** of $\mathbb{R}$ says that for each positive real number α there exists an integer n such that $0 < 1/n < \alpha$.

If a set of real numbers has an **upper (lower) bound** then it has infinitely many such. It also has a *unique* **least upper bound (greatest lower bound)** – **lub** and **glb**. If a set of rationals has an upper bound it will have infinitely many *rational* upper bounds but its unique lub in $\mathbb{R}$ may well be a *non-rational* real number. This **completeness property** of $\mathbb{R}$ distinguishes the set of real numbers from $\mathbb{Q}$.

CHAPTER 10

The Real Numbers and the Complex Numbers

Happiness is an imaginary condition.   THOMAS SZASZ

This chapter introduces the complex numbers in an informal way. Only their simpler properties are proved; nevertheless some of the proofs are quite interesting. Complex numbers can profitably be employed in solving problems concerning nothing but real numbers (see Examples 5, 6, 7). Even so it is perhaps a touch surprising that they are a fundamental tool in certain branches of applied mathematics, physics and engineering.

Introduction

As early as the sixteenth century, mathematicians thought that it would be nice to be able to say that every quadratic equation has two roots, every cubic three roots, and so on. In his *Ars Magna* of 1545, Cardano asked for two 'numbers' whose sum is 10 and product is 40. He gave the answer $5 + \sqrt{-15}$ and $5 - \sqrt{-15}$ (these numbers being the roots of the equation $x(10 - x) = 40$.

If we introduce the symbol 'i' to represent the 'number' $\sqrt{-1}$ then the above numbers may be written $5 + i\sqrt{15}$ and $5 - i\sqrt{15}$.[1] Such numbers were used (in private) by many mathematicians but they were not given greater mathematical

---

[1] $\sqrt{-1}$ was first abbreviated to i by Euler in 1777.

credence until around 1800 when Argand and, prior to him, Wessel and Gauss interpreted them as points in a two-dimensional plane (cf. the representation of real numbers along a (one-dimensional) line).

JEAN ROBERT ARGAND	CASPAR WESSEL
(18 July 1768–13 August 1822)	(8 June 1745–25 March 1818)

The history of the discovery of the geometric representation of complex numbers is not unlike that of some other discoveries/inventions in mathematics. Once an idea is well established and named after its 'founder' it is noticed that others had similar (though often less well developed) ideas earlier. In this case the name that stands out is that of Argand, although it later transpired that Wessel had published similar ideas seven years earlier. Unfortunately for Wessel his sole mathematical publication was written in Danish in the annals of the Royal Danish Academy and was overlooked for 98 years.

After attending the University of Copenhagen, Wessel, who was born in Vestby, Norway, became a cartographer with the Danish Survey Commission. It is not surprising that Wessel was interested in the problem of representing the length and direction of a line analytically. Wessel defined 'units' $1, \varepsilon, -1, -\varepsilon$ along what we would call the x- and y-axes and showed that each line can be represented by expressions $a + \varepsilon b$ and $r(\cos \vartheta + \varepsilon \sin \vartheta)$. He then explained how such expressions should be added and multiplied and observed that 'none of the common rules of operation is contravened'.

The work of Argand, who was born in Geneva, almost met the same fate. He published his ideas anonymously in 1806 but admitted his authorship following a request to do so by J.-F. Français who learned of these concepts from a letter written by Legendre which he found in his dead brother's papers. Unlike Wessel, Argand published other mathematical works.

Wessel was not the first to attempt to represent complex numbers geometrically. An unsuccessful attempt was made by Wallis in the seventeeth century. Furthermore, if Wessel can be said to have anticipated Argand, perhaps Gauss anticipated both. As he observed in his first published treatment (1831) of the subject, he used ideas implicit in his doctoral dissertation of 1799. Indeed, it was Gauss's paper of 1831 which drew attention to these ideas and, belatedly for Argand, and very belatedly for Wessel, gained them the notice they deserved.

An informal definition ... causes problems

Informally we define a **complex number** to be an object of the form $a + ib$ where a and b are real numbers and i represents a quantity whose square is -1. The number a is called the **real part** of $a + ib$ and (the *real number*) b (*not* ib) is called the **imaginary part** of $a + ib$.[2] We sometimes write $a = \mathcal{R}e(a + ib)$ and $b = \mathcal{I}m(a + ib)$. (In particular, each real number a may be regarded as a complex number, namely $a + i0$. A complex number of the form $0 + ib$ is called **wholly imaginary**.)

From a logical point of view the 'definition' causes a problem since we really

---

[2] Descartes, 1637, introduced this term for numbers of the form $a \pm \sqrt{-b}$.

want to say, for instance, that $(a + ib) + (c + id) = (a + c) + i(b + d)$ and that $(a + ib) \cdot (c + id)$ $[= ac + aid + ibc + ibid] = (ac - bd) + i(bc + ad)$. This requires, amongst other things, that $ai = ia$ and $bi = ib$. Is this fair? After all, i is a 'new' object: is it even a 'number'? Hamilton[3] got rid of the problem by defining complex numbers as ordered pairs (a, b) of real numbers. We shall leave Hamilton's approach to others and accept the justifiable fact that defining complex numbers, their sums and their products as we have done creates no difficulties.

EXERCISE 1   Write the following in the form $a + ib$ (a and b being real numbers). Note that we often write $a + bi$ instead of $a + ib$.

(a) $(3 + 2i) + (5 - 3i)$
(b) $(3 + 2i) \cdot (5 - 3i)$
(c) $(1 + i)^2$
(d) $(1 + i)^3$
(e) $(1 + i)^4$
(f) $(a - ib) \cdot (c - id)$ (Compare this with $(a + ib) \cdot (c + id)$.)

Dealing with fractions

Example 1
Write $(4 + 5i)/(2 - 3i)$ in the form $a + ib$.

DISCUSSION
Naively we might put

$$\frac{4 + 5i}{2 - 3i} = x + iy$$

and multiply up to get

$$4 + 5i = (2 - 3i)(x + iy) = (2x + 3y) + i(-3x + 2y)$$

We then only have to solve the system of simultaneous (linear) equations

$$2x + 3y = 4$$
$$-3x + 2y = 5$$

However, a clever move (which you and I may never have thought of for ourselves but which, now we have seen it, we will never forget) is to write

$$\frac{4 + 5i}{2 - 3i} = \frac{4 + 5i}{2 - 3i} \cdot \frac{2 + 3i}{2 + 3i}$$

the point being that the new denominator, $(2 - 3i)(2 + 3i) = 13$, is a number *with zero imaginary part*. So we have:

---

[3] Sir William Rowan Hamilton (4 August 1805–2 September 1865).

SOLUTION

$$\frac{4 + 5i}{2 - 3i} = \frac{4 + 5i}{2 - 3i} \cdot \frac{2 + 3i}{2 + 3i} = \frac{(8 - 15) + i(10 + 12)}{13} = \frac{-7 + 22i}{13}$$

EXERCISE 2 Write $(3 + 7i)/(2 - 3i)$ in the form $a + ib$. Can you now *guess* the $a + ib$ form for $(3 - 7i)/(2 + 3i)$ (without doing any work)?

Complex numbers and the rules of arithmetic

At this late stage in the proceedings it seems rather pedantic to observe that complex numbers satisfy the usual rules of arithmetic. For example, if α, β, γ are complex numbers then

(Ia) $\alpha + \beta = \beta + \alpha$
(Im) $\alpha\beta = \beta\alpha$
(IIa) $(\alpha + \beta) + \gamma = \alpha + (\beta + \gamma)$
(IIm) $(\alpha\beta)\gamma = \alpha(\beta\gamma)$
(III) $\alpha(\beta + \gamma) = \alpha\beta + \alpha\gamma$, etc.

Do note, however, that there is no trichotomy law, as there is for the real numbers. If there *were*, we would (since $i \neq 0$) have either $i > 0$ or $i < 0$, i.e. $-i > 0$. But then one could, by squaring up, presumably 'prove' that $i^2 > 0$, i.e. $-1 > 0$, or $(-i)^2 > 0$, i.e. $-1 > 0$.

The Fundamental Theorem of Algebra

That the complex numbers are quite remarkable is emphasized by the **Fundamental Theorem of Algebra**. Let

$$f(z) = c_n z^n + c_{n-1} z^{n-1} + \dots + c_1 z + c_0$$

be a polynomial (in z) with complex coefficients $c_n, c_{n-1}, \dots, c_0$. Then, $f(z)$ factorizes into a product $c_n(z - \alpha_1)(z - \alpha_2)\dots(z - \alpha_n)$ of n linear factors (i.e. factors of degree 1) for suitable complex numbers $\alpha_1, \alpha_2, \dots, \alpha_n$. In particular, $\alpha_1, \alpha_2, \dots, \alpha_n$ (which may not all be distinct) are the n roots of $f(z)$.

Example 2

(a) $z^4 + 2z^2 + 1 = (z - i)(z - i)(z + i)(z + i)$
(b) $z^3 - 7z^2 + 25z - 39 = (z - 3)(z - \{2 - 3i\})(z - \{2 + 3i\})$
(c) $z^3 - (2 + 2i)z^2 + (-2 + 3i)z - (5 + 5i) = (z - \{3 + i\})(z - \{-1 + 2i\})(z + i)$

The Fundamental Theorem of Algebra is too tricky to prove here. The first adequate proof was given by Gauss when aged about 22, Euler and others having previously failed in the attempt. Note that the corresponding result for real polynomials is *false*: polynomials with *real* coefficients need *not* factorize into products of linear factors $x - \alpha_i$ with *real* numbers α_i. Indeed, there are no *real* numbers α_1, α_2 such that the polynomial $x^2 + 1$ factorizes as $(x - \alpha_1)(x - \alpha_2)$. Put another way: to factorize a real polynomial into linear factors we may have to call upon complex numbers. Therefore, to factorize a polynomial with complex coefficients, it would not seem unreasonable to expect to have to call upon some new 'supercomplex' numbers. The remarkable fact is that we don't need to 'invent' any new numbers to allow the factorization to go ahead: the complex numbers we already have to hand will suffice.

According to the Fundamental Theorem of Algebra, the polynomial $z^2 - (1 + i)$ will have two roots, namely the (two) square roots of $1 + i$. How can we find these?

DISCUSSION
We *could* suppose that $1 + i = (x + iy)^2 = x^2 - y^2 + 2ixy$ and then solve the pair of equations

$$x^2 - y^2 = 1$$
$$2xy = 1$$

To do this, note that

$$(x^2 + y^2)^2 = (x^2 - y^2)^2 + 4x^2y^2$$

Thus

$$(x^2 + y^2)^2 = 1^2 + 1^2 = 2$$

Hence

$$x^2 + y^2 = +\sqrt{2}$$

(It can't be $-\sqrt{2}$. Why not?) Therefore $2x^2 = 1 + \sqrt{2}$ (so that $2y^2 = -1 + \sqrt{2}$). Consequently

$$x = \pm\{\sqrt{(1 + \sqrt{2})}\}/\sqrt{2}$$
$$y = \pm\{\sqrt{(-1 + \sqrt{2})}\}/\sqrt{2}$$

So there are *four* values for $x + iy$. *What's this?* (Only *two* square roots were expected!) This reminds me of the problem of the cubic with six roots (in Chapter 1).

EXERCISE 3 Try to resolve the above four roots/two roots difficulty.

EXERCISE 4 Find the square roots of $3 - 4i$.

COMMENT We shall see an alternative (possibly better) way to proceed with finding certain roots a little later.

Equations with real coefficients

The Fundamental Theorem of Algebra says something special about polynomial equations *with real coefficients*. To see what, we need to introduce complex conjugates.

Definition 1
The **complex conjugate** of the complex number $a + ib$ is the complex number $a - ib$. It is usual to denote it by $\overline{a + ib}$.

We have already used this idea in Example 1. Complex conjugates appear here in a result (Theorem 1) due to Euler.

Theorem 1
If $f(x) = a_n x^n + a_{n-1} x^{n-1} + \ldots + a_1 x + a_0$ is a polynomial with *real* coefficients, and if $a + ib$ is a complex root of $f(x)$, then so is $a - ib$. (In other words, *complex roots of a real polynomial occur in pairs of complex conjugates.*)

To prove this we need some simple properties of complex conjugates.

Lemma 1[4]
Let $\alpha = a + ib$ and $\beta = c + id$ be complex numbers. Then

(i) $\overline{\alpha + \beta} = \bar{\alpha} + \bar{\beta}$
(ii) $\overline{\alpha \cdot \beta} = \bar{\alpha} \cdot \bar{\beta}$
(iii) $\alpha + \bar{\alpha} = 2a$ is real
(iv) $\alpha - \bar{\alpha} = 2ib$ is wholly imaginary
(v) $\alpha = \bar{\alpha}$ iff α is real; $\alpha = -\bar{\alpha}$ iff α is wholly imaginary
(vi) $\bar{\bar{\alpha}} = \alpha$ for *every* α

Proof of Lemma 1   This is straightforward and is left to you.

COMMENT   Parts (i) and (ii) of Lemma 1 extend readily to sums and products of many numbers. For example, $\overline{\alpha_1 \alpha_2 \ldots \alpha_n} = \bar{\alpha}_1 \bar{\alpha}_2 \ldots \bar{\alpha}_n$. In particular, for each complex number α and each integer n, we have $\overline{\alpha^n} = (\bar{\alpha})^n$. (What method might you use to prove that?)

We can now prove Theorem 1.

Proof of Theorem 1   Suppose that the complex number α is a root of $a_n x^n + a_{n-1} x^{n-1} + \ldots + a_1 x + a_0$. Then

$$a_n \alpha^n + a_{n-1} \alpha^{n-1} + \ldots + a_1 \alpha + a_0 = 0$$

It follows that

$$\overline{a_n \alpha^n + a_{n-1} \alpha^{n-1} + \ldots + a_1 \alpha + a_0} = \bar{0} = 0$$

Thus

$$0 = \overline{a_n \alpha^n + a_{n-1} \alpha^{n-1} + \ldots + a_1 \alpha + a_0}$$

$$= \overline{a_n \alpha^n} + \overline{a_{n-1} \alpha^{n-1}} + \ldots + \overline{a_1 \alpha} + \overline{a_0} \text{ (using the above Comment)}$$

$$= a_n \bar{\alpha}^n + a_{n-1} \bar{\alpha}^{n-1} + \ldots + a_1 \bar{\alpha} + a_0 \text{ (since } \bar{a}_i = a_i, \text{ etc., because each } a_i \text{ is real)}$$

But this says that $\bar{\alpha}$ is (also) a root of $a_n x^n + a_{n-1} x^{n-1} + \ldots + a_1 x + a_0$.

---

[4] A **lemma** is usually a result which helps prove a theorem but isn't of sufficient interest or depth to stand as a theorem on its own.

Example 3

Given that the polynomial

$$p(z) = z^4 - 6z^3 + 26z^2 - 46z + 65$$

has $1 + 2i$ as a root, find the other three roots.

SOLUTION   Since $1 + 2i$ is a root of $p(z)$, so is $1 - 2i$. This means that both $z - \{1 + 2i\}$ and $z - \{1 - 2i\}$ *and, hence, their product* are factors of $p(z)$. Now

$$
\begin{aligned}
(z - \{1 + 2i\})(z - \{1 - 2i\}) &= (\{z - 1\} - 2i)(\{z - 1\} + 2i) \\
&= \{z - 1\}^2 - (2i)^2 \qquad \text{(I)} \\
&= z^2 - 2z + 5
\end{aligned}
$$

Simply dividing $z^2 - 2z + 5$ into $p(z)$ gives

$$z^4 - 6z^3 + 26z^2 - 46z + 65 = (z^2 - 2z + 5)(z^2 - 4z + 13)$$

This second quadratic has (using the usual formula) roots

$$z = \frac{-(-4) \pm \sqrt{\{(-4)^2 - 4.1.13\}}}{2.1}$$

that is, $z = \frac{1}{2}(4 \pm \sqrt{-36}) = 2 \pm 3i$. Thus the (four) roots of the given quartic equation are $1 + 2i$, $1 - 2i$, $2 + 3i$ and $2 - 3i$.

(Note the little trick of rewriting in 'difference of two squares' form in (I).)

EXERCISE 5   Given that $7 - 2i$ is a root of $z^4 - 21z^3 + 73z^2 + 721z - 4134$, show that the quartic has two real roots.

EXERCISE 6   Given that 3 is a root of the polynomial $z^4 - 2z^3 - 5z^2 - 2z + 24$, I claim that its complex conjugate $\overline{3}$ $(= 3)$ is *another* root. That is, I claim that $(z - 3)$ will appear *twice* as a factor of the polynomial. Identify where, in the proof of Theorem 1, my claim would falter.

EXERCISE 7   Why does a proof of the Fundamental Theorem of Algebra for polynomials with real coefficients not follow immediately from Theorem 1?

Modulus–argument form

As noted above, complex numbers can be associated with points in the x–y plane. We make the complex number $a + ib$ correspond to the point whose (Cartesian) coordinates are (a, b). Thought of in this way, the real plane is sometimes called the **Argand diagram**. However, the plane can be coordinatized in many ways, one of which is by using *polar coordinates*. A diagram should help.

From Figure 10.1 we see that, if a point has Cartesian coordinates (x, y) and polar coordinates (r, ϑ), then $x = r\cos\vartheta$ and $y = r\sin\vartheta$. It is easily checked that $r = +\sqrt{(x^2 + y^2)}$ whilst if ϑ is chosen so that $-\pi < \vartheta \leqslant \pi$ then ϑ is one of the *two* possible angles for which $\tan\vartheta = y/x$. Hence if α is the complex number $x + iy$ then $\alpha = r(\cos\vartheta + i\sin\vartheta)$. This is called the **polar form** of α. The value r is called the

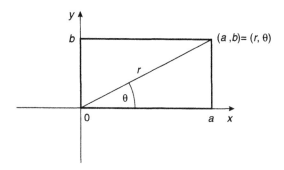

Figure 10.1

modulus of α and we write $r = |\alpha|$. As drawn, $-\pi < \vartheta \leqslant \pi$ and as such ϑ is called the **principal argument** or **principal amplitude** of α, written $\vartheta = \text{Arg}(\alpha)$ with a capital 'A'. However, since, for all integers n,

$$\cos(\vartheta + 2n\pi) = \cos\vartheta$$

and

$$\sin(\vartheta + 2n\pi) = \sin\vartheta$$

it is often convenient to regard any one (or indeed *all*) of these angles $\vartheta + 2n\pi$ as *an* **argument** (or **amplitude**) of α. We then write $\vartheta = \arg(\alpha)$ – using a small 'a'. The expression $r(\cos\vartheta + i\sin\vartheta)$ is often abbreviated to $r\,\text{cis}\,\vartheta$.

EXERCISE 8 For each of the following complex numbers α, find the modulus $|\alpha|$ and the principal value $\text{Arg}(\alpha)$ of the argument. Then write α in polar form:

(a) 2   (g) $1 + i$
(b) -3 (h) $1 - i$
(c) $\cos\vartheta + i\sin\vartheta$ (i) $3 - 4i$
(d) $\cos\vartheta - i\sin\vartheta$ (j) $11 + 60i$
(e) i (k) $\sin 73° + i\cos 73°$
(f) $-i$

EXERCISE 9 Write the following complex numbers in Cartesian form: (a) $\text{cis}\,\frac{\pi}{6}$; (b) $\text{cis}\,\frac{5\pi}{6}$; (c) $4\,\text{cis}\,\pi$; (d) $\sqrt{2}\,\text{cis}\,\frac{\pi}{4}$.

EXERCISE 10 For complex numbers α $(=x + iy)$ and β show that:

(a) $|\bar{\alpha}| = |\alpha|$
(b) $\alpha\bar{\alpha} = |\alpha|^2 \,(= x^2 + y^2)$
(c) $|\alpha\beta| = |\alpha||\beta|$
(d) **By** choosing α and β (almost) at random, exhibit complex numbers α and β such that $|\alpha + \beta| \neq |\alpha| + |\beta|$
(e) If α is real (i.e. $\mathscr{I}m(\alpha) = 0$), then $|\alpha|$ coincides with value $|x|$ introduced in Chapter 4
(f) $\mathscr{R}e\,(\alpha) \leqslant |\alpha|$

(*Hint*: For (c), first prove that $|\alpha\beta|^2 = \alpha\beta\overline{\alpha\beta} = \dots = |\alpha|^2|\beta|^2$ and then take (positive) square roots. (Why 'positive'? Is that fair?).)

Next, a couple of results which *we* shall not call upon but which appear time and again in mathematics – so much so that the first has been given a name.

Theorem 2
For all complex numbers α, β, γ and δ

(i) $|\alpha + \beta| \leqslant |\alpha| + |\beta|$
(ii) $|\gamma - \delta| \geqslant ||\gamma| - |\delta||$

> DISCUSSION
> Taking a hint from the previous Comment and Exercise 10, we can try to prove $|\alpha + \beta| = |\alpha| + |\beta|$ by proving the 'same' inequality holds between their squares and then taking (positive) square roots.

Proof of Theorem 2   On the one hand

$$|\alpha + \beta|^2 = (\alpha + \beta)\overline{(\alpha + \beta)} = \alpha\bar{\alpha} + \alpha\bar{\beta} + \beta\bar{\alpha} + \beta\bar{\beta} = |\alpha|^2 + \alpha\bar{\beta} + \beta\bar{\alpha} + |\beta|^2 \qquad \text{(II)}$$

On the other hand

$$(|\alpha| + |\beta|)^2 = |\alpha|^2 + 2|\alpha||\beta| + |\beta|^2 \qquad \text{(III)}$$

Now $\alpha\bar{\beta} + \beta\bar{\alpha} = \alpha\bar{\beta} + \overline{\alpha\bar{\beta}}$. Hence, by Lemma 1(iii)

$$\alpha\bar{\beta} + \beta\bar{\alpha} = 2\mathscr{R}e(\alpha\bar{\beta})$$

Consequently, by Exercise 10(f) and (c)

$$\alpha\bar{\beta} + \beta\bar{\alpha} = 2\,\mathscr{R}e(\alpha\bar{\beta}) \leqslant 2|\alpha\bar{\beta}| = 2|\alpha||\beta|$$

Comparing (II) and (III) establishes (i).
    To prove (ii) replace α by $\gamma - \delta$ and β by δ in (i). We obtain:

$$|\{\gamma - \delta\} + \delta| \leqslant |\{\gamma - \delta\}| + |\delta|$$

that is

$$|\gamma| \leqslant |\gamma - \delta| + |\delta|$$

and hence

$$|\gamma| - |\delta| \leqslant |\gamma - \delta|$$

Repeat this with γ and δ interchanged. We get

$$|\delta| - |\gamma| \leqslant |\delta - \gamma| \{= |\gamma - \delta|\}$$

We have therefore shown that the *real number* $|\gamma| - |\delta|$ and its negative $-(|\gamma| - |\delta|)$ are both less than or equal to $|\gamma - \delta|$. We deduce, immediately, that $||\gamma| - |\delta|| \leqslant |\gamma - \delta|$, as claimed.

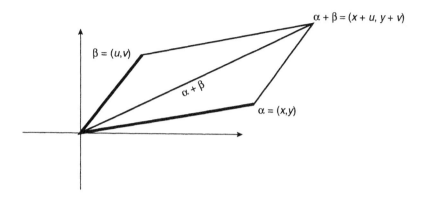

Figure 10.2

COMMENT   The inequality (i) is called the **triangle inequality**. The reason lies in Figure 10.2. Note that, if $\alpha = x + iy$ and if $\beta = u + iv$ then, pictorially, their sum is given by the parallelogram of vector addition.

EXERCISE 11   Prove, using Theorem 2 that, for all complex numbers α, β and γ,

(a) $|\alpha + \beta + \gamma| \leqslant |\alpha| + |\beta| + |\gamma|$
(b) $|\alpha - \beta| + |\beta - \gamma| \geqslant |\alpha - \gamma|$

Polar form of complex multiplication

We have seen that complex number multiplication (in Cartesian form) is a bit ugly. On the other hand, in polar form it is very pretty.

Theorem 3
Let $\alpha = r(\text{cis } \vartheta)$ and $\beta = s(\text{cis } \phi)$ be complex numbers in polar form. Then

$$\alpha\beta = rs(\text{cis}\{\vartheta + \phi\})$$

DISCUSSION
Clearly, the only problem is, is $(\text{cis }\vartheta)(\text{cis }\phi) = \text{cis}\{\vartheta + \phi\}$?

Proof of Theorem 3   Now[5]

$$(\cos\vartheta + i\sin\vartheta)(\cos\phi + i\sin\phi) = \cos\vartheta\cos\phi + i\cos\vartheta\sin\phi + i\sin\vartheta\cos\phi - \sin\vartheta\sin\phi$$
$$= \cos\vartheta\cos\phi - \sin\vartheta\sin\phi + i(\cos\vartheta\sin\phi + \sin\vartheta\cos\phi)$$
$$= \cos\{\vartheta + \phi\} + i\sin\{\vartheta + \phi\}$$

Thus $\alpha\beta = rs(\text{cis}\{\vartheta + \phi\})$, as claimed.

[5] We use the well-known identities:

$$\sin(\vartheta + \phi) = \sin\vartheta\cos\phi + \cos\vartheta\sin\phi \quad \text{and} \quad \cos(\vartheta + \phi) = \cos\vartheta\cos\phi - \sin\vartheta\sin\phi$$

COMMENT Here $\vartheta + \phi$ may well lie outside the interval $-\pi$ to π. However, by our definition of argument (with a small 'a') we are entitled to say that

$$\arg(\alpha \cdot \beta) = \arg(\alpha) + \arg(\beta)$$

rather than determine the precise number of 2π by which $\vartheta + \phi$ exceeds π or falls short of $-\pi$.

With this proviso we may remember Theorem 3 by saying that the *moduli of* α *and* β *are multiplied together* whilst the *arguments of* α *and* β *are added*.

Figure 10.3 shows the effect of multiplication by $2\operatorname{cis}\phi$.

EXERCISE 12 Draw the line joining the complex numbers $\alpha = 1 + 3i$ and $\beta = 2 + 5i$. Now draw the lines joining (a) the pair $i\alpha$ and $i\beta$; (b) the pair $(1 - i)\alpha$ and $(1 - i)\beta$.

De Moivre's theorem

Putting $\phi = \vartheta$ (and $r = s = 1$) in Theorem 3 yields

$$(\cos\vartheta + i\sin\vartheta)^2 = \cos 2\vartheta + i\sin 2\vartheta$$

Immediately one *must* ask oneself, might $(\cos\vartheta + i\sin\vartheta)^n = \cos n\vartheta + i\sin n\vartheta$ hold *for all integers n?* We could try to prove it. By mathematical induction?

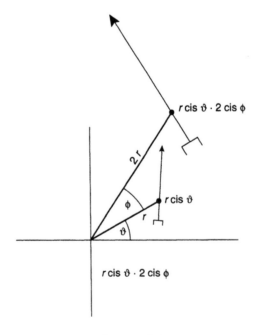

Figure 10.3

Theorem 4 (De Moivre's theorem)
For all integers n,

$$(\cos \vartheta + i \sin \vartheta)^n = \cos n\vartheta + i \sin n\vartheta$$

DISCUSSION

All integers? Let us at least try to prove the result claimed for all positive integers – that suggests we use mathematical induction – and then try to see how the negative n might relate to the positive n.

Proof of Theorem 4   For each positive integer n, let S(n) be

$$(\cos \vartheta + i \sin \vartheta)^n = \cos n\vartheta + i \sin n\vartheta$$

Since $(\cos \vartheta + i \sin \vartheta)^1 = \cos 1\vartheta + i \sin 1\vartheta$, our induction proof begins. Now suppose that

$$(\cos \vartheta + i \sin \vartheta)^k = \cos k\vartheta + i \sin k\vartheta$$

Then

$$
\begin{aligned}
(\cos \vartheta + i \sin \vartheta)^{k+1} &= (\cos \vartheta + i \sin \vartheta)^k (\cos \vartheta + i \sin \vartheta) \\
&= (\cos k\vartheta + i \sin k\vartheta)(\cos \vartheta + i \sin \vartheta) \\
&= \cos k\vartheta \cos \vartheta - \sin k\vartheta \sin \vartheta + i(\cos k\vartheta \sin \vartheta + \sin k\vartheta \cos \vartheta) \\
&\qquad\qquad\qquad\qquad\qquad\qquad\qquad\qquad\qquad\text{(cf. Theorem 3)} \\
&= \cos \{k + 1\} \vartheta + i \sin \{k + 1\} \vartheta
\end{aligned}
$$

as required. Thus the theorem is true for all $n \geqslant 1$.

MORE DISCUSSION

The case $n = 0$ is easily dealt with. What if n is negative? Then $n = -m$ where m is positive. Can we use the first part of the proof on m?

More proof   The case $n = 0$ requires that $(\cos \vartheta + i \sin \vartheta)^0 = \cos 0 + i \sin 0$, which is true. Finally, if n is negative, we write $n = -m$ *where m is positive*. We therefore know from what we have just proved that

$$(\cos \vartheta + i \sin \vartheta)^m = \cos m\vartheta + i \sin m\vartheta$$

Hence

$$
\begin{aligned}
(\cos \vartheta + i \sin \vartheta)^n &= (\cos \vartheta + i \sin \vartheta)^{-m} \\
&= \frac{1}{\cos m\vartheta + i \sin m\vartheta} \\
&\overset{6}{=} \frac{1}{(\cos m\vartheta + i \sin m\vartheta)} \frac{(\cos m\vartheta - i \sin m\vartheta)}{(\cos m\vartheta - i \sin m\vartheta)} \\
&= \cos m\vartheta - i \sin m\vartheta \quad \text{(since the denominator is } \cos^2 m\vartheta + \sin^2 m\vartheta = 1)
\end{aligned}
$$

---

[6] Notice the use of the standard 'trick' introduced in Example 1 on page 166.

Therefore

$$(\cos \vartheta + i \sin \vartheta)^n = (\cos \vartheta + i \sin \vartheta)^{-m}$$

$$= \cos m\vartheta - i \sin m\vartheta = \cos(-n)\vartheta - i \sin(-n)\vartheta$$

$$= \cos n\vartheta + i \sin n\vartheta$$

(since $\cos(-x) = \cos x$ and $\sin(-x) = -\sin x$ for all x)

ABRAHAM DE MOIVRE
(26 May 1667–27 November 1754)

Born in 1667 and coming from a Protestant family, De Moivre was one of forty thousand Hugenots who emigrated to Britain following the revocation, by Louis XIV in 1685, of the Edict of Nantes. In France he had attended a Protestant academy where he read Huygens's book *The Mathematics of Chance*.

In England he made a modest living as a private tutor in mathematics, reading Newton's *Principia* whilst journeying from one pupil to another, and by solving problems in probability for well-to-do patrons who wished to improve their chances of winning at gambling.

Despite never obtaining a professorship in a British university, De Moivre's talents were recognized with a fellowship of the Royal Society in 1697 and, in 1719, he was made a member of a commission to settle the Leibniz/Newton dispute regarding priority in the invention of the calculus.

In 1718 there appeared his best-known book *The Doctrine of Chance*. In this work he obtains what is now known as Stirling's formula approximating $n!$ by the expression $cn^{n+1/2}e^{-n}$ where c is a constant determined, by Stirling, to be equal to $\sqrt{(2\pi)}$. Also in this text is the first use of the concept of *generating function* which is fundamental in the theory of probability – and was later used by Euler to solve problems arising from difference equations (such as Fibonacci's).

Although listed in the *Dictionary of Scientific Biography*, as a probabalist, De Moivre published some papers in other areas of mathematics; indeed his first paper, in 1692, concerned Newton's 'fluxions'. Of these other results, the best known is the theorem, first stated in 1722, which bears his name.

Aged 87 he succumbed to lethargy, sleeping 20 hours each day. The cause of his death was given as 'somnolence'.

EXERCISE 13   Write in the form $a + ib$ the following complex numbers.

(a) $\left(\cos \dfrac{3\pi}{28} + i \sin \dfrac{3\pi}{28} \right)^7$

(b) $\left(\cos \dfrac{\pi}{4} + i \sin \dfrac{\pi}{4} \right)^9$

(c) $\left(\sin \dfrac{\pi}{4} + i \cos \dfrac{\pi}{4} \right)^{11}$

(d) $\left(\cos \dfrac{2\pi}{3} - i \sin \dfrac{2\pi}{3} \right)^{11}$

(e) $(\sqrt{3} + i)^{201}$ $\left(Hint: \sqrt{3} + i = 2 \left(\dfrac{\sqrt{3}}{2} + \dfrac{1}{2}i \right). \right)$

Uses of De Moivre's theorem 1: finding roots

De Moivre's theorem can help in finding the nth roots of certain complex numbers and in establishing trigonometric identities. Three examples follow. The first depends on a consequence of Theorem 4.

Corollary 1
The k kth roots of $r \operatorname{cis} \vartheta$ are $r^{1/k}$ $[\operatorname{cis} \{(\vartheta + 2t\pi)/k\}]$ where t takes the values $0, 1, 2, \ldots, k - 1$ and $r^{1/k}$ denotes the (unique) positive real kth root of the positive real number r.

Proof of Corollary 1   Using De Moivre's theorem we see that the kth power of $r^{1/k} [\operatorname{cis} \{(\vartheta + 2t\pi)/k\}]$ is $r \operatorname{cis}(\vartheta + 2t\pi)$, which is $r \operatorname{cis} \vartheta$. Further, no two of these k roots are equal. For if

$$\cos \left\{ \frac{\vartheta + 2s\pi}{k} \right\} = \cos \left\{ \frac{\vartheta + 2t\pi}{k} \right\}$$

and

$$\sin \left\{ \frac{\vartheta + 2s\pi}{k} \right\} = \sin \left\{ \frac{\vartheta + 2t\pi}{k} \right\}$$

then $\{(\vartheta + 2s\pi)/k\}$ and $\{(\vartheta + 2t\pi)/k\}$ must differ by an integer multiple of 2π. But if $0 \leqslant s, t < k$ this is impossible unless $s = t$.

Example 4

(a) Find the (five) fifth roots of 1 and plot them on the Argand diagram.
(b) Find the (seven) seventh roots of $\alpha = 3 - 4i$.

SOLUTION

(a) Since $1 = 1 (\operatorname{cis} 0)$ the five roots are given by $1^{1/5} [\operatorname{cis} \{(0 + 2t\pi)/5\}]$ where $t = 0, 1, 2, 3, 4$. Thus the roots are $\operatorname{cis}(0\pi/5)$, $\operatorname{cis}(2\pi/5)$, $\operatorname{cis}(4\pi/5)$, $\operatorname{cis}(6\pi/5)$, $\operatorname{cis}(8\pi/5)$. If necessary, these can be written as $1, a + ib, c + id, c - id, a - ib$ where

$$a = \frac{-1 + \sqrt{5}}{4}, \quad b = \frac{\sqrt{(10 + 2\sqrt{5})}}{4}, \quad c = \frac{-1 - \sqrt{5}}{4}, \quad d = \frac{\sqrt{(10 - 2\sqrt{5})}}{4}$$

How do we determine a, b, c, d? See Exercise 18.
   We can represent these roots pictorially as the five vertices of the regular pentagon inscribed in a circle of unit radius (Figure 10.4).

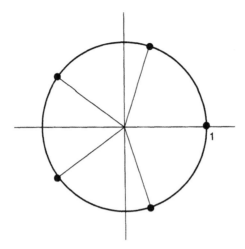

Figure 10.4

(b) Here $|\alpha| = +\sqrt{\{3^2 + (-4)^2\}} = 5$, so write $\alpha = 5(\operatorname{cis}\vartheta)$ where $\cos\vartheta = \frac{3}{5}$ and $\sin\vartheta = -\frac{4}{5}$. Hence the required roots are $5^{1/7}[\operatorname{cis}\{(\vartheta + 2t\pi)/7\}]$ where t runs from 0 to 6 inclusive and ϑ is any angle such that $\cos\vartheta = \frac{3}{5}$ and $\sin\vartheta = -\frac{4}{5}$.

COMMENT   There seems little point in trying to 'simplify' the given solution since $\cos\{(\vartheta + 2t\pi)/7\}$ and $\sin\{(\vartheta + 2t\pi)/7\}$ are not readily evaluated.

EXERCISE 14   Find (simplifying the cos and sin values if they are readily evaluated) all the (complex number) solutions of the following equations:

(a) $z^2 = i$ (e) $z^2 + 2z + 4 = 0$
(b) $z^6 = -1$ (f) $z^4 + 2z^2 + 4 = 0$
(c) $z^3 = 2i$ (g) $z^5 - (1 + i) = 0$
(d) $z^3 = -2i$

EXERCISE 15   Plot, as in Example 4: (a) the three cube roots of 1; (b) the three cube roots of -1; (c) the six sixth roots of -1; (d) the four fourth roots of i.

EXERCISE 16   Find the square roots of $3 - 4i$ by De Moivre's theorem. Is the solution any clearer than in Exercise 4?

Uses of De Moivre's theorem 2: trigonometrical identities

The usefulness of De Moivre's theorem in finding (or even remembering) certain trigonometric identities is shown in the following example.

Example 5
Show that $\cos 4\vartheta = 8\cos^4\vartheta - 8\cos^2\vartheta + 1$.

SOLUTION $\cos 4\vartheta$ is the real part of $\text{cis}\, 4\vartheta$ which is equal to $(\text{cis}\,\vartheta)^4$ by De Moivre's theorem. However, by the binomial theorem

$$(\text{cis}\,\vartheta)^4 = (\cos\vartheta + i\sin\vartheta)^4$$
$$= \cos^4\vartheta + 4\cos^3\vartheta\,(i\sin\vartheta) + 6\cos^2\vartheta\,(i\sin\vartheta)^2 + 4\cos\vartheta\,(i\sin\vartheta)^3 + (i\sin\vartheta)^4$$

the real part being $\cos^4\vartheta - 6\cos^2\vartheta\sin^2\vartheta + \sin^4\vartheta$. If we now replace $\sin^2\vartheta$ by $1 - \cos^2\vartheta$ (and $\sin^4\vartheta$ by $(1 - \cos^2\vartheta)^2$) we get the result claimed.

EXERCISE 17

(a) Find a formula for $\sin 3\vartheta$ in terms of powers of $\sin\vartheta$.
(b) Find a formula for $\cos 4\vartheta$ in terms of powers of $\sin\vartheta$.
(c) Can you find a formula for $\sin 4\vartheta$ in terms of powers of $\cos\vartheta$?

EXERCISE 18   Setting $c = \cos\vartheta$, $s = \sin\vartheta$, prove that

$$\sin 5\vartheta = s(5c^4 - 10c^2s^2 + s^4)$$

Now put $\vartheta = 36°$ and use the identity $\cos 2\vartheta = 2\cos^2\vartheta - 1$ to obtain the solutions given in Example 4(a) above.

Powers in terms of multiple angles – by Euler's formula

It is possible to reverse the above procedure and to convert powers such as $\sin^7\vartheta$ into sums involving multiple angles (but *no* powers > 1). This relies upon *Euler's formula* as given below.
    You probably know that $\sin x$, $\cos x$ and e^x have 'series representations'

$$\sin x = \frac{x}{1!} - \frac{x^3}{3!} + \frac{x^5}{5!} - \frac{x^7}{7!} + \dots$$

$$\cos x = 1 - \frac{x^2}{2!} + \frac{x^4}{4!} - \frac{x^6}{6!} + \dots$$

$$e^x = 1 + \frac{x}{1!} + \frac{x^2}{2!} + \frac{x^3}{3!} + \frac{x^4}{4!} + \frac{x^5}{5!} + \frac{x^6}{6!} + \dots$$

These expansions can be shown to be meaningful (and valid) for all real values of x – indeed for all complex values of x. In particular, replacing each x by $i\alpha$, where α denotes any real or complex number, we obtain:

$$e^{i\alpha} = 1 + \frac{(i\alpha)}{1!} + \frac{(i\alpha)^2}{2!} + \frac{(i\alpha)^3}{3!} + \frac{(i\alpha)^4}{4!} + \frac{(i\alpha)^5}{5!} + \frac{(i\alpha)^6}{6!} + \dots$$

$$= 1 \qquad -\frac{\alpha^2}{2!} \qquad +\frac{\alpha^4}{4!} \qquad -\frac{\alpha^6}{6!} \qquad + \dots$$

$$+ i\left\{\frac{\alpha}{1!} \qquad -\frac{\alpha^3}{3!} \qquad +\frac{\alpha^5}{5!} \qquad + \dots\right\}$$

We immediately obtain **Euler's formula**:

$$e^{i\alpha} = \cos\alpha + i\sin\alpha$$

COMMENT   Several deductions follow. For example,

1. Setting $\alpha = \pi$ we obtain the remarkable formula $e^{i\pi} = -1$ connecting the four basic mathematical constants, 1, π, e and i via the most basic relationship, that of equality.
2. Next, replacing α first by $n\vartheta$, then by $-n\vartheta$ and then by (i) adding and (ii) subtracting the results we obtain

$$\cos n\vartheta = \frac{1}{2}(e^{in\vartheta} + e^{-in\vartheta})$$

and

$$\sin n\vartheta = \frac{1}{2i}(e^{in\vartheta} - e^{-in\vartheta})$$

It is these two latter formulae which enable us to reverse the process in Example 5, as demonstrated in the next example.

Example 6
Write $\sin^3\vartheta$ in terms of sines/cosines of multiple angles.

SOLUTION   Put $v = e^{i\vartheta}$ so that

$$\cos n\vartheta = \left\{\frac{1}{2}(v^n + v^{-n})\right\} \quad\text{and}\quad \sin n\vartheta = \left\{\frac{1}{2i}(v^n - v^{-n})\right\}$$

Then

$$\sin^3\vartheta = \left\{\frac{1}{2i}(v - v^{-1})\right\}^3 = -\frac{1}{8i}(v^3 - 3v^2v^{-1} + 3vv^{-2} - v^{-3})$$

$$= -\frac{1}{8i}(v^3 - v^{-3}) + \frac{3}{8i}(v - v^{-1})$$

$$= -\frac{1}{4}\left\{\frac{1}{2i}(v^3 - v^{-3})\right\} + \frac{3}{4}\left\{\frac{1}{2i}(v - v^{-1})\right\}$$

$$= -\frac{1}{4}\sin 3\vartheta + \frac{3}{4}\sin\vartheta$$

EXERCISE 19

(a) Write $\sin^4\vartheta$ in terms of cosines of multiple angles.
(b) Can you think why no similar formula in terms of sines of multiple angles is available? (*Hint:* Assume $\sin^4\vartheta = a_0 + a_1\sin\vartheta + a_2\sin 2\vartheta + \ldots$. Now replace ϑ by $-\vartheta$.)

One final example illustrating how the use of complex numbers can solve a problem concerning only real numbers is given here.

Example 7

Find a formula for $C = 1 + \cos \vartheta + \cos 2\vartheta + \dots + \cos n\vartheta$.

DISCUSSION

Here is a super method which it is useful to remember because it appears again in several other contexts.

Define s to be $\sin 0\vartheta + \sin \vartheta + \sin 2\vartheta + \dots + \sin n\vartheta$ and form the quantity

$$C + iS = \operatorname{cis} 0\vartheta + \operatorname{cis} \vartheta + \operatorname{cis} 2\vartheta + \dots + \operatorname{cis} n\vartheta$$
$$= (\operatorname{cis} \vartheta)^0 + (\operatorname{cis} \vartheta) + (\operatorname{cis} \vartheta)^2 + \dots + (\operatorname{cis} \vartheta)^n$$

This can be summed easily since it is a geometric progression. We then take its real part, which is, of course, C.

SOLUTION   Put $S = \sin \vartheta + \sin 2\vartheta + \dots + \sin n\vartheta$. Then, by De Moivre's theorem,

$$C + iS = 1 + (\operatorname{cis} \vartheta) + (\operatorname{cis} \vartheta)^2 + \dots + (\operatorname{cis} \vartheta)^n$$
$$= \frac{1 - (\cos \vartheta + i \sin \vartheta)^{n+1}}{1 - (\cos \vartheta + i \sin \vartheta)}$$
$$= \frac{1 - \cos \{n + 1\}\vartheta - i \sin \{n + 1\}\vartheta}{1 - \cos \vartheta \qquad - i \sin \vartheta}$$

Multiplying numerator and denominator by $1 - \cos \vartheta + i \sin \vartheta$ and tidying up, this becomes $(X + iY)/Z$ where

$$X = \{(1 - \cos (n + 1)\vartheta\}(1 - \cos \vartheta) + \sin \{n + 1\}\vartheta \sin \vartheta$$
$$= 1 - \cos \vartheta + \cos n\vartheta - \cos \{n + 1\}\vartheta$$

and

$$Z = (1 - \cos \vartheta)^2 + (\sin \vartheta)^2 = 2 - 2\cos \vartheta$$

Finally we note that

$$1 - \cos \vartheta = 2 \sin^2 \left\{ \frac{\vartheta}{2} \right\}$$

whilst

$$\cos n\vartheta - \cos (n + 1)\vartheta = 2 \sin \{(n + \tfrac{1}{2})\vartheta\} \sin \left\{ \frac{\vartheta}{2} \right\}$$

so that

$$X = 2 \sin^2 \left\{ \frac{\vartheta}{2} \right\} + 2 \sin \left\{ \frac{\vartheta}{2} \right\} \sin \{(n + \tfrac{1}{2})\vartheta\} = 2 \sin \left\{ \frac{\vartheta}{2} \right\} \left[\sin \left\{ \frac{\vartheta}{2} \right\} + \sin \{(n + \tfrac{1}{2})\vartheta\} \right]$$
$$= 2 \sin \left\{ \frac{\vartheta}{2} \right\} \left[2 \sin \left\{ \left(\frac{n + 1}{2} \right)\vartheta \right\} \sin \left\{ \frac{n\vartheta}{2} \right\} \right]$$

and

$$\frac{X}{Z} = \frac{\sin\left\{\left(\dfrac{n+1}{2}\right)\vartheta\right\}\sin\left(\dfrac{n\vartheta}{2}\right)}{\sin\left\{\dfrac{\vartheta}{2}\right\}}$$

The above is only the tip of the iceberg. We can go on to consider general functions of a complex variable, their derivatives and their integrals over contours in the complex plane. Such activity has applications in many areas of applied mathematics – even in civil engineering (where it is important in studying fluid seepage under dams). Back in pure mathematics, an estimate for the number of primes less than the integer n can be obtained via the *Prime Number Theorem* which may be proved by considering complex integrals and logarithms. But these are other stories.

Summary

Following their introduction by means of a 'problem', **complex numbers** were used informally by many mathematicians before it was deemed necessary to try to put them on a firmer footing. Simply defining a complex number to be 'something of the form $a + ib$ (or $a + bi$) where a and b are real numbers and i represents $\sqrt{-1}$' can lead to (temporary) difficulties – which can be eliminated by using Hamilton's definition of complex numbers using ordered pairs.

Sticking to the $a + ib$ notation we illustrated the 'trick' for simplifying a fraction of two complex numbers, and observed that the complex numbers satisfy the basic rules of arithmetic. We then stated the very important **Fundamental Theorem of Algebra** and deduced that, in each polynomial equation with *real* coefficients, the roots appear in **complex conjugate** pairs. The important concepts of **modulus** and **argument** were introduced via **polar representation** of complex numbers, and their more prominent properties – in particular the **triangle inequality** – were established. The neat polar form of complex multiplication leads, naturally, to **De Moivre's theorem** with its applications to finding roots and obtaining trigonometrical identities. Complex numbers, in particular the theory of functions of a complex variable, are important in many branches of mathematics, physics and engineering where complex functions are able to express, mathematically, the idea of flows and potentials.

CHAPTER 11

Guessing, Analogy and Transformation

It's better only guessing: analogies decide nothing, but they can make one feel more at home.

BLAKE

Those who transform the world are not the statesmen but the scientists.

AUDEN

In this chapter and the next we look at some more examples of problems and theorems whose proofs employ methods named in these two chapters' titles. We do this to emphasize the importance of these methods not only for the research mathematician but also for the undergraduate for whom coursework problems, can, after all, present a corresponding challenge.

Introduction

The words in the title name three important techniques which can fruitfully be used in tackling problems and attempting proofs. We have already seen examples of all three, but here are some more nice examples which emphasize these techniques.

Guessing: to find the answer

Apart from making conjectures, which may be described more as 'fairly confident' guesses it is not so clear that guessing can be of help in the precise area of pure mathematics. But it can. Indeed, even at a very primitive level you can find yourself guessing whenever you perform a division of, say, 1043 by 178 by hand ('Does 178 go into 1043 six times – or is it only five?'). Of course, making a guess without incorporating a subsequent check is useless.

EXERCISE 1 (From a turn-of-the-century common arithmetic book for young schoolchildren.) A total of £1500 is loaned to two men. At the end of eight months £936 is repaid by one man; at the end of ten months £630 is repaid by the other man. Assuming they both paid at the same rate of simple interest, find the amount loaned to each and the common rate of interest. (*Hint:* Go on, guess! But *check your guess* to see if it is correct.)

You might also find yourself guessing if you are asked to find the function $y(x)$ which solves the differential equation

$$\frac{d^2 y}{dx^2} + 3\frac{dy}{dx} + 2y = 0 \tag{I}$$

given that $y(x)$ must also satisfy the *boundary conditions* that $y = 1$ when $x = 0$ and $y = 2$ when $x = 2\pi$. If you had never seen anything like this before, what could you do? Perhaps you should first concentrate on trying to solve the equation and, having obtained its most general solution, fit this solution to the boundary conditions. How do you find this general solution, though? You may try guessing a solution, say $y(x) = ax^3 + bx^2 + cx + d$. (You might choose a polynomial solution because polynomials are such 'nice' functions.) On reflection, a polynomial is a poor choice. Why?

If, at this point, you are stuck, *see if you can solve a simpler but related problem.* It can't do any harm and it might help. What about

$$3\frac{dy}{dx} + 2y = 0?$$

Can you solve *that*? Of course you can. From

$$\int\frac{dy}{y} = -\frac{2}{3}\int dx$$

you obtain

$$\log_e y = -\frac{2}{3}x + c$$

Consequently

$$y = e^{-\frac{2}{3}x + c} = Ae^{-\frac{2}{3}x}$$

where $A = e^c$ is some constant.

Returning to equation (I), perhaps you should try $y(x) = Ae^{\alpha x}$ and see what happens. Well, replacing y in (I) by $Ae^{\alpha x}$, you obtain

$$\alpha^2 Ae^{\alpha x} + 3\alpha Ae^{\alpha x} + 2Ae^{\alpha x} = 0$$

Now, since $e^{\alpha x}$ is never 0 and $A = 0$ gives an uninteresting answer, you can divide through this last equation by $Ae^{\alpha x}$ to obtain $\alpha^2 + 3\alpha + 2 = 0$, which implies that $\alpha = -1$ or -2. Which is it? And what about the value of A?

Actually you are not limited to choosing one value of α to the exclusion of the other since if Ae^{-x} and Be^{-2x} are solutions of (I) then so is their sum. (Note also that the (so far) arbitrary constants A and B need not be equal.)

You may readily check that $y(x) = Ae^{-x} + Be^{-2x}$ does indeed satisfy (I). To make it satisfy the boundary conditions you will need to observe that

if $x = 0$ then $1 = y(0) = (Ae^{-0} + Be^{-0})$
if $x = 2\pi$ then $2 = y(2\pi) = Ae^{-2\pi} + Be^{-4\pi}$

You can now find A and B by solving these two equations simultaneously.

EXERCISE 2   You are asked to find a function $y(x)$ satisfying the differential equation

$$\frac{d^2y}{dx^2} + 3\frac{dy}{dx} + 2y = f(x)$$

Which of the functions

(i) $ax + b$
(ii) ae^x
(iii) $ax^3 + bx^2 + cx + d$
(iv) $a\sin x$
(v) $a\sin x + b\cos x$

would you try for y if

(a) $f(x) = 7x^3$
(b) $f(x) = 4\sin x$
(c) $f(x) = 7x^4 + 5\cos x$?

Guessing has long been made use of. The Egyptian mathematicians of c. 1650 BC used a kind of guessing to solve problems such as 'A quantity together with a seventh of the quantity is 31. What is the quantity?' Of course, with the advantage of good *notation* we can transform the problem to the trivial one of 'given $x + \frac{x}{7} = 31$, find x'. The Egyptian solution? 'Try $x = 7$; then $x + \frac{x}{7} = 8$. This is not the required answer.' After multiplying 8 by $2, 1, \frac{1}{2}, \frac{1}{4}$ and $\frac{1}{8}$, and adding (to get 31), they did the same to 7 to find the appropriate value for x. This procedure was called the 'Rule of False Position' (see Bunt *et al.* (1988) and Burton (1995)).

EXERCISE 3   Three men A, B and C share $1100. B has $4\frac{1}{2}$ times as much as A, and C has as much as A and B put together. Find how much each receives (a) by algebra, (b) by false position, by assuming A gets $2 (chosen so that B's share is a whole number). Then scale your answer as appropriate.

Guessing is, of course, not always safe. Witness the real number $e^{\pi\sqrt{163}}$ mentioned in the preface which is so nearly an integer. Even Euler wasn't immune. Noting that $x^3 + y^3 = z^3$ has no solution with positive integers x, y, z, Euler speculated that, for each $n \geqslant 3$, no nth power could be the sum of $n - 1$ positive nth powers. But in 1988 Elkies found that $2\,682\,440^4 + 15\,365\,639^4 + 18\,796\,760^4 = 20\,615\,673^4$. Later Frye found a smaller solution.

Some more examples

We now look at two more examples.

Example 1
Find an integer n such that, for certain, there is a root of the polynomial $p(x) = 3x^5 - 71x^4 + x^2 - 2x - 2$ between n and $n + 1$.

> DISCUSSION
> One *could* look to see if $n = 0$, $n = 1$, $n = 2$, etc. would do. But, noticing that, for large x, the two leading terms dominate the given polynomial, I might begin my searches with that non-zero integer x which comes nearest to solving the equation $3x^5 - 71x^4 = 0$, i.e. $x = 24$.
>   So we look to see if the polynomial changes sign as we pass from $x = 23$ to $x = 24$.

SOLUTION   Since $3 \cdot (23)^5 = 19\,309\,029$ and $71 \cdot (23)^4 = 19\,868\,711$ it is clear that, $p(23)$ will be negative. Since $3 \cdot (24)^5 = 23\,887\,872$ and $71 \cdot (24)^4 = 23\,556\,096$ it is clear that $p(24)$ will be positive. It is immediate, by the intermediate value theorem of Chapter 6, that $p(x)$ will have a root between 23 and 24.

COMMENT   A lucky first guess!

EXERCISE 4   The equation $n^3 + 125n + 1001 = 129\,618\,467$ has an integer solution. Find it using only pencil and paper. (Approximately which n do you try first?)

EXERCISE 5   Find, by repeated approximation, a root 2 and 3 and correct to 3 decimal places of $x^3 - 9x + 9$.

Example 2
What is the least value of the integer n such that $n!$ ends in 1000 zeros?

> DISCUSSION
> According to Exercise 42 in Chapter 4, we have to find the least n such that
>
> $$\left[\frac{n}{5}\right] + \left[\frac{n}{25}\right] + \left[\frac{n}{125}\right] + \left[\frac{n}{625}\right] + \left[\frac{n}{3125}\right] + \ldots = 1000$$
>
> if indeed this is possible. But which n to look at? Perhaps solving
>
> $$\frac{n}{5} + \frac{n}{25} + \frac{n}{125} + \frac{n}{625} + \frac{n}{3125} + \ldots = 1000$$

would put us in the right area? Now, the sum of the infinite geometric series

$$\frac{1}{5} + \frac{1}{25} + \frac{1}{125} + \frac{1}{625} + \frac{1}{3125} + \dots \text{ is } `\frac{a}{1-r}`$$

which, in this case is

$$\frac{\frac{1}{5}}{1 - \frac{1}{5}} = \frac{1}{4}$$

Perhaps we should begin looking around $n = 4000$?

SOLUTION

$$\left[\frac{4000}{5}\right] + \left[\frac{4000}{25}\right] + \left[\frac{4000}{125}\right] + \left[\frac{4000}{625}\right] + \left[\frac{4000}{3125}\right]$$
$$+ \dots = 800 + 160 + 32 + 6 + 1 = 999$$

Hence we should take $n = 4005$.

EXERCISE 6   Show that there exists no integer n such that $n!$ ends in precisely 123 zeros.

Guessing: to discover or prove the theorem

Guessing is certainly very important in *discovering* theorems. Researchers guess frequently. As we have said, anyone formulating a conjecture is making a guess. Guessing is also used in *proving* theorems when we have to decide which approach is best suited to achieve our aim. Often our first attempts fail. For example, in the proof of Statement 10 we wrongly guessed that, for composite n, $2^n - 1$ is composite because it is a multiple of 3. This sort of guessing bears some resemblance to the problem of choosing (guessing) a suitable method – out of many available – to evaluate an indefinite integral.

EXERCISE 7   A rectangular sheet of 200 stamps (10 × 20) is to be separated into 200 separate stamps by tearing down entire rows or columns of perforations as shown for the 5 × 4 case in Figure 11.1. Guess the least number of tears that must be made. (*Hint:* First try the 2 × 2 and 3 × 2 cases, make a conjecture and try to find a proof of your conjecture by... what *method*? Have a guess!)

Analogy

It seems to me that there is quite a close connection between analogy and guessing, in that the latter can often depend on the former. You say to yourself, 'I've seen something like this before. I wonder if a similar solution/proof will work here. I can't see any reason why not – so I shall *guess* that it does and try to establish the

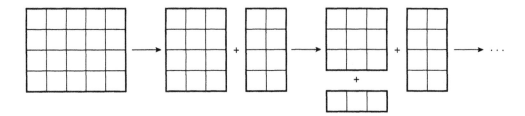

Figure 11.1

required result by an analogous solution/proof.' Sometimes the problem has *already been solved*, but in another guise.

Example 3
I invited six people to a party. Some pairs shook hands. Show that amongst the six there must have been either (a) at least three, all of whom shook hands, or (b) at least three, no two of whom shook hands.

> DISCUSSION
> This reminds me of something from an earlier chapter. If I had asked each pair who shook hands to hold opposite ends of a piece of red string and each pair who didn't shake hands to hold opposite ends of a piece of blue string.... Need I say more?

EXERCISE 8   If $f(x)$ is the product function $g(x)h(x)$ then

$$\frac{df}{dx} = \frac{dg}{dx}h + g\frac{dh}{dx}$$

$$\frac{d^2f}{dx^2} = \frac{d^2g}{dx^2}h + 2\frac{dg}{dx}\frac{dh}{dx} + g\frac{d^2h}{dx^2}$$

$$\frac{d^3f}{dx^3} = \frac{d^3g}{dx^3}h + 3\frac{d^2g}{dx^2}\frac{dh}{dx} + 3\frac{dg}{dx}\frac{d^2h}{dx^2} + g\frac{d^3h}{dx^3}$$

Do these coefficients remind you of anything? Write down the full expansion of d^5f/dx^5. (Notice that you are not asked to prove anything, merely to guess.) Conjecture a general result. By what method might you prove it? Take an educated guess.

Here is another example of 'I've seen something like this before'.

Example 4
Consider the sequence beginning

$$\qquad\qquad 3 \quad 11 \quad 25 \quad 45 \quad 71 \quad 103 \quad 141 \quad \ldots \qquad\qquad\text{(II)}$$

with first differences   8   14   20   26   32   38 ...

and second differences   6   6   6   6   6 ...

Is there a formula generating the terms of the given sequence?

DISCUSSION
Is there an analogy here with differentiation where, if f is a function whose nth derivative is a constant, then f is a polynomial of degree n? Let us *guess* that that is the case and assume that the integers in the sequence (II) are generated by a *quadratic polynomial* $an^2 + bn + c$.

We then want to find a, b and c such that $a \cdot 1^2 + b \cdot 1 + c = 3, a \cdot 2^2 + b \cdot 2 + c = 11$ and $a \cdot 3^2 + b \cdot 3 + c = 25$. We leave the finding of a, b and c to you but ask, 'is our guess valid?'

EXERCISE 9   Is there any harm in assuming that the sequence above is generated by the *cubic* polynomial $an^3 + bn^2 + cn + d$? Would you guess 'yes' or 'no'? What would you (confidently?) *expect* the value of a to be?

As a really impressive use of analogy, we shall derive (rather than merely confirm) the remarkable formula in Statement 16.5. We begin with an example.

Example 5
The sequence $2, 1, -7, 19, -43, 91, \ldots$ is obtained from the *initial conditions* $a_1 = 2$ and $a_2 = 1$ and, for all integers $t \geqslant 3$, from the *recurrence relation* $a_t + 3a_{t-1} + 2a_{t-2} = 0$. We are to find a formula for the *general term* a_n.

DISCUSSION
If, as in the case of the differential equation considered earlier, we have no idea what to do, we can likewise begin with a simpler problem: we ignore the initial conditions and try to solve the simpler (but related?) problem

$$3a_{t-1} + 2a_{t-2} = 0$$

that is,

$$a_{t-1} = -\tfrac{2}{3}a_{t-2}$$

A solution to this is immediate. Taking $a_1 = A$ (some randomly chosen real number) we see that, for each integer $n \geqslant 1$

$$a_n = A\left(-\tfrac{2}{3}\right)^{n-1}$$

Continuing as above, we first attempt to solve $a_t + 3a_{t-1} + 2a_{t-2} = 0$ by trying $a_t = A\alpha^{t-1}$. We therefore have

$$A\alpha^{t-1} + 3A\alpha^{t-2} + 2A\alpha^{t-3} = 0$$

Neglecting the uninteresting cases $A = 0$ and $\alpha = 0$ we obtain

$$\alpha^2 + 3\alpha + 2 = 0$$

from which we deduce that $\alpha = -1$ or -2. Continuing with the analogy with differential equations we find that, for real numbers A and B, $a_t = A(-1)^{t-1}$, $a_t = B(-2)^{t-1}$ and also

$$a_t = A(-1)^{t-1} + B(-2)^{t-1} \tag{III}$$

are all solutions of $a_t + 3a_{t-1} + 2a_{t-2} = 0$. We can now determine A and B in (III) by using the initial conditions. We obtain

$$2 = a_1 = A(-1)^0 + B(-2)^0$$

and

$$1 = a_2 = A(-1)^1 + B(-2)^1$$

from which we find $A = 5$ and $B = -3$. Thus the general term a_n is given by

$$a_n = 5(-1)^{n-1} - 3(-2)^{n-1}$$

We can now give the formula for the nth term of the Fibonacci sequence.

Example 6

The nth term $f(n)$ of the Fibonacci sequence is given by

$$f(n) = \frac{1}{\sqrt{5}}\left(\frac{1+\sqrt{5}}{2}\right)^n - \frac{1}{\sqrt{5}}\left(\frac{1-\sqrt{5}}{2}\right)^n$$

SOLUTION   To solve the recurrence relation, namely, $f(n) = f(n-1) + f(n-2)$, we try putting $f(n) = \alpha^n$.[1] We obtain

$$\alpha^n - \alpha^{n-1} - \alpha^{n-2} = 0$$

which, on cancelling by α^{n-2}, gives rise to the equation

$$\alpha^2 - \alpha - 1 = 0$$

This has roots $\alpha_1 = (1+\sqrt{5})/2$ and $\alpha_2 = (1-\sqrt{5})/2$. We then have, for each n, $f(n) = c_1\alpha_1{}^n + c_2\alpha_2{}^n$, where c_1 and c_2 are constants which we may find from the initial conditions.

We have:

$$f(1) = 1 = c_1\left(\frac{1+\sqrt{5}}{2}\right) + c_2\left(\frac{1-\sqrt{5}}{2}\right)$$

and

$$f(2) = 1 = c_1\left(\frac{1+\sqrt{5}}{2}\right)^2 + c_2\left(\frac{1-\sqrt{5}}{2}\right)^2$$

Solving these simultaneous equations we find $c_1 = -c_2 = 1/\sqrt{5}$.

EXERCISE 10   To help determine if a number of the form $2^n - 1$ is prime, Lucas[2] introduced the sequence $L_1, L_2, L_3, \ldots$ given by $L_1 = 1$, $L_2 = 3$ and, for each $n \geqslant 3$, $L_n = L_{n-1} + L_{n-2}$. Find a formula giving the value of L_n.

---

[1] There is no harm done in omitting the constant A (since we shall cancel it) and taking $f(n)$ as α^n instead of α^{n-1} (merely for aesthetical reasons).

[2] François-Edouard-Anatole Lucas, inventor of games and recreations including the Tower of Hanoi, died 3 October 1891. See the Comment on p. 104 in Chapter 6.

EXERCISE 11 As we saw in Chapter 5, it is easy to prove that, for all real numbers a and b, $a^2 + b^2 \geqslant 2ab$. Which, if any, of the following is the 'correct' analogue for three numbers? (i) $a^2 + b^2 + c^2 \geqslant 2(ab + bc + ca)$; (ii) $a^2 + b^2 + c^2 \geqslant 2abc$; (iii) $a^2 + b^2 + c^2 \geqslant 3abc$; (iv) $a^3 + b^3 + c^3 \geqslant 3abc$.

Finally, let us examine an attempt to prove a *result* analogous to Statement 14 by using an analogous *method*. Let us consider the *odd* primes arranged as follows:

$$3 \quad 7 \quad 11 \quad * \quad 19 \quad 23 \quad * \quad 31 \quad * \quad * \quad 43 \quad ...$$
$$5 \quad * \quad 13 \quad 17 \quad * \quad * \quad 29 \quad * \quad 37 \quad 41 \quad ...$$

We see that those in the upper list are all of the form $4k + 3$ whereas those in the lower list are all of the form $4k + 1$. One may ask if there are infinitely many odd primes in each list, or are almost all the primes in one of the lists with only a finite number in the other?

As we have in our repertoire a method for proving the set of all primes to be infinite, we naturally try to follow a line of argument similar to that taken in the proof of Statement 14. Let us see where this gets us. We shall try to prove the following theorem.

Theorem 1
There are infinitely many primes of the form $4k + 3$.

DISCUSSION
We aim to follow the proof of Statement 14 as closely as possible. We supposed that there are only finitely many, $p_1, p_2, ..., p_t$, say, and we let q be a prime divisor of $N = p_1 \cdot p_2 \cdot ... \cdot p_t + 1$. Since $q = p_j$ for some j ($1 \leqslant j \leqslant t$) we deduced that q divided N and $N - 1$ and hence their difference, 1; a contradiction.

Let us try the same moves here. We put $N = p_1 \cdot p_2 \cdot ... \cdot p_t + 1$ where the p_i are now the (supposedly finitely many) primes of the form $4k + 3$. Now, unfortunately, if q is a prime dividing N, there seems no obvious reason why q must be one of the p_i (since q might be a prime of the form $4k + 1$ – indeed it might even be equal to 2.)

At this point one just has to fish around trying this and that until a workable idea arises. It turns out that a good move is to take N not as above but as $N = 4 \cdot p_1 \cdot p_2 \cdot ... \cdot p_t - 1$ and to note that, if N factorizes as a product $q_1 \cdot q_2 \cdot ... \cdot q_s$ of primes, then we shall be able to get a contradiction as in Statement 14 *if any one of the q_j is equal to one of the p_i*. Thus we want to show that one of the q_j is of the form $4k + 3$. Can we achieve this?

Now, N is clearly an odd integer – so that, at least, no q_j is equal to 2. Consequently, each q_j is (a prime) of the form $4k + 1$ or of the form $4k + 3$. So the question is, must it be the case that at least one of the q_j be of the form $4k + 3$? Let us suppose not. Then all the q_j are of the form $4k + 1$. Can this be? If they *all* are, then their product is also of the form $4k + 1$ (see Exercise 12) whereas (and this explains the choice of N as $4 \cdot p_1 \cdot p_2 \cdot ... \cdot p_t - 1$ as distinct from $4 \cdot p_1 \cdot p_2 \cdot ... \cdot p_t + 1$) N is of the form $4k + 3$ (since each integer of the form $4S - 1$ is also of the form $4(S - 1) + 3$). We therefore have our proof (cf. the proof of Statement 14).

Proof of Theorem 1   Suppose, to the contrary, that there are only a finite number of primes of the form $4k + 3$. Let these primes be $p_1 = 3$, $p_2 = 7, \ldots, p_t$. Now let N denote the integer $4 \cdot p_1 \cdot p_2 \cdot \ldots \cdot p_t - 1$. Write $N = q_1 \cdot q_2 \cdot \ldots \cdot q_s$ where the q_i are primes (not necessarily distinct from each other). Now, not *all* the q_j are of the form $4k + 1$ since, if they were, so too would be their product $q_1 \cdot q_2 \cdot \ldots \cdot q_s$ (see Exercise 12) whereas N is of the form $4k - 1$ and hence of the form $4k + 3$. This shows that *at least one* of the q_j is of the form $4k + 3$ and, therefore, *must be one of the p_i*. Indeed, suppose that $q_m = p_n$. Then $q_m | N$ and $(q_m =) p_n | (N - 1)$. Hence $q_m | N - (N - 1)$, that is, $q_m | 1$. But this is impossible since q_m is prime. It follows that our supposition is wrong. Hence the number of primes of the form $4k + 3$ is infinite.

COMMENT   We made quite heavy use of *reductio* in this proof.

EXERCISE 12   Show that if a and b are integers of the form $4k + 1$ then so is their product. (*Hint:* Let $a = 4u + 1$ and $b = 4v + 1$. Show that $ab = 4w + 1$ for suitable integer w.) Now prove that the product of *any number* of integers of the form $4k + 1$ is of the form $4k + 1$. (Which method do you expect to use?)

EXERCISE 13   Try to prove that there are infinitely many primes of the form $4k + 1$ by copying the proof of Theorem 1, including taking $N = 4 \cdot p_1 \cdot p_2 \cdot \ldots \cdot p_t + 1$ where $p_1, p_2, \ldots, p_t$ are the supposed finitely many primes of the form $4k + 1$.) Identify the point at which your attempt breaks down.

EXERCISE 14   By copying the proof of Theorem 1 whilst making the obvious changes, prove that there are infinitely many primes of the form $6k + 5$. (*Hint:* In the proof of Theorem 1 replace N by $6 \cdot p_1 \cdot p_2 \cdot \ldots \cdot p_t - 1$ where the p_i are the (supposed finitely many) primes of the form $6k + 5$.

COMMENT   It is a famous result of Dirichlet's that *if a and b are positive integers such that $(a, b) = 1$ then, in the arithmetic progression $a, a + b, a + 2b, a + 3b, \ldots$ there are infinitely many primes.* The proof is hard.

EXERCISE 15   Show that if $(a, b) > 1$ then the above arithmetic progression has at most one prime.

The Division Theorem for Polynomials

An instance of an important result in one area being *directly suggested* by an analogous and equally important result in another area is demonstrated by Theorem 2.

Theorem 2 (The Division Theorem for Polynomials)
Let

$$a(x) = a_m x^m + a_{m-1} x^{m-1} + \ldots + a_1 x + a_0$$

and

$$b(x) = b_n x^n + b_{n-1} x^{n-1} + \ldots + b_1 x + b_0 \quad \text{(not the zero polynomial)}$$

be polynomials in x with real coefficients a_i $(0 \leqslant i \leqslant m)$ and b_j $(0 \leqslant j \leqslant n)$. Then there exists polynomials $q(x)$ and $r(x)$ (in x with real coefficients) such that $a(x) = q(x) \cdot b(x) + r(x)$ and the degree of $r(x)$ is less than the degree of $b(x)$. (*Please read Comment 2 below.*)

COMMENTS

1. Compare the statement of this theorem with that of Theorem 2 in Chapter 8. (We use $a = qb + r$ here rather than $a = mb + r$ as we did there because we wish to use the letter m for the degree of the polynomial $a(x)$.)
2. This theorem is actually *false* as stated. You will learn of the trivial amendment which corrects it in the answer to Exercise 3 at the end of Chapter 13 (page 221). However, see if you can spot the 'mistake' *now*.

Proof of Theorem 2   There is no harm in supposing that $a_m \neq 0$ and $b_n \neq 0$, so that $a(x)$ and $b(x)$ have degrees m and n respectively.

1. If $m < n$ then we take $q(x) = 0$, the zero polynomial, and $r(x) = a(x)$. We then have

$$a(x) = q(x) \cdot b(x) + r(x)$$

with $\deg\{r(x)\} < \deg\{b(x)\}$, as required.

2. If $m \geqslant n$ we prove the theorem by induction on the difference $m - n$, starting with $m - n = 0$, that is with $m = n$.

     The case $m - n = 0$. If we take $q(x)$ to be the constant polynomial a_m/b_n $(= a_m/b_m)$ we see that $r(x)$ $(= a(x) - q(x) \cdot b(x))$ has lesser degree than $b(x)$. Therefore, in this case the claimed result is proved.

     The cases $m - n > 0$. We now assume the result proved for all cases where $m - n \leqslant k$ and establish it for all cases $m - n = k + 1$. We begin by forming a new polynomial

$$f(x) = a(x) - (a_m/b_n)x^{m-n} \cdot b(x)$$

which has degree $d \leqslant m - 1$ (since the subtracted term exactly cancels out the leading term of $a(x)$). Since

$$d - n \leqslant (m - 1) - n = k$$

we may apply the induction hypothesis to the pair of polynomials $f(x)$ and $b(x)$. Hence we may assume that there are polynomials $q_1(x)$ and $r_1(x)$ such that

$$f(x) = q_1(x) \cdot b(x) + r_1(x)$$

with the degree of $r_1(x)$ less than the degree of $b(x)$. We now see that

$$a(x) = f(x) + (a_m/b_n)x^{m-n} \cdot b(x)$$
$$= \{(a_m/b_n)x^{m-n} + q_1(x)\} \cdot b(x) + r_1(x)$$

which completes our proof by induction.

EXERCISE 16   Why, in proving the case $m - n = k + 1$, do we need to call upon the induction hypothesis corresponding to *all* cases $m - n \leqslant k$ rather than just the single case $m - n = k$?

EXERCISE 17   Find $q(x)$ and $r(x)$ as above if

$$a(x) = 3x^5 + x^3 - x^2 + \tfrac{1}{4}x + \tfrac{1}{2}$$
$$b(x) = x^2 + \tfrac{2}{3}x + \tfrac{3}{4}$$

EXERCISE 18   Let $a(x) = x + 1$ and $b(x) = 2x$. Show that there *do not* exist polynomials $q(x)$ and $r(x)$ with coefficients in $\mathbb{Z}$ (and degree $r(x) <$ degree $b(x)$) such that

$$a(x) = m(x)b(x) + r(x)$$

(This shows that Theorem 2 may fail if only integer coefficients are permitted.)

The Factor Theorem

As an important consequence of Theorem 2 we have the Factor Theorem.

Theorem 3 (The Factor Theorem)
Let $p(x)$ be a polynomial with real coefficients. Then α is a **root** or **zero** of $p(x)$ (that is, $p(\alpha) = 0$) iff $x - \alpha$ is a **factor** of $p(x)$ (that is, $p(x) = (x - \alpha) \cdot h(x)$ for some real polynomial $h(x)$).

Proof ('If': $\leftarrow$) From $p(x) = (x - \alpha) \cdot h(x)$ we obtain

$$p(\alpha) = (\alpha - \alpha) \cdot h(\alpha) = 0$$

('Only if': $\rightarrow$) By Theorem 2 we can write

$$p(x) = q(x) \cdot (x - \alpha) + r(x)$$

where $r(x)$ is a constant c, say, since degree $(x - \alpha) = 1$. Since $p(\alpha) = 0$ we find

$$0 = p(\alpha) = q(\alpha) \cdot (\alpha - \alpha) + c$$

Hence $c = 0$. Consequently

$$p(x) = q(x) \cdot (x - \alpha)$$

as claimed.

EXERCISE 19

(a) Show, using the corollary, that $3\tfrac{1}{2}$ is a root of the polynomial

$$6x^5 - 19x^4 - 19x^3 + 46x^2 - 22x + 28$$

(*Hint*: Show, by division, that $(2x - 7)$ is a factor.)
(b) By replacing x by -1, in the polynomial

$$p(x) = 17x^5 + 60x^4 - 22x^3 - 47x^2 + 7x - 11$$

show that $x + 1$ is a factor of $p(x)$.

COMMENTS

1. Just as the Division Theorem for $\mathbb{Z}$ enables us, through the Euclidean Algorithm, to find the greatest common divisor of two given integers (not both zero) so does Theorem 2 give rise to a similar theory. One consequence is that polynomials with real (respectively, rational) coefficients also factorize (essentially) uniquely into products of 'prime' polynomials with real (respectively, rational) coefficients. Amongst other things, this allows us to prove correct the rules for finding partial fractions.

2. We record here, without proof, an *analogue* of Theorem 1 of Chapter 10:

if $P(x) = a'_n x^n + a_{n-1} x^{n-1} + \ldots + a_1 x + a_0$ is a polynomial with rational coefficients, if u, v and $d \in \mathbb{Q}$ where $\sqrt{d}$ is irrational and if $u + v\sqrt{d}$ is a root of $P(x)$ then so too is $u - v\sqrt{d}$.

For a proof see Hall and Knight (1887, p. 148).

Transformation

Sometimes it is possible to change a problem which appears to be difficult into a form in which it can be solved much more readily. Sometimes the change itself requires a stroke of genius. Sometimes a clever method, once learnt, can be used time and again.

First a splendid transformation from the past. In 1736, Euler resolved the question as to whether or not it was possible to walk through the town of Konigsberg in such a way that one could cross each of the town's seven bridges exactly once and return to one's starting point.

Figure 11.2 shows how the town's bridges were placed. Euler represented each landmass by a dot and each bridge by a line joining the two appropriate dots. See Figure 11.3. Because there is an odd number of different bridges ending at each dot,

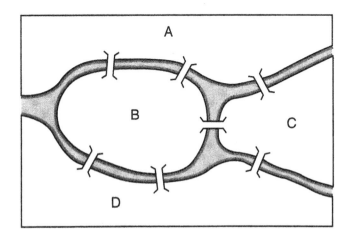

Figure 11.2

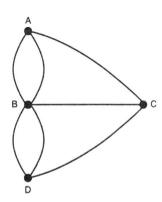

Figure 11.3

it is then fairly clear that a walk of the desired kind leaving, say, dot A on one bridge must, eventually return to A over a second bridge and then leave A on the third bridge – and be unable to get back to A without crossing one bridge for a second time. The same argument applies to any route starting at dot C and dot D and a similar argument applies at dot B where five bridges meet.

More familiar examples

1. Perhaps the example with which readers will be most familiar is that of changing an integral, by substitution, into one which is easier to do.
2. Another familiar example is that of solving a system of simultaneous linear equations by changing it into 'echelon form'. For example, we solve

$$1x + 2y = 3$$
$$4x + 5y = 6$$

by subtracting 4 times equation 1 from equation 2 in order to transform it to the system

$$1x + 2y = 3$$
$$-3y = -6$$

which is more easily solved.

3. Here is an example where a subtle change of emphasis is helpful.

   CLAIM   In any triangle the three angle bisectors are concurrent. (Figure 11.4 provocatively attempts to show them to be non-concurrent.) How do you prove the claim? One idea might be to try to prove that the area of the central triangle is zero; but it seems difficult to see what first step one can take to achieve that. An alternative approach is indicated in Figure 11.5: draw the angle bisectors at B and C to meet at O. Draw AO. *Then try to prove that AO bisects angle BAC.*

EXERCISE 20   Explain why, in (2) above, the solution of the second system of equations is the same as the solution of the first.

EXERCISE 21   Prove the claim made in (3) above. (*Hint:* Use Figure 11.5.)

EXERCISE 22   Show, likewise, that the medians in any triangle are concurrent. (*Hint:* Use Figure 11.6.)

Analogy with (3) above suggests the following *problem*. You may take a *guess* at the answer.

EXERCISE 23   *Problem:* If the angle *bisectors* meet in a point, do the angle *trisectors* do anything special? Using a protractor, draw the angle trisectors of several triangles,

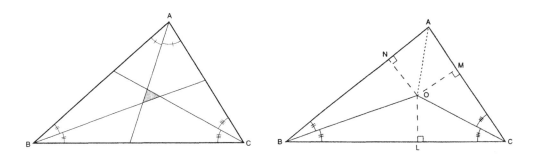

Figure 11.4   Figure 11.5

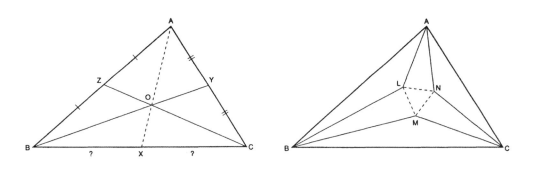

Figure 11.6   Figure 11.7

as in Figure 11.7. From the evidence do you suspect anything concerning their points of meeting L, M and N?

It can also be useful to interpret an algebraic problem as a geometrical one. (The turning of geometrical problems in to algebraic ones has been much employed since Descartes (and Fermat) founded algebraic geometry.) Consider, for instance, the first system of equations in (2) above. If we interpret each equation as a line which can be drawn in the plane then it is clear that (i) the two lines may cross each other at a single point, or (ii) the lines may be distinct but parallel, or (iii) the lines may coincide. These are clearly the only possibilities. Their algebraic reinterpretation says: each pair of simultaneous linear equations has (i) a unique solution or (ii) no solution or (iii) infinitely many solutions. (Cf. Theorem 6 in Chapter 12.)

Some delightful examples

Here are some more examples of transforming what appear to be tricky problems into trivialities.

Example 7
An organizer of a knock-out tennis tournament with 113 entrants wants to know for how many days he should book hotel rooms for the players. He knows that, on average, he can count on getting through 12 games per day. (There is no interruption for bad weather since the tournament is indoors.)

COMMENT   What the organizer really wants to know is, what is the total number of matches that will have to be played, and can he arrange a system of 'byes' which will minimize this number of games? Of course, one can just count $[113/2] = 56$ games (with 1 player left over) $+ [57/2] = 28$ games (with one player left over) $+ [29/2] = 14$ games, etc. This gives *an* answer, but is it the only way? Yes, for the reason explained below.

SOLUTION   The number of games is 112. (Because there is exactly one tournament winner, there must be 112 losers. Each loser loses in only one game.)

Next, a problem I first saw when I was about 7 years old. (You will be pleased to hear that I am not one of those who can brag 'And I did it then.')

Example 8
A fly (f) is stuck at a point 1 m from the floor and 1 m from the end of a sidewall (wall A) of a room and a spider (s) is positioned 1 m from the ceiling and 1 m from the end of the backwall (wall B) of the same room (Figure 11.8). The room is rectangular and is $6 \times 4 \times 3$ m (in length, depth and height, respectively). If the spider takes the shortest route to the fly, how far does she have to walk?

SOLUTION   Draw an opened-out view of the room as in Figure 11.9. Then the answer to the question is immediate. The spider uses Pythagoras' theorem to calculate the distance via the floor that she has to travel as $\sqrt{(5^2 + 6^2)} = \sqrt{61}$ m.

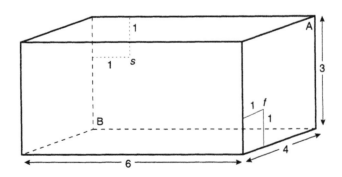

Figure 11.8

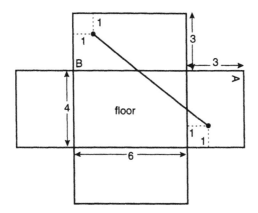

Figure 11.9

EXERCISE 24   Is the corresponding 'ceiling route' shorter?

Here is another question about a fly! One story attached to it demonstrates the
mental powers of the Hungarian/American mathematician John von Neumann.

Example 9
Two cyclists, starting 15 miles apart, simultaneously begin cycling towards one
another each at a speed of 10 miles per hour. At the moment of departure a fly
flies, at 12 miles per hour, from the nose of one cyclist until it lands on the nose of
the second cyclist. The fly then flies back, at 12 miles per hour until it reaches the
nose of the first cyclist again. This procedure continues until the cyclists' noses meet.
How far has the fly flown in that time?

> DISCUSSION
> It seems that we have to determine the lengths of each of the infinitely many shorter
> and shorter trips that the fly makes back and forth between the cyclists and then
> add these up. Looking at the problem from a different angle, though, we may ask,
> how long does the fly fly for? That time, multiplied by 12 will give the required
> distance in miles.

SOLUTION   The cyclists approach each other at a combined speed of 20 mph.
Starting 15 miles apart they therefore meet after three-quarters of an hour. In that
time the fly has travelled 9 miles.

The story told about von Neumann is that he was asked this problem at a party.
Almost instantly came the reply '9 miles'. 'Aha,' said the questionner, 'you spotted
the quick way of doing it. Most people try to do it by summing an infinite series.'
'But that's the way I *did* do it', replied von Neumann.

JOHN LOUIS VON NEUMANN
(28 December 1903–8 February 1957)

John Louis von Neumann was born in Budapest on 28 December 1903, the eldest of three sons of a banker. His exceptional mathematical ability, recognized early, was nurtured by private tuition; his first research paper was published at age 18.

After spells at the Universities of Berlin and Hamburg, von Neumann joined Princeton University in 1930.

Von Neumann's mathematical achievements were very wide ranging, encompassing several areas of pure and applied mathematics and theoretical physics. It was natural that his expertise was called upon in connection with the Los Alamos project to construct the atomic bomb.

As Example 9 in the text makes clear, he was a phenomenally quick thinker. It is said that he could talk faster in any one of several languages than most people can in their own.

In pure mathematics he made fundamental contributions in logic and set theory, in measure theory, in Lie group theory (where he answered one of the 23 difficult questions which Hilbert had, in 1900, urged the mathematical community to try to solve). He was, for twenty years, the leading researcher in the field of operators on Hilbert space and founded the field of rings of operators.

In applied mathematics his most famous work, *The Mathematical Foundations of Quantum Mechanics* (1932), used Hilbert space theory to make the foundations of quantum mechanics secure.

The need for numerical calculations to be performed more rapidly took von Neumann in the direction of computing machines and, after the Second World War, he both devised methods of computer programming and supervised the building of a computer at Princeton University.

Amongst all his other interests, von Neumann also found time to found the Theory of Games in which the concept of 'strategy' in an economical, sociological, political or simply 'games' setting was subjected to mathematical analysis.

> Recognizing the possibility of applications, von Neumann and Oscar Morgenstern
> produced their book *The Theory of Games and Economic Behaviour* in 1927.
> Von Neumann died of cancer, in Washington, at the age of 53.

Finally, a transformation I really like. The problem is easy enough to state.

Example 10

In how many different ways can the equation $x + y + z + t = 37$ be solved in positive integers?

DISCUSSION

We are looking for solutions such as $x = 17$, $y = 2$, $z = 17$, $t = 1$, which is, of course, different from $x = 17$, $y = 17$, $z = 1$, $t = 2$. What alternative do we have to listing all possible solutions? Answer: look at it as follows. (The method may *never* have occurred to you, but when you see it you won't forget it.)

SOLUTION   (Replacing 37 by 14 merely for convenience of explanation.) First, solving $x + y + z + t = 14$ with each of x, y, z and $t \geqslant 1$ is the same as solving $X + Y + Z + T = 10$ with each of X, Y, Z and $T \geqslant 0$ (why?)
   With each solution X, Y, Z, T we associate a self-explanatory diagram:

   with solution 1, 7, 0, 2 we associate the diagram * / ******* / / **
   with solution 1, 4, 2, 3 we associate the diagram * / **** / ** / ***
   with diagram **** / * / ***** / we associate the solution 4, 1, 5, 0

(Notice that, to split the stars into four groups we need just three dividers.) The answer to our problem is therefore just the number of ways of choosing where to put the three dividers in the 13 $(= 10 + 3)$ places available. The answer, then, is the binomial coefficient $\binom{13}{3} = 286$.

EXERCISE 25   Explain why, from the diagrams in Example 10, the answer can also be seen to be $\binom{13}{10}$.

EXERCISE 26   How many distinct solutions are there in integers $\geqslant 0$ of:

(a) $x + y + z + t + w = 99$
(b) $x + y + z \leqslant 20$

(*Hint:* For (b), match each solution of $x + y + z \leqslant 20$ with a solution of $x + y + z + t = 20$ (with $t \geqslant 0$).)

CHAPTER 12

Generalization and Specialization

All generalizations are dangerous – even this one. DUMAS *fils*

His specialization is omniscience.
SHERLOCK HOLMES (talking about Mycroft)

Generalizing is a natural activity for mathematicians: the more general the result, the wider its field of application. Generalization can help understanding since it tends to remove what proves to be irrelevant detail from an argument. However, specialization can be used to check that general statements are 'probably' valid. Furthermore, the 'general case' is sometimes little more than several copies of the special case superimposed on one another.

Introduction

Generalization is exactly what it says. Results already obtained are extended so that more cases are covered, the original ones becoming *special cases* of the generalization. In this respect generalization differs from *analogy* in that, in the latter, new results, whose statements or proofs *resemble* the originals, are obtained but they do *not* include the originals as special cases. For example, *generalization* is the step in which you pass from noting that $1^3 = 1^2$, $1^3 + 2^3 = 3^2$, $1^3 + 2^3 + 3^3 = 6^2$ and $1^3 + 2^3 + 3^3 + 4^3 = 10^2$ to proving the (general) statement that $1^3 + 2^3 + \dots + n^3 = [n(n + 1)/2]^2$ *for all n*. On the other hand, *analogy* is drawn when, after noting that $1 + 2 + ... + n (= n(n + 1)/2)$ is of degree 2 in n and that $1^2 + 2^2 + ... + n^2$ $(= n(n + 1)(2n + 1)/6)$ is of degree 3 in n, you say: 'I expect that $1^4 + 2^4 + ... + n^4$ is of degree 5 in n'. Then, of course, the claim that, *for each positive integer k,* $1^k + 2^k + ... + n^k$ is of degree $k + 1$ in n is another example of generalization.

Specialization is, by its name, the opposite of generalization. It seems to be useful in two main directions. First, if a result which is claimed to hold generally fails to hold in a special, perhaps extreme, case then the claimed general result is surely invalid. Thus specialization can detect weak spots in general theories. Second, if your aim is to prove some general result or theorem then there would seem little chance of success if there are special cases which you are unable to resolve. So it is often profitable to try special cases first. Furthermore, a successful attack on a special case of your problem may show the way to resolving the general case. Sometimes the general case is merely several copies of the special case superimposed on one another.

Generalization

More examples

As just indicated, one of the *signs* by which one can recognize that some form of generalization may have taken place is the preponderence, in statements and formulae of various kinds, of *letters* rather than specific objects (for example, numbers) often accompanied by the words 'for all'. (The mathematicians of long ago did not possess the notation that we take for granted today and had to describe what were, in fact, widely applicable methods by means of giving many specific examples. Thus the Babylonians gave specific numeral examples showing the method of solving quadratic equations. (See Boyer (1968, p. 34).) How much easier it is to have available the *general* formula $x = [-b \pm \sqrt{(b^2 - 4ac)}]/2a$ for giving the roots of *every* quadratic equation.)

There are several reasons why we seek to generalize. One is natural curiosity: how much further can I push this result? It is a desire to aim for the widest possible application as well as a desire to see *what*, if anything, *stops* us from going even further. Another reason is a desire to show that two, hitherto unconnected, results or theories are just particular instances (*specializations*) of a single, much more general case. One of the great quests at present in physics is the search for a unified field theory which will explain the four natural forces (strong nuclear, weak nuclear, electromagnetic and gravitational) in terms of a single unified force.

In this book we operate under the first of these reasons and see where we get to. Accordingly, I shall pick out some of the results we have obtained to date and see what we might learn by attempting an extra 'push'.

As a very simple beginning we generalize Statement 1.

Theorem 1
No integer n which ends in a 2, 3, 7 or 8 can be a perfect square.

Proof of Theorem 1 (Cf. the proof of Statement 1.) Suppose that the given integer n is the square of some integer t, say. Now $t = 10m + r$ for some suitable integer

m and some suitable value for r where $0 \leqslant r \leqslant 9$. But then $t^2 = (10m + r)^2$ and so the last digit of n $(= t^2)$ is the same as the last digit of r^2, which is, therefore, $0, 1, 4, 9, 6$ or 5. Thus, if n's final digit is a 2 or 3 or 7 or 8, n cannot be a perfect square.

COMMENT This theorem clearly has Statement 1 as a (very special) case.

EXERCISE 1 Show that none of

$1! + 2! + 3! + 4!$
$1! + 2! + 3! + 4! + 5!$
$1! + 2! + 3! + 4! + 5! + 6!$

is a perfect square. Find and prove a generalization concerning $1! + 2! + ... + n!$.

Another rather simple act of generalization results in the following theorem.

Theorem 2
The sum of the interior angles of an n-sided polygon is $(n - 2)\pi$ radians.

COMMENT We shall omit a formal proof but the *method* one can employ is noteworthy. The general case, as Figure 12.1 demonstrates, is proved by 'adding together' a number of special cases each corresponding to that of a triangle where $n = 3$. This adding together is the method of *superposition of special cases*. You have used this method many times, for example in finding the derivative (or integral) of a sum $f(x) + g(x)$ of two functions by differentiating (or integrating) each of $f(x)$ and $g(x)$ separately and then adding the results. We have already used the method in this book in the solution of the differential equation (I) at the start of Chapter 11 as well as in Examples 5 and 6 on difference equations in that chapter. It is also used in Theorem 6 and Example 3 later in this chapter.

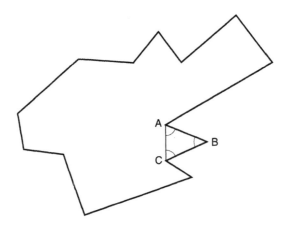

Figure 12.1

Next we generalize Statement 12 substantially.

Theorem 3

Let z be an integer which is not a perfect square. The $\sqrt{z}$ is not a rational number.

DISCUSSION

We might think of copying the proof for $\sqrt{2}$ (Statement 12) as far as possible. Alternatively we may feel that the Fundamental Theorem of Arithmetic might be of use. Whichever way we choose (let's go for the latter), it seems reasonable to try the contrapositive approach and attempt to show that if $\sqrt{z}$ is a rational number then z is a perfect square.

Proof of Theorem 3   Suppose that $\sqrt{z} = a/b$. The $b^2 \cdot z = a^2$. Now let p be a prime[1] dividing z. Suppose that p^α, p^β and p^γ are the largest powers of p which divide a, b and z respectively. Using the Fundamental Theorem of Arithmetic several times we can complete the proof as follows. $p^{2\alpha}$ and $p^{2\beta}$ are the exact powers of p which divide a^2 and b^2. Hence, comparing the powers of p on each side of the equality $b^2 \cdot z = a^2$ we see that the exact power of p which divides z is $p^{2\alpha - 2\beta}$, i.e. $p^{2(\alpha - \beta)}$. We deduce that, in the prime factorization of z, each prime divisor is raised to an even power. This means that z is a perfect square.

In Theorem 5 in Chapter 8 we showed that the gcd (a, b) of two integers (not both zero) is expressible as a 'linear combination' $sa + tb$. Analogy would suggest that this might extend to three integers.

Theorem 4

Every three integers a, b and c (not all zero) have a gcd which can be expressed as a linear combination $ra + sb + tc$.

(Perhaps even the proof is analogous to that for pairs of integers?)

COMMENT   In fact this *analogy* can be regarded as a case of *generalization* since on *specializing* c to be 0 we deduce, since the gcd of a, b and 0 is the same as the gcd of a and b (is it?), that

$$(a, b) = ra + sb + t0 = ra + sb$$

DISCUSSION

How might we prove this? If we knew that the gcd of a, b and c (which we denote by (a, b, c)) is equal to the gcd of (a, b) and c then we could first say (why?) that there exist integers δ and γ such that

$$(a, b, c) = ((a, b), c) = \delta(a, b) + \gamma c$$

We could also claim that there exist integers α and β such that $(a, b) = \alpha a + \beta b$. But then we should have

$$(a, b, c) = \delta\alpha a + \delta\beta b + \gamma c$$

which is of the required form.

---

[1] Considering the powers of just one prime is simpler than comparing the complete prime factorizations of $b^2 \cdot z$ and a^2.

An alternative approach is to use the well-ordering property of the integers. As this approach is not one which naturally suggests itself (until you have seen how useful it can be) we shall merely give the following proof without trying, by further discussion, to motivate the method.

Proof of Theorem 4   Given integers a, b, c (not all zero – why not?) consider the subset P, say, of positive integers in the set S where

$$S = \{ra + sb + tc : r \in \mathbb{Z}, s \in \mathbb{Z}, t \in \mathbb{Z}\}$$

P is not empty (why not?) and so has a least positive member d, say. We express d as $r_0 a + s_0 b + t_0 c$ and claim that d is the required gcd of a, b and c.

To this end choose $z (= r_1 a + s_1 b + t_1 c)$ arbitrarily from S and, using the Division Theorem for $\mathbb{Z}$, write $z = md + e$ where $m, e \in \mathbb{Z}$ and $0 \leqslant e < d$. Then

$$e = z - md$$

$$= (r_1 - mr_0)a + (s_1 - ms_0)b + (t_1 - mt_0)c$$

If e were positive then e would belong to P (since it is of the appropriate form). But by choice of d, e cannot be positive. Hence $e = 0$. Consequently, $z = md$. This means that *every* element in S is an integer multiple of d. In particular $d|a$, $d|b$ and $d|c$ so that d *is a common divisor of a, b and c*. However, if f is a common divisor of a, b and c then f divides $r_0 a + s_0 b + t_0 c$, that is, $f|d$. Thus the positive integer d is a common divisor of a, b and c which is greater than or equal to every common divisor of a, b and c. It is therefore the *greatest common divisor* of a, b and c . Since d is a linear combination of a, b and c the proof is complete.

COMMENT   Theorem 4 can be *generalized* so as to apply to any number of given integers, not just three.

EXERCISE 2   Explain why the set P described in the above proof is not empty.

EXERCISE 3   Find integers r, s and t such that

$$(66, 110, 165) = 66r + 110s + 165t$$

(*Hint:* Write $(66, 110)$ in the form $66x + 110y$ then $((66, 110), 165)$ in the form $(66, 110)z + 165t$.)

The following looks like a candidate for proof by mathematical induction.

Theorem 5
Let $a_1, a_2, \ldots, a_m$ be any m positive real numbers. Then

$$\frac{a_1 + a_2 + \ldots + a_m}{m} \geqslant \sqrt[m]{(a_1 a_2 \ldots a_m)}$$

DISCUSSION

How might we go from the case of $m = 2$ to $m = 3$? A natural approach would be to group two of the numbers together writing

$$\frac{a_1 + a_2 + a_3}{3} = \frac{2}{3} \cdot \frac{a_1 + (a_2 + a_3)}{2} \geqslant \frac{2}{3} \cdot \sqrt{\{a_1(a_2 + a_3)\}}$$

which doesn't look hopeful. We can, however, show that the case $m = 3$ actually follows from the case $m = 4$. Indeed, assuming the case $m = 4$, we merely take a_4 to be $\sqrt[3]{(a_1 a_2 a_3)}$.

We then get

$$\frac{a_1 + a_2 + a_3 + \sqrt[3]{(a_1 a_2 a_3)}}{4} \geqslant \sqrt[4]{\{a_1 a_2 a_3 \sqrt[3]{(a_1 a_2 a_3)}\}} = \sqrt[3]{(a_1 a_2 a_3)} \quad \text{(why?)}$$

Subtracting $\sqrt[3]{(a_1 a_2 a_3)}/4$ from each side we get

$$\frac{a_1 + a_2 + a_3}{4} = \frac{3}{4}(\sqrt[3]{a_1 a_2 a_3})$$

But how can we prove (especially by induction) the case $m = 4$ *before* the case $m = 3$? Answer: as follows.

Proof of Theorem 5   Let $S(n)$ be: for each integer n,

$$\frac{1}{2^n}(a_1 + a_2 + \ldots + a_{2^n}) \geqslant \sqrt[2^n]{(a_1 a_2 \ldots a_{2^n})}$$

$S(0)$ claims that $a_1 \geqslant a_1$ and so $S(0)$ is trivially true. Assume that $S(k)$ is true and look at the sum $A + B$ where

$$A = \frac{1}{2^k}(a_1 + a_2 + \ldots + a_{2^k})$$

and

$$B = \frac{1}{2^k}(a_{2^k + 1} + a_{2^k + 2} + \ldots + a_{2^{k+1}}).$$

We first use the inequality $(A + B)/2 \geqslant \sqrt{(AB)}$ and then, inside the square root, we use the inequalities

$$A \geqslant \sqrt[2^k]{(a_1 a_2 \ldots a_{2^k})} \quad \text{and} \quad B \geqslant \sqrt[2^k]{(a_{2^k + 1} a_{2^k + 2} \ldots a_{2^{k+1}})}$$

(which we obtain from the induction assumption $S(k)$) to deduce that $S(k + 1)$ is valid.

(As is typical in many mathematics books, we 'leave the reader to complete the proof' in Exercise 4.)

EXERCISE 4   Complete the proof of Theorem 5. (*Hint:* To prove the result for an integer m other than a power of 2 (which we have just done) choose t so that $2^{t-1} < m < 2^t$. In the inequality

$$\frac{1}{2^t}(a_1 + a_2 + \ldots + a_{2^t}) \geqslant \sqrt[2^t]{(a_1 a_2 \ldots a_{2^t})}$$

replace each of the $a_{m+1}, a_{m+2}, \ldots, a_{2^t}$ by $\sqrt[m]{(a_1 a_2 \ldots a_m)}$ and proceed as in the above Discussion.)

EXERCISE 5

(a) Show that, for positive real numbers a, b and c,

$$27a^2 b^2 c^2 \leqslant (ab + bc + ca)^3$$

(b) Show that, for (positive) integers $a_1, a_2, \ldots, a_m$, we have

$$a_1^m + a_2^m + \ldots + a_m^m \geqslant m \cdot a_1 \cdot a_2 \cdot \ldots \cdot a_m$$

In Chapter 11 we saw (pictorially) that a system of two (simultaneous) linear equations in two unknowns has either (i) no solutions, (ii) a unique solution or (iii) infinitely many solutions. The proof extends (i.e. generalizes) easily not only to systems of n equations in n unknowns but also to the following.

Theorem 6
Let S be a system of m equations in n unknowns where m and n are not necessarily equal. The S has (A) no solution or (B) a unique solution or (C) infinitely many solutions.

DISCUSSION
As it would be impossible even to discuss this theorem meaningfully by use of pictures, for its proof we must resort to use of algebra. If we let X stand for, 'S is a system of m equations, etc.' then the theorem is of the type IF X THEN (A or B or C). So to prove it we may show that C follows from the supposition of (X and $\neg$A and $\neg$B). We therefore suppose that our given system of equations *has* a solution but (and!) that it is not unique, and we try to show that there must be infinitely many. There is then a great temptation to see what the difference of two of these solutions yields.

Proof of Theorem 6 Let the system of equations be

$$a_{11}x_1 + a_{12}x_2 + \ldots + a_{1n}x_n = b_1$$
$$a_{21}x_1 + a_{22}x_2 + \ldots + a_{2n}x_n = b_2 \tag{I}$$
$$\vdots \qquad \vdots \qquad \vdots \qquad \vdots$$
$$a_{m1}x_1 + a_{m2}x_2 + \quad + a_{mn}x_n = b_m$$

Suppose that $x_1 = u_1, x_2 = u_2, \ldots, x_n = u_n$ is one solution of this system of equations and that $x_1 = v_1, x_2 = v_2, \ldots, x_n = v_n$ is a second. Then

$$a_{11}u_1 + a_{12}u_2 + \ldots + a_{1n}u_n = b_1 \qquad a_{11}v_1 + a_{12}v_2 + \ldots + a_{1n}v_n = b_1$$
$$a_{21}u_1 + a_{22}u_2 + \ldots + a_{2n}u_n = b_2 \text{ and } a_{21}v_1 + a_{22}v_2 + \ldots + a_{2n}v_n = b_2$$
$$\vdots \quad \vdots \quad \vdots \quad \vdots \qquad\qquad \vdots \quad \vdots \quad \vdots$$
$$a_{m1}u_1 + a_{m2}u_2 + \quad + a_{mn}u_n = b_m \qquad a_{m1}v_1 + a_{m2}v_2 + \quad + a_{mn}v_n = b_m$$

By subtraction we see that $x_1 = u_1 - v_1, x_2 = u_2 - v_2, \ldots, x_n = u_n - v_n$ is a *non-zero* solution of the system

$$a_{11}x_1 + a_{12}x_2 + \ldots + a_{1n}x_n = 0$$
$$a_{21}x_1 + a_{22}x_2 + \ldots + a_{2n}x_n = 0 \qquad \text{(II)}$$
$$\vdots \qquad \vdots \qquad \vdots \qquad \vdots$$
$$a_{m1}x_1 + a_{m2}x_2 + \quad + a_{mn}x_n = 0$$

It then follows easily that $x_1 = r(u_1 - v_1)$, $x_2 = r(u_2 - v_2), \ldots, x = r(u_n - v_n)$ is, for each (non-zero) real number r, also a (non-zero) solution of (II). Adding this solution of (II) to the solution $x_1 = u_1, x_2 = u_2, \ldots, x_n = u_n$ of (I) we see that, for each r, $x_1 = u_1 + r(u_1 - v_1)$, $x_2 = u_2 + r(u_2 - v_2), \ldots, x_n = u_n + r(u_n - v_n)$ is a solution of (I). This completes the proof.

EXERCISE 6   Given that $(-3, 0, 1)$ and $(8, -3, 3)$ are both solutions (x, y, z) of the simultaneous equations

$$-x - 3y + z = \quad 4$$
$$2x + 8y + z = -5$$

Show that $(8 - \{-3\}, -3 - 0, 3 - 1)$ is a solution of

$$-x - 3y + z = 0$$
$$2x + 8y + z = 0$$

and that the general solution of the given system of equations is $(x, y, z) = (-3 + 11a, -3a, 1 + 2a)$, a being any real number. (But isn't $(8 + 11a, -3 - 3a, 3 + 2a)$ the general solution? Explain!)

In Exercise 27 in Chapter 8 we gave you, albeit in mixed-up order, the proof of Fermat's Little Theorem. Later we stated Euler's *generalization*.

Theorem 7
Let a and m be positive integers with $(a, m) = 1$. Then $a^{\phi(m)} \equiv 1 \pmod{m}$.

COMMENT   We shall not prove this here. We merely note that if $m = p$ is prime, then $\phi(m) = p - 1$ and so Fermat's Little Theorem is a *special case*. As well as having serious application to the construction of *public key encryption systems*, we have reintroduced the theorem here because it helps to prove the following fun result:

Let n be a positive integer. Then there is a multiple of n which uses only the digits 0 and 1.

DISCUSSION
Suppose that n is divisible by 2^a and by 5^b (and no higher powers of 2 and 5). Let $N = n/(2^a 5^b)$. Then $(N, 10) = 1$. By Theorem 7 we have $N \mid (10^{\phi(N)} - 1)$, an integer comprising $\phi(N)$ digits, all 9s. If $3 \nmid N$ then $N \mid 111 \ldots 1$ where there are $\phi(N)$ ones. Then, if c is the larger of a and b, we see that $2^c 5^c N$ is a multiple of n which comprises $\phi(N)$ ones followed by c zeros.

EXERCISE 7 Write out a proof of this result after deciding what you must do if the number *n* *is* divisible by some power of 3.

Specialization

Extreme special cases

Do you have trouble remembering which of the infinite series

$$x - \frac{x^3}{3!} + \frac{x^5}{5!} - \frac{x^7}{7!} + \dots \quad \text{or} \quad 1 - \frac{x^2}{2!} + \frac{x^4}{4!} - \frac{x^6}{6!} + \frac{x^8}{8!} - \dots$$

is that for sin *x* and which for cos *x*? If so, just consider the special case $x = 0$. You will then readily recall which is which – provided you also remember that $\sin 0 = 0$ and/or $\cos 0 = 1$.

Consider, now, the following example.

Example 1

Yesterday you forgot the formula for the surface area of a cone. Your friend suggested it is $2\pi rh$ where *r* and *h* are as in Figure 12.2(a). Is he right? You can check – how?

SOLUTION   If the formula is as claimed then the surface area of the frustum in Figure 12.2(b) will be $2\pi rh - 2\pi RH$. Then the surface area of the cylinder in Figure 12.2(c) will be $2\pi r(h - H)$, which is correct. So was the formula correctly remembered? (See Exercise 8.)

EXERCISE 8   Let the height *h* of the cone in Figure 12.2(a) gradually decrease towards 0. If *h* *were* 0 the cone would be a disc of area $2\pi r^2$. So what?

EXERCISE 9   A plane can fly at 400 kph in still air. Each day the plane flies in a straight line from city A to city B and back. On a day where there is a 50 kph wind

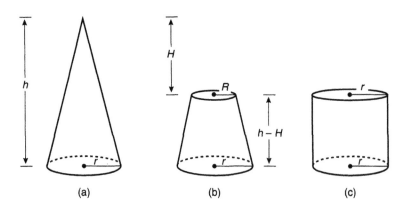

(a)   (b)   (c)

Figure 12.2

blowing directly from B to A all day, will the total flight time from A to B and back take (i) more; (ii) less; (iii) the same time as on a windless day? (*Hint:* What if the wind were 400 kph?)

EXERCISE 10   For the area of a quadrilateral the Egyptians used the formula $\frac{1}{2}(a + c) \cdot \frac{1}{2}(b + d)$ where a, b, c, d are the lengths of the sides taken in order. Show that this formula is incorrect by considering a quadrilateral with one side taken to be very small – even zero, if you wish.

The philosophy behind Exercises 8 and 10 and Example 1 is: a universal theorem must remain valid even in *extreme cases*. Extreme cases may also show the way to establish the general case, as in the following old problem.

Example 2
A square board is divided into a $2n \times 2n$ grid of idential squares and then the two opposing corners are cut off. (Figure 12.3(a) shows the case $n = 4$.) By a 'tile' we mean a rectangular shape of size 1×2. Can the remaining $4n^2 - 2$ squares be covered, in a non-overlapping way, by $2n^2 - 1$ tiles? (*Hint:* Can you see the *real* reason why the answer is 'no' if $n = 1$? Figure 12.3(b).)

Special cases show the way

There is no hope of proving a particular result true for, say, all triangles if you cannot prove it for, say, all *isosceles* triangles. Furthermore, if the result *is* true for all triangles, a proof of it for isosceles triangles may at least *suggest* an approach to proving the general case. The idea of *working towards* the proof of a general case by first examining (and, perhaps, even proving) special cases, has already been remarked upon. Here is another example, where superposition of special cases comes to the rescue.

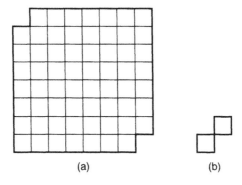

(a)   (b)

Figure 12.3

Example 3
You are asked to find a polynomial $p(x)$, say, in x and with real coefficients which is such that $p(2) = 2$, $p(4) = 4$, $p(5) = 5$ and $p(11) = 12$. How can you find it?

DISCUSSION
Can you find it? It is not immediately clear that there *is* such a polynomial – and if there are many which should we choose? It certainly seems clear from plotting the points $(2, 2)$, $(4, 4)$, $(5, 5)$ and $(11, 12)$ that neither a straight line $(y = mx + c)$ nor a quadratic curve $(y = ax^2 + bx + c)$ will pass through these points. Perhaps we should try for a cubic – or a quartic – or, to be on the safe side a polynomial of degree n? It might be nice to go for a polynomial of least possible degree. Since, given *two* points, we could fit a unique straight line through them, I could *generalize* with a *guess* that, given four points we shall be able to find a unique(?) cubic which will do the trick. This seems reasonable since, *assuming* such a cubic, $p(x) = ax^3 + bx^2 + cx + d$, say, exists, we will obtain, on substituting $2, 4, 5$ and 11 in turn for x, a system of four simultaneous equations in four unknowns which should yield the required values for a, b, c and d.

We shall leave such calculations to you and show you, admittedly using a scarcely more general example, a method which is equally applicable to all problems of this type and which allows you to write down the desired polynomial immediately. Again we have to admit that the method leaves a certain amount of tidying up to be done at the end.

Let us start again. If, given, $p(x_1) = y_1$, $p(x_2) = y_2$, $p(x_3) = y_3$ and $p(x_4) = y_4$, how can we find the desired polynomial? If we feel stuck, can we do a special case? Which special case? Are there some 'easy' special cases? Yes! For example, when all the y_i are zero. Unfortunately, in this case the zero function $p(x) = 0$, whose graph is a straight line, will solve this easy case. This is clearly *too special*.

What if *only three of the* y_i (say y_2, y_3, y_4) are zero? Since we want $p(x_2) = 0$, $p(x_3) = 0$ and $p(x_4) = 0$, we might choose

$$p(x) = (x - x_2)(x - x_3)(x - x_4)$$

Then, certainly, $p(x_2) = 0$, $p(x_3) = 0$ and $p(x_4) = 0$, whilst

$$p(x_1) = (x_1 - x_2)(x_1 - x_3)(x_1 - x_4)$$

This may not be equal to y_1, but, by replacing $p(x)$ by

$$P_1(x) = \frac{p(x)}{(x_1 - x_2)(x_1 - x_3)(x_1 - x_4)} y_1$$

we obtain a function, $p_1(x)$ which is such that $P_1(x_1) = y_1$ whilst $P_1(x_2) = P_1(x_3) = P_1(x_4) = 0$.

Now, maybe, you can see the light. We, likewise, define functions $P_2(x)$, $P_3(x)$ and $P_4(x)$ such that $P_2(x_2) = y_2$, $P_3(x_3) = y_3$ and $P_4(x_4) = y_4$ with all other $P_2(x_i)$, $P_3(x_i)$ and $P_4(x_i)$ equal to 0. Then the sum

$$P_1(x) + P_2(x) + P_3(x) + P_4(x)$$

is easily seen to be a cubic polynomial satisfying the required conditions. Furthermore it is unique, as you will prove in Exercise 11.

SOLUTION   Define

$$P_1(x) = \frac{(x-4)(x-5)(x-11)}{(2-4)(2-5)(2-11)} \cdot 2$$

$$P_2(x) = \frac{(x-2)(x-5)(x-11)}{(4-2)(4-5)(4-11)} \cdot 4$$

$$P_3(x) = \frac{(x-2)(x-4)(x-11)}{(5-2)(5-4)(5-11)} \cdot 5$$

$$P_4(x) = \frac{(x-2)(x-4)(x-5)}{(11-2)(11-4)(11-5)} \cdot 12$$

The required polynomial is their sum.

EXERCISE 11

(a) Calculate the cubic polynomial $P_1(x) + P_2(x) + P_3(x) + P_4(x)$.
(b) Use the Fundamental Theorem of Algebra to prove that the above cubic polynomial is unique. (*Hint:* If $a_3x^3 + a_2x^2 + a_1x + {}^\cdot a_0$ and $b_3x^3 + b_2x^2 + b_1x + b_0$ are two polynomials which solve this problem, then their difference $(a_3 - b_3)x^3 + (a_2 - b_2)x^2 + (a_1 - b_1)x + (a_0 - b_0)$ is a cubic with *four* different roots.)

Finally we mention, without proof, a nice geometrical example in which the 'general case' is *equivalent* to a very special case.

Example 4
Consider six points $A_1, B_1, C_1, C_2, B_2, A_2$ drawn in that order on an ellipse. Pascal proved that the three intersections (i) of A_1B_2 with B_1A_2; (ii) of B_1C_2 with C_1B_2 and (iii) of C_1A_2 with A_1C_2 lie on a straight line. (Draw yourself a picture now!) How to prove this? I suggest that if you cannot prove it for the special case of a *circle* then the chances of proving it for an ellipse are nil. In fact the result is true for a circle (one uses Menelaus' theorem – whatever that may be). But the point I wish to make is that the corresponding result for an ellipse then follows immediately. The reason is that, by definition, every ellipse is obtainable by projecting a circle onto a plane placed at a suitable angle. Now, since such a projection clearly sends straight lines to straight lines, the result for every ellipse follows immediately. In a nutshell, the proof for the circle *is the proof for the general ellipse* – of which the circle is a special case.

Reduction to a previous case

This last example illustrates the technique of *reduction to a previous case*. In this instance one proves the desired result by making it depend wholly on the special case of the circle. Theorem 2 gives a better example: there the proof for the general polygon depends directly on the same result for a triangle, which is proved first. To some extent a proof by induction is also of this type: the quest for the truth of $S(k + 1)$ being made to depend on the truth of $S(k)$. Mathematicians' use of this method suggests the final self-deprecating joke in this book. It goes (against the mathematician) as follows:

A psychologist, wishing to compare the thought processes of a physicist and a mathematician, invites the former to move a matchbox from a chair to a bookcase. This the physicist achieves with ease. After restoring the matchbox to the chair, the psychologist then invites the mathematician to do the same. (Each is, of course, out of the room whilst the other is tested so that neither can see the other's solution.) He too accomplishes the task without difficulty. 'Now', says the psychologist to the physicist, 'I make the problem more difficult. Your task now is to move the matchbox from the *table* to the bookcase.' The physicist performs this task with aplomb. Restoring the matchbox to the table, the psychologist now invites the mathematician into the room to attempt the same task. 'Easy', says the mathematician. He takes the matchbox off the table and places it on the *chair*, saying as he does so, 'Now I have reduced our task to that of the previous problem.'

CHAPTER 13

Fallacies and Paradoxes
– and Mistakes

A fallacy, like a witness committing perjury, is far more convincing than the truth.

THORNTON JAMES

How quaint the ways of paradox.

WILLIAM SCHWENK GILBERT

> This chapter presents a collection of fallacies (results which appear to be true because of convincing supporting proofs which are, in fact, unsound) and paradoxes (results which are valid even though they may seem difficult to believe). Readers are invited to test their critical faculties by examining the following proofs and saying whether or not they are sound.

Introduction

I have long thought that one learns a lot more from trying to find the *mistakes* in a specious argument than one does in merely checking through an argument which one is convinced is correct. Indeed, the main reason for asking you, in Exercise 4(b) of Chapter 6, to try to prove $\sqrt{4}$ irrational by copying the proof that $\sqrt{2}$ is irrational, was to show you where, in the $\sqrt{2}$ proof, the *real action* takes place. Many of the steps in that proof do not depend on which integer is having its square root examined, but there must be at least one step which works for $\sqrt{2}$ but fails for $\sqrt{4}$. Such steps form the *real core* of the $\sqrt{2}$ proof and, having identified them, one can better see what really makes that proof 'tick'.

Professional mathematicians often have to read over proofs of results they 'know' (that is, feel 'in their bones') are true, just to see if the argument given is sufficient to establish a result they, in any case, believe. (A good technique for doing this is to disbelieve every bit of what is written, even going so far as to attempt to see if a counterexample springs to mind, until one feels obliged to admit its truth.) This can be quite taxing. How much more motivating it is to try to find the error, especially if it is subtle, in an argument which claims to prove something that no-one could possibly believe, for example that there exists some positive real number which is equal to zero. This latter is what two of our examples will show. (To those who are already shouting, 'Thou shalt not divide by zero', let me say that it is not that easy.)

Bogus proofs

Here is a selection of bogus proofs and doubtful results. Your job is to identify where the 'lapses in thinking' (if any) occur. We begin with an argument due to Zeno[1] which is almost 2500 years old.

Example 1
The story of Achilles and the tortoise.

Achilles and a tortoise agreed to have a 100-metre race. Achilles, confident of winning, magnanimously gave the tortoise a 10-metre start. Mrs Achilles had her doubts, though. 'Ach,' she said, 'I think you made an error in giving the tortoise a start, because, no matter how fast you run, it will take you some time to get to where the tortoise started from. In this time the tortoise will have moved on, not very far, perhaps, but you will still have to make up this extra distance, during which time the tortoise will have moved on further. I don't think you can ever catch him up, my dear. By the way, is your tendon better?'

Surely, Mrs Achilles is wrong, but *how do you explain that*?

EXERCISE 1   Starting from 1 o'clock, the minute hand on a clock will in due course lie *exactly* over the hour hand. At what time (exactly)? (Note that the Achilles argument shows that the minute hand will *never* catch the hour hand.)

EXERCISE 2   Assume that Achilles runs at a constant speed equal to 100 times that of the tortoise. After how many metres does Achilles draw level with the tortoise?

Next, an integral offers a surprise.

Example 2
$\pi = 0$.

[1] Zeno of Elea, 5th century BC.

'Proof'

$$\int_{-1}^{1} \frac{dx}{1 + x^2} = \left[\tan^{-1}x \right]_{-1}^{1} = \frac{\pi}{4} - \left(-\frac{\pi}{4} \right) = \frac{\pi}{2}$$

We now make the substitution $x = 1/u$, so that $dx = -du/u^2$. Also, when $x = 1$ then $u = 1$ and when $x = -1$ then $u = -1$. Hence

$$\int_{-1}^{1} \frac{dx}{1 + x^2} = \int_{-1}^{1} \frac{-du/u^2}{1 + (1/u)^2} = -\int_{-1}^{1} \frac{du}{u^2 + 1} = -\frac{\pi}{2}$$

since the last integral is the negative of the given one. Consequently $\pi/2 = -\pi/2$ from which we deduce that $\pi = 0$.

   How do you explain that?

COMMENT   We *must* identify what has gone wrong here. We cannot afford to abandon the method of substitution altogether, just because of one 'bad egg'.

EXERCISE 3

$$\int_{-1}^{1} \frac{dx}{x^2} = \left[-\frac{1}{x} \right]_{-1}^{1} = -1 - (1) = -2$$

shows that an area *above* the x-axis (between -1 and 1) may be negative. Criticize.

EXERCISE 4

$$\int \frac{dx}{x} = \text{(by 'parts')} \; x \cdot \frac{1}{x} - \int x \cdot \left(\frac{-1}{x^2} \right) dx$$

Subtracting $\int dx/x$ from each side yields $0 = 1$. Criticize.

We have previously seen that $\frac{1}{2} = 0$ (Chapter 1) and $\pi = 0$ and $1 = 0$ (above). So is $\log_e 2$.

Example 3
There exists a positive number, $\log_e 2$, which is equal to 0.

'Proof 1'   Let

$$S = 1 + \frac{1}{2} + \frac{1}{3} + \frac{1}{4} + \frac{1}{5} + \frac{1}{6} + \frac{1}{7} + \ldots$$

Then

$$\frac{1}{2}S = \frac{1}{2} + \frac{1}{4} + \frac{1}{6} + \frac{1}{8} + \ldots$$

Consequently

$$0 = S - \frac{1}{2}S - \frac{1}{2}S =$$

$$\left\{ 1 + \frac{1}{2} + \frac{1}{3} + \frac{1}{4} + \frac{1}{5} + \frac{1}{6} + \frac{1}{7} + \ldots - \left(\frac{1}{2} + \frac{1}{4} + \frac{1}{6} + \frac{1}{8} + \ldots \right) \right\} - \left(\frac{1}{2} + \frac{1}{4} + \frac{1}{6} + \frac{1}{8} + \ldots \right)$$

$$\left\{ 1 \quad\;\; + \frac{1}{3} \quad\;\; + \frac{1}{5} \quad\;\; + \frac{1}{7} + \ldots \qquad\qquad\qquad\qquad\; \right\} - \left(\frac{1}{2} + \frac{1}{4} + \frac{1}{6} + \frac{1}{8} + \ldots \right)$$

$$= 1 - \frac{1}{2} + \frac{1}{3} - \frac{1}{4} + \frac{1}{5} - \frac{1}{6} + \frac{1}{7} - \ldots$$

But this number (which is known to equal $\log_e 2$: see Phillips (1950, p. 344)) is positive since clearly

$$\left(1 - \frac{1}{2} \right) + \left(\frac{1}{3} - \frac{1}{4} \right) + \left(\frac{1}{5} - \frac{1}{6} \right) + \left(\frac{1}{7} - \frac{1}{8} \right) + \ldots > 0$$

How do you explain that?

COMMENT   If you think you can explain it by mumbling something about $-\frac{1}{2}$ being moved past infinitely many other summands in order to cancel out the $\frac{1}{2}$ etc., what about the following 'proof' of the same result?

'Proof 2'   Let $T = \log_e 2$. Then

$$2T = 2 - \frac{2}{2} + \frac{2}{3} - \frac{2}{4} + \frac{2}{5} - \frac{2}{6} + \frac{2}{7} - \ldots$$

Now for each fraction $2/n$ with an odd denominator (e.g. $+\frac{2}{3}$, $+\frac{2}{37}$) there exists a fraction $-2/2n$ (e.g. $-\frac{2}{6}$, $-\frac{2}{74}$) with denominator of double the size. The sum of this pair is $1/n$. On the other hand, each of the fractions $-2/4n$ can be rewritten as $-1/2n$. It therefore appears that $2T = 1 - \frac{1}{2} + \frac{1}{3} - \frac{1}{4} + \frac{1}{5} - \frac{1}{6} + \frac{1}{7} - \ldots$, that is, $2T = T$. Hence $T = 0$ even though we know $(1 - \frac{1}{2}) + (\frac{1}{3} - \frac{1}{4}) + (\frac{1}{5} - \frac{1}{6}) + (\frac{1}{7} - \ldots$ is positive.
How do you explain that?

COMMENT   The explanation is quite involved (see Hirst (1995, p. 158)) so we won't give it here.

EXERCISE 5   Criticize the following proof that $\infty = -1$. Set $U = 1 + 2 + 4 + 8 + \ldots$, then

$$2U = 2 + 4 + 8 + \ldots = U - 1$$

Hence

$$U = -1$$

Now for something a little easier.

Example 4
The length of the hypotenuse of a right-angled triangle is equal to the sum of the lengths of the other two sides.

'Proof'   Split the line AC in Figure 13.1 into n equal parts and draw the jagged line J as indicated. Clearly, for each n, the length of J is equal to the sum of the lengths of its vertical pieces and its horizontal pieces, and this is equal to the sum of the lengths of the sides AB and BC. Now let n increase indefinitely. Clearly J tends more and more into coincidence with the line AC. Thus AC has length as claimed.
   How do you explain that?

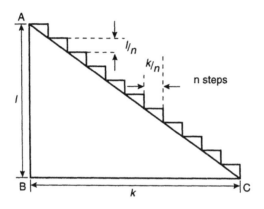

Figure 13.1

COMMENT   I hope that you are not tempted to say, 'Clearly it is wrong since the length of the hypotenuse cannot be the sum of the lengths of the other two sides.' Such a response is of *no use whatsoever*. We *must* find out where we have gone wrong since we don't want to make the same mistake again. In a more subtle problem we might not recognize an answer as being absurd, as we do here. So *why* is it wrong? Identify the point(s) with which you disagree and say why you disagree with them.

Next, a puzzling example due, some say, to Aristotle.

Example 5
Every two circles have the same diameter.

'Proof'   Imagine a railway engine wheel of radius r, as in Figure 13.2, with its (larger) flange of radius R (which helps to keep the engine on the rails). As the point A on the wheel completes one revolution, so does the point B on the flange. Now A travels a distance $2\pi r$ along the rail, but B, on the flange, also travels a distance $2\pi r$ parallel to the rail. Hence $2\pi r = 2\pi R$. Therefore $r = R$.
   How do you explain that?

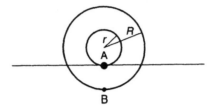

Figure 13.2

Example 6

The volume V and the surface area S of a (*regular*) cone of height h and radius r are given by $V = \frac{1}{3}\pi r^2 h$ and $S = \pi rh$.

'Proof' Imagine the cone split into n parts of equal height as shown in Figure 13.3. It is easy to check that the radii of the discs, starting with the bottom one, are

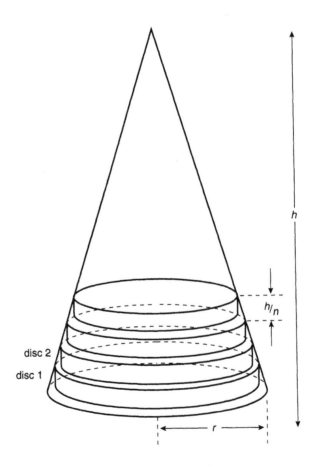

Figure 13.3

$r(1 - 1/n), r(1 - 2/n), \ldots, r\{1 - (n - 1/n)\}$, each one having height h/n. Consequently the total volume and vertical surface area of these discs are, respectively:

$$\sum_{k=1}^{n-1} \pi r^2 \left(1 - \frac{k}{n}\right)^2 \cdot \frac{h}{n} = \pi r^2 \cdot \frac{h}{n} \sum_{k=1}^{n-1} \left(1 - \frac{k}{n}\right)^2$$

and

$$\sum_{k=1}^{n-1} 2\pi r \left(1 - \frac{k}{n}\right) \cdot \frac{h}{n} = 2\pi r \cdot \frac{h}{n} \sum_{k=1}^{n-1} \left(1 - \frac{k}{n}\right)$$

These can be rewritten, respectively, as

$$\pi r^2 \cdot \frac{h}{n^3} \sum_{k=1}^{n-1} (n - k)^2 = \pi r^2 \cdot \frac{h}{n^3} \cdot \frac{(n-1)n(2n-1)}{6} \tag{I}$$

$$2\pi r \cdot \frac{h}{n^2} \sum_{k=1}^{n-1} (n - k) = 2\pi r \cdot \frac{h}{n^2} \cdot \frac{(n-1)n}{2} \tag{II}$$

If we now let n increase indefinitely, so that more (thinner) discs more and more nearly fill the cone, we see from (I) that V must be $\pi r^2 h/3$ and from (II) that S must be πrh. *How do you explain that?*

COMMENT   What is the problem? In Chapter 12 we saw, by 'flattening out' the cone to get the extreme case of a disc, that $2\pi rh$ was not the correct formula for the surface area of a cone as drawn here – and here we have the same intrusive h which, if reduced to 0, gives a zero value for the area contained inside a circle. Hence, if the surface area formula is wrong, what reliance can you place on the volume formula *obtained from the same method*? (Actually *that* formula *is* correct.)

Perhaps our 'flatten the cone' argument gives us at least a hint for the correct formula for the cone's surface area. On flattening the cone towards becoming a disc, we really want h to change not to 0 but to r (because the area inside a circle of radius r is πr^2, not $\pi r \cdot 0$. This would have come about if the area of the cone had been found to be πrH where H is the 'slant height' of the cone.

So I *conjecture* that the surface area of the cone is πrH, as above, and you will *prove* it!

EXERCISE 6   Prove it. Now!

Next a pair of simultaneous equations:

Example 7
To solve the pair of simultaneous equations

$$\begin{cases} 2x^2 - 3x - 4 = 0 & \text{(III)} \\ 3x^2 - 6x - 8 = 0 & \text{(IV)} \end{cases}$$

SOLUTIONS

1. Forming (IV) − (III) gives $x^2 - 3x - 4 = 0$ so that $x = 4$ or -1.
2. Forming $2 \cdot$(IV) − $3 \cdot$(III) gives $-3x - 4 = 0$ so that $x = -\frac{3}{4}$.

How do you explain that?

Now for some number theory. The statement is correct – it is Theorem 10 of Chapter 8 – but this new 'proof' is inadequate. Why?

Example 8
Let p be a prime. Then for each integer r such that $1 \leqslant r \leqslant p - 1$ the binomial coefficient $\binom{p}{r}$ is an integer which is divisible by p.

'Proof' From Chapter 8,

$$\binom{p}{r} = p \cdot \frac{(p - 1)(p - 2)\ldots(p - r + 1)}{1 \cdot 2 \cdot 3 \cdot \ldots \cdot r}$$

Hence it is immediate that $\binom{p}{r}$ is a multiple of p.
  What is wrong with this argument?

Here are two bogus bits of mathematics admitted to by lecturers. In the first, the result claimed is false – and yet surely the proof is correct?

Example 9
If A, B, C and D are sets with $A \times B = C \times D$ then $A = B$ and $C = D$.

'Proof' Let $a \in A$. Then there exists $b \in B$ (indeed possibly many such b) such that $(a, b) \in A \times B$. But $A \times B = C \times D$ and so $(a, b) \in C \times D$. Hence, immediately, $a \in C$. This proves that $A \subseteq C$. In a similar way we can show that $C \subseteq A$ and, likewise, that $B \subseteq D$ and $D \subseteq B$. All this proves that $A = C$ and $B = D$.
  What is wrong with this argument?

In the second, the statement *is* correct (it is Theorem 2(i) of Chapter 10) but the proof is insufficient. Why? Fortunately, the lecturer, whose name must remain a mystery, discovered the mistake before giving the lecture. However, its accidental occurrence has provided a nice example for a certain book about proofs.

Example 10
Each pair α, β of complex numbers, satisfies the inequality

$$|\alpha + \beta| \leqslant |\alpha| + |\beta|$$

'Proof' On the one hand

$$|\alpha + \beta|^2 = (\alpha + \beta)\overline{(\alpha + \beta)} = \alpha\bar{\alpha} + \alpha\bar{\beta} + \beta\bar{\alpha} + \beta\bar{\beta}$$

$$= |\alpha|^2 + \alpha\bar{\beta} + \beta\bar{\alpha} + |\beta|^2 \tag{V}$$

On the other hand

$$(|\alpha| + |\beta|)^2 = |\alpha|^2 + 2|\alpha||\beta| + |\beta|^2 \tag{VI}$$

Now, $\alpha\bar{\beta} + \beta\bar{\alpha} = \alpha\bar{\beta} + \overline{\alpha\bar{\beta}}$ is real. Hence $\alpha\bar{\beta} + \beta\bar{\alpha} = |\alpha\bar{\beta} + \beta\bar{\alpha}|$. But

$$|\alpha\bar{\beta} + \beta\bar{\alpha}| \leqslant |\alpha\bar{\beta}| + |\beta\bar{\alpha}| = |\alpha||\bar{\beta}| + |\beta||\bar{\alpha}| = 2|\alpha||\beta|$$

Comparing (V) and (VI) establishes the result claimed.
    What is wrong with this argument?

COMMENT   The following moral emerges: trust no-one where mathematical proofs are concerned. Always *retain a healthy scepticism* where statements and proofs of statements are concerned until you really are fully satisfied as to their truth (or falsity).

To redress the balance, there now follow some proofs 'donated' by students. The first one attempts a direct proof of the 'only if' part of Statement 11.

Example 11
If x is an integer such that x^2 is odd then x is odd.

'Proof' If x were odd then $x^2 - x$ would be the difference of two odd integers and would, therefore, be even. But $x^2 - x = x(x - 1)$ is a product of consecutive integers and so it *certainly is* even. Therefore x *is* odd.
    What is wrong with this argument?

Example 12
Let s and t be integers. Prove that if $(s, t) = 1$ then $(s + 2t, 3s + 4t) = 1$.

'Proof' Taking $s = 3$ and $t = 8$ we see that $(s, t) = 1$. Now $(s + 2t, 3s + 4t) = (19, 41)$ which is also equal to 1.
    What is wrong with this argument?

Example 13
Show that, for each integer n, $7\!\!\not|\,(n^2 + 4)$.

'Proof' Now $n = 10k + r$ where $r = 0, 1, 2, ..., 9$. Thus n^2 ends in a $0, 1, 4, 9, 16, 25, 36, 49, 64$ or 81. Hence $n^2 + 4$ ends in $4, 5, 8, 13, 20, 29, 40, 53, 68$ or 85. Since none of these integers is divisible by 7, $n^2 + 4$ can never be divisible by 7.
    So, what is wrong with this argument?

Yet another – but is this one sound?

Example 14
If a and b are positive integers with $a^2 = b^3$ then a is a perfect cube.

'Proof' $a = (\sqrt{b})^3$. Suppose that $\sqrt{b} \notin \mathbb{Z}$. Now $(\sqrt{b})^2 \cdot \sqrt{b} = a \in \mathbb{Z}$. Hence $\sqrt{b} \in \mathbb{Z}$. Consequently a is the cube of some integer (namely $\sqrt{b}$.)
    Is anything wrong with this argument?

What about this one?

Example 15
To show that, if a, b are integers such that $a \cdot b$ is odd then both a and b are odd.

'Proof' Suppose that $a \cdot b = m^2 - 1$ where m is even. Then

$$a \cdot b = (m - 1) \cdot (m + 1)$$

Hence $a = m - 1$ and $b = m + 1$ which are both odd since m is even.
    Is anything wrong with this argument?

Example 16
To generate an infinite sequence of primes, take the integers $p_1, p_2, p_3, p_4, \ldots$, where $p_i = 2^{p_{i-1}} - 1$. Starting with $p_1 = 2$ we get $2, 3, 7, 127, 2^{127} - 1, \ldots$, all of which are prime.
    Is anything wrong with this argument?

Example 17
Is the statement $(\exists y \in \mathbb{Z})(\forall x \in \mathbb{Z})[x < y]$ true or false?

'Proof' The statement $(\exists y \in \mathbb{Z})(\forall x \in \mathbb{Z})[x < y]$ is false because, taking $y = 1$ we could take $x = 2$.
    Is anything wrong with this argument?

One final student contribution.

Example 18
As the positive integer n increases indefinitely the quantity $(1 + 1/n)^n$ approaches the value 1.

'Proof' As n increases, $1 + 1/n$ clearly approaches the value 1. But, for all n, $1^n = 1$. That proves it!
    Is anything wrong with this argument?

COMMENT   You may say that, for any particular n, $1 + 1/n > 1$ and so certainly $(1 + 1/n)^n$ is greater than 1. True, but so what? A sequence of numbers all greater than 1 can approach the value 1. Indeed the same remarks apply to $1 + 1/n$ itself – and that seems to strengthen my case!

Perhaps the following is easier to swallow. The concept of integral taken over an infinite range is easy to understand.

Example 19
Consider the infinite 'trumpet' in Figure 13.4 constructed by rotating the graph of $y = 1/x$, drawn from 1 to ∞, about the x-axis.
    By first evaluating the surface area and the volume of the standard (shaded) 'element' of the trumpet, we see that the surface area and the volume of the trumpet

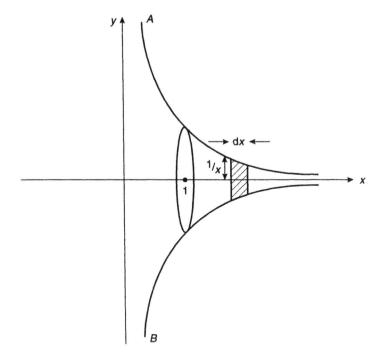

Figure 13.4

from $x = 1$ to $x = A$ are given, respectively, by

$$\int_1^A 2\pi\left(\frac{1}{x}\right) dx = 2\pi\{\log_e A\}$$

and

$$\int_1^A \pi\left(\frac{1}{x}\right)^2 dx = \pi\left\{\frac{1}{1} - \frac{1}{A}\right\}$$

As A increases indefinitely we see that the first integral takes on larger and larger values (since $\log_e A$ increases as A does) but the second integral approaches the value π (since $1/A$ approaches the value 0 as A increases indefinitely). This shows that, as we take more and more trumpet, the volume gets nearer and nearer to π but the surface area increases without bounds.

How do you explain that?

Example 20
Tonpres-Dehy was an unpopular swot! He passed all his examinations with an average mark of 71.27%. Ellafmas was popular and nearly as clever, taking fewer examinations but averaging 70.37%. Each was taking one final examination. The school prize would go to the one who had the higher average. 'And the final result:

Ellafmas 95%.' Cries of 'Hooray for Ellers!' 'For Tonpres-Dehy, 98%.' Groans all round as the headmaster rose to give the prize to T-D. Only Archie Newgauss, the mathematics student, could see that the headmaster might be about to award the prize to the wrong boy.

How do you explain that?

Example 21

Here is another proof showing that case $n = 4$ implies case $n = 3$ in Theorem 5 in Chapter 12.

'Proof' We are given that

$$\frac{a + b + c + d}{4} \geqslant \sqrt[4]{(abcd)}$$

for all sets of four positive reals a, b, c and d. Setting $d = (a + b + c)/3$ we obtain

$$\frac{a + b + c + \{(a + b + c)/3\}}{4} \geqslant \sqrt[4]{[abc \cdot \{(a + b + c)/3\}]}$$

But the left-hand side of this inequality is just $(a + b + c)/3$ and so, since we also have

$$\sqrt[4]{[abc\{(a + b + c)/3\}]} \geqslant \sqrt[4]{\{abc \sqrt[3]{(abc)}\}}$$

which is equal to $\sqrt[3]{(abc)}$, the proof is complete.

What, if anything, is wrong with this argument?

Example 22

Every positive integer is odd.

'Proof' Let $S(n)$ be the assertion: each positive integer $\leqslant n$ is odd. Clearly $S(1)$ is true. Now suppose $S(k)$ is true and consider the integer $k + 1$. By the induction assumption $k - 1$ is odd. Hence $k + 1$ $(= \{k - 1\} + 2)$ is *certainly* odd. This completes the proof.

How do you explain that?

Finally, an example from the fairground.

Example 23

A fairground stallholder advertised: 'Choose an integer in the range 1 to 6 (inclusive). Throw three dice. If one of your dice shows your number then you get 10 pence, if two show it you get 20 pence, if all three show it you get 30 pence. If none of the three shows your number then you give me 10 pence. My dice are guaranteed fair by the circus strong man. All complaints to him!'

The budding mathematics student thought about this and reasoned as follows. 'If, say, I choose number 4 then each of the three dice has one chance in six of showing a 4. So, on each throw of three dice I have three chances in six of throwing at least one 4 – that is, I have an even chance of winning and losing. But sometimes

– when a throw shows either two or three 4s – I will win more than 10 pence. Consequently, if I play long enough, I'll make a fortune. It can't fail.'

The next day the student received a letter telling him he had been rejected by all the university mathematics departments he had applied to through UCAS and he also saw the stallholder getting out of his Rolls-Royce and going into the bank carrying a very large bag of money.

How do you explain that?

'I don't believe it!' (Meldrew)

1. Which of the following equalities are true for *all* pairs of positive real numbers *a* and *b*? (Each of them has been assumed to be true, by different students in examinations across the years.) For each 'equality' which does not always hold give values of *a* and *b* which prove your claim.

   (i) $\dfrac{1}{a} + \dfrac{1}{b} = \dfrac{1}{a+b}$

   (ii) $\sqrt{(a+b)} = \sqrt{a} + \sqrt{b}$

   (iii) $(a+b)^2 = a^2 + b^2$

   (iv) $|a+b| = |a| + |b|$

   (v) $\log_e(a+b) = \log_e a + \log_e b$

   (vi) $\log_e ab = \log_e a \cdot \log_e b$

   (vii) $e^a \cdot e^b = e^{ab}$

   Here are some more contentious statements. Do you think they are correct? If so, try to confirm them, if not, give counterexamples showing them to be untrue.

2. If a function $f(x)$ is such that, for *every* $n \geq 0$, its *n*th derivative $f^n(x)$ is equal to 0 at the origin, i.e. $f^n(0) = 0$ for all integers $n \geq 0$, then $f(x)$ must be the *zero function*, i.e. $f(r) = 0$ for all real numbers *r*.

3. The degree of the zero polynomial, 0, must be zero. (After all, polynomials such as $ax^2 + bx + c$ have degree 2 (if $a \neq 0$, of course) whilst all those of the form $ax + b$ have degree 1 and all constant polynomials, including the zero polynomial have degree zero. I mean, what *else* could it be?)

4. Fil once said to me that, in a lottery, I would stand a better chance of winning if I bought the ten tickets numbered $1, 2, 3, 4, 5, 6, 7, 8, 9, 10$ rather than the ten tickets numbered $1, 3, 5, 7, 9, \ldots, 19$ because, in the latter case, the winning lottery number may fall into one of the even numbered gaps.

5. A marathon is a race run over a distance of 26 miles and 385 yards. I was told that, in a recent race, over *every* distance of one mile (and by that I mean from *every* point A on the course to that point B, 1 mile farther down the road from A, no matter which point is chosen for A) Barbara ran faster than Julie – and yet Julie won. (I can't believe this can be true. Can you?)

6. 'The simultaneous equations

$$2.91x + 1.33y = 4.82$$
$$4.21x + 1.93y = 6.98$$

do have solutions $x = 1.1294...$ and $y = 1.1529...$, sir', said the new recruit, Matthew Matticks to his boss, the innumerate horseshoe magnate Ivor Fortune. 'Ah', said Fortune, 'the figure 2.91 should have read 2.90. Sorry! That change won't make much difference to your x and y will it?' 'No, sir,' said Matticks. 'Both x and y will certainly remain greater than the 1.00 reading necessary for the project to succeed.' Is Matticks correct? (In working this out retain *all* decimal places.)

CHAPTER 14

A Mixed Bag

A veritable rag-bag of... fireworks.   N.F. SIMPSON

This chapter comprises a collection of problems. Its purpose is threefold. First, it gives you some examples on which to test your mathematical skills. Second, the (curt but complete) solutions given in the Appendix can be used as a check against your solutions and/or as exercises in reading and understanding proofs. Third, the problems are meant to be fun, to be a source of enjoyment. Some are extensions of or variations on results already seen in the text. Others are 'new' and may require some thought and ingenuity. All I ask is that you have the fortitude not to peek at their solutions until you have had an honest go at them. To encourage you to refrain as long as possible from looking at the solutions a list of hints follows that of the problems.

PROBLEM 1   Show the analogue of Exercise 1 in Chapter 12, namely that, for each $n > 1$, the sum $\sum_{k=1}^{n} n!$ is not a perfect cube. Generalize this by showing that, for all integers $t \geqslant 3$, it is not a perfect tth power.

PROBLEM 2   Find a formula for the sum $1 \cdot n + 2 \cdot (n-1) + 3 \cdot (n-2) + ... + (n-1) \cdot 2 + n \cdot 1$.

PROBLEM 3   From computer evidence I conjectured that all of the numbers 211, 322 111, 4 332 221 111, 544 333 222 211 111,..., etc. are square-free. In fact, I was wrong. Show this using only pencil and paper.

PROBLEM 4 (The anti-Goldbach Theorem – Just (1973)) Prove that 38 is the largest integer which *cannot* be expressed as a sum of two *odd* composite numbers.

PROBLEM 5 Show, for $n \in \mathbb{Z}^+$, that

$$\left(1 + \frac{1}{n}\right)^n < 3$$

and that

$$\left(1 + \frac{1}{n}\right)^n < \left(1 + \frac{1}{n+1}\right)^{n+1}$$

PROBLEM 6

(a) Show that if x and y are integers with $3 \leqslant x < y$ then $x^y > y^x$.
(b) Show that if x and y are positive integers such that $x < y$ and $x^y = y^x$, then $x = 2$ and $y = 4$.

PROBLEM 7 Prove that there is *no* polynomial

$$p(x) = a_n x^n + a_{n-1} x^{n-1} + \dots + a_1 x + a_0$$

with integer coefficients a_i which is prime for *every* positive integer value of x.

PROBLEM 8 If

$$ax + by = e$$
$$cx + dy = f$$

(where $a, b, c, d, e, f \in \mathbb{Q}$) can be solved using *real* numbers x, y, then need it have a solution with *rational* numbers x, y?

PROBLEM 9

(a) Show that there exist integers x, y, z and t such that
$$23 = x^3 + y^3 + z^3 + t^3$$
(b) Show that 23 is not the sum of *three* (positive or negative) cubes.

PROBLEM 10 Show that if $u^2 - 2v^2 = 1$ then
$$(u^2 + 2v^2)^2 - 2(2uv)^2 = 1$$
Use this to obtain better and better rational approximations to $\sqrt{2}$.

PROBLEM 11 We know that $\sum_{k=1}^{n} k^3$ is *always* a square (Statement 16.1, Chapter 7). This suggests that we ask:

(a) Is $\sum_{k=1}^{n} k$ ever a square if $n > 8$?

(b) Is $\sum_{k=1}^{n} k^2$ ever a square if $n > 1$?

PROBLEM 12 Find the least integer whose cube ends in 111.

PROBLEM 13 Are there no integers a and b such that $a^2 + b^2 = 1\,234\,567$?

PROBLEM 14 Prove that there is no example of a sum of *four* consecutive integers whose sum is a perfect square.

PROBLEM 15 Prove that for all positive integers n neither of the following is a perfect square:

(a) $n(n + 1)(n + 2)$
(b) $n(n + 1)(n + 2)(n + 3)$

PROBLEM 16 Show that, for each real number $x > 1$, $\log_e x < x - 1$.

PROBLEM 17 Show that, if p is a prime dividing $4 \cdot (5 \cdot 13 \cdot 17 \cdot ...)^2 + 1$, then p is of the form $4k + 1$. Use this to prove that there are infinitely many primes of the form $4k + 1$.

PROBLEM 18 Show that a perfect square can end in *three* 4s but not *four* 4s.

PROBLEM 19 Show that, if $h > 0$ and $n \geqslant 2$, then

$$(1 + h)^n \geqslant \frac{n(n-1)}{2} \cdot h^2$$

Deduce that if $t > 1$ is given then there exists an integer n such that

$$0 < (n)^{1/n} < t$$

PROBLEM 20 Show that, in the infinite sequence 31, 331, 3331, 33 331,... not all pairs of integers are coprime.

PROBLEM 21 Let k be a positive odd integer. Show that not all the numbers in the infinite sequence $k, k + 2, k + 2^2, k + 2^3, k + 2^4, ...$ can be prime.

PROBLEM 22 Show that, if $2^n + 1$ is prime then n is a power of 2.

PROBLEM 23 Given a fixed perimeter, the rectangle with largest area is a square. Generalize this to: given that $a_1, a_2, ..., a_n$ are variable positive real numbers whose sum $a_1 + a_2 + ... + a_n$ is the fixed quantity t, show that their *product* is maximum when each a_i is equal to t/n.

PROBLEM 24 At a party many people shake hands. Show that the *number* of people who shake hands an odd number of times is even.

PROBLEM 25 Without using Oresme's result (see the proof of Statement 18 on page 121) try to discover a value of t (depending on r) so that the sum

$$\frac{1}{r} + \frac{1}{r+1} + \frac{1}{r+2} + \frac{1}{r+3} + \dots + \frac{1}{t(r)}$$

is greater than or equal to 1. Confirm your findings with a proof.

PROBLEM 26 An *autosquare* is a positive integer which is equal to the sum of the squares of its digits. Show that 1 is the only autosquare.

PROBLEM 27 For each $n \in Z^+$, define sos(n) to be the *sum of the squares of the digits of* n. For each $t \in Z^+$, define $\text{sos}^t(n) = \text{sos}\{\text{sos}^{t-1}(n)\}$. Prove that, for each $n \in Z^+$, there exists $k \in Z^+$ such that either $\text{sos}^k(n) = 1$ or $\text{sos}^k(n) = 4$.

PROBLEM 28 Show that, if the sets $\{\{a\}, \{a, b\}\}$ and $\{\{c\}, \{c, d\}\}$ are equal then $a = c$ and $b = d$. (Thus $\{\{a\}, \{a, b\}\}$ may be taken as the *definition of ordered pair* (a, b).)

PROBLEM 29 *Wilson's theorem* states that:

If p is a prime then $(p - 1)! + 1$ is a multiple of p.

Prove that the converse of this is true, namely that *if n is an integer greater than 1 such that $n \mid \{(n - 1)! + 1\}$ then n is prime.*

PROBLEM 30 Show that, if $2^k - 1$ is prime, then $2^{k-1}(2^k - 1)$ is a perfect number. Find five perfect numbers.

PROBLEM 31 Show that each positive integer is the sum of distinct powers of 2.

PROBLEM 32 Prove by mathematical induction that, for each $n \in Z^+$,

$$(x + y)^n = x^n + \binom{n}{1}x^{n-1}y + \binom{n}{2}x^{n-2}y^2 + \dots + \binom{n}{n-1}xy^{n-1} + y^n$$

PROBLEM 33 The statement $(\exists x \in Z^+)[x \text{ is prime and } x^2 + 200 \text{ is prime}]$ is false. Prove it.

PROBLEM 34 Write (25 273, 7363) in the form $25\,273s + 7363t$ with $s > 0$. Show that it is impossible to write (25 273, 7363) as $25\,273s + 7363t$ with $0 \leqslant t \leqslant 100$.

PROBLEM 35 Let $f(k)$ denote the kth term of the Fibonacci sequence. Show that, for $m, n \in Z^+$

$$f(m + n) = f(m - 1) \cdot f(n) + f(m) \cdot f(n + 1)$$

PROBLEM 36 With $f(k)$, m and n as in Problem 35, show that if $m \mid n$ then $f(m) \mid f(n)$.

PROBLEM 37   Consider the sequence 1, 2, 4, 8, 16, 23, 28, 38, 49,... where each term after the first is the sum of the previous term and the sum of its digits, (e.g. $28 = 23 + 2 + 3$). Show that no term in this infinite sequence is divisible by 9.

PROBLEM 38   In contrast to the claim of Final Exercise 2(b) in Chapter 1, show that there exist infinitely many $k \in \mathbb{Z}^+$ for which both $6k - 1$ and $6k + 1$ are *composite*.

PROBLEM 39   Prove that $\sqrt{2} + \sqrt{3} + \sqrt{5}$ is irrational.

PROBLEM 40   Show that the equation $x^2 - 2y^2 = 5$ has no solution in $\mathbb{Q}$.

PROBLEM 41   Is either the difference or the sum of 2.345 678 910 111 213... and 1.234 567 891 011 121... a rational number?

PROBLEM 42   I met a man who said that it was possible to place 45 balls into 12 bags so that there is an odd number of balls in each bag. I don't believe him. Do you?

PROBLEM 43   Someone once phoned a colleague for advice. He had to complete the following table so that the entries in the rows and in the columns add up to the numbers shown. *I* can't find a solution. Can you?

					Row totals
?	83	?	?	?	429
68	?	?	?	65	424
?	?	59	?	?	478
?	?	83	71	?	422
Column totals 344	378	366	323	332	

PROBLEM 44   A problem in an arithmetic book asked: 'A storekeeper had a stock of over 200 paper hats but couldn't sell any at 50 pence each. So he lowered the price and sold the lot in a day. His takings were £78.07. How many hats did the storekeeper sell at what price?' (Assume that all hats were sold at the same price.)

PROBLEM 45   (A newspaper's example of the type of maths problem which defeated most school-leavers!) Four points in the plane are such that AB = AC, AD = BD, angle DAC = 39°. Find angle BAD.

PROBLEM 46   Do white sheep eat more than black sheep?

For your computer

PROBLEM 47   Show that the integer 169 can be expressed as a sum of n squares of *positive* integers for each n with $1 \leqslant n \leqslant 5$. Now use the theorem (see Chapter 1) that every positive integer m is expressible as a sum of at most four positive integer squares to prove that *each integer* $\geqslant 34$ *is expressible as a sum of exactly five positive integer squares*.

PROBLEM 48 *Conjecture (of Goldbach, 1752)*: Each (positive) integer n is expressible in the form $q + 2a^2$ where (i) q is either 1 or prime and (ii) $a \geqslant 0$. Test this conjecture up to $10\,000$.

PROBLEM 49 It was claimed (by von Sterneck, 1903) that if x, y, z and t are positive integers such that $x^3 + y^3 + z^3 = 3t^3$ then, necessarily, $x = y = z$. Fleck (1906) stated that this was not true. Who is correct?

PROBLEM 50 Is every odd integer $\geqslant 3$ either (i) a prime or (ii) a sum of a prime and a power of 2?

Hints

This list is compressed purposely to make it more difficult to read so that you may think twice before seeking help from it!

1. An integer divisible by 3 but not by 27 cannot be a cube – nor any higher power. **2.** Evaluate the sum for $n = 1, 2, 3, 4, \ldots$. Does taking first differences help you to see a pattern? **3.** What is the square of the smallest odd prime? **4.** Write (for example) $10k + 6 = 21 + (10k - 15)$. **5.** Use the binomial expansion. In the first case compare with $1 + \dfrac{1}{2^0} + \dfrac{1}{2^1} + \ldots + \dfrac{1}{2^n}$. In the second case compare ith term of first expansion with ith of second. **6.** Let $y = x + t$ and compare x^{y-x} with $\left(\dfrac{x+t}{x}\right)^x = \left(1 + \dfrac{1}{x+t-1}\right)^x \left(1 + \dfrac{1}{x+t-2}\right)^x \ldots \left(1 + \dfrac{1}{x}\right)^x$. **7.** For each integer t, $p(ta_0)$ is a multiple of a_0. **8.** Two cases: $ad - bc \neq 0$; $ad - bc = 0$. **9.** (b) For $x \in \mathbb{Z}$ show $x^3 \equiv 0$ or 1 or -1 (mod 9) (using binomial theorem); 23 is of the form $9k - 4$. **10.** If $u^2 - 2v^2 = 1$ then $(u/v)^2 = 2 + (1/v)^2$ so, for large u and v, u/v is, approximately...? **11.** Find *infinitely many* such as follows: if we set $n = 2v^2$ and $n + 1 = u^2$ then $n(n + 1)/2 = (uv)^2$. Now use the result in Problem 10. **12.** Assume the number is $\ldots d \cdot 10^3 + c \cdot 10^2 + b \cdot 10 + a$. Find, successively, $a, b, c, \ldots$. **13.** An odd sum of two squares must be the sum of an odd square and an even square. **14.** $(n - 1) + n + (n + 1) + (n + 2)$ is of the form...? **15.** (a) $(m - 1)m(m + 1) = m(m^2 - 1)$, a product of coprime numbers. Hence each is a ...? (b) Evaluate for $n = 1, 2, 3, 4$. Notice a pattern. **16.** Look at $f(1)$ and $f'(x)$ where $f(x) = \log_e x - (x - 1)$. **17.** Suppose that p is a prime of the form $4k + 3$. If $a^2 \equiv -1$ (mod p) then $a^{4k+2} \equiv -1$ (mod p). But $a^{p-1} \equiv 1$ (mod p). **18.** Can $(100a + b)^2$ (where $0 \leqslant b \leqslant 99$) end in four 4s? **19.** Use $(1 + h)^n \geqslant 1 + nh + \frac{1}{2}n(n - 1)h^2 + \ldots > \frac{1}{2}n(n - 1)h^2$. For second part set $n^{1/2} = 1 + k(n)$ (where $k(n) > 0$). **20.** Is any later term of the sequence divisible by 31? Yes, *if* there is an integer of the form $33\ldots300$ which is divisible by 31. **21.** $p | 2^p - 2$ and so if $p | k + 2$.... **22.** If r is odd then $y^r + 1 = (y + 1)(y^{r-1} - y^{r-2} + y^{r-3} - \ldots - y + 1)$. Hence if $n = 2^\alpha \cdot r$.... **23.** $t = (a_1 + a_2 + \ldots + a_n)/n \geqslant \sqrt[n]{(a_1 a_2 \ldots a_n)}$. But there can be equality. When? **24.** When x and y shake hands write down the ordered pairs (x, y) and (y, x). How many x can there be for which the set of all (x, y) is odd? An *even* number of odds is required to make an even. **25.** Try pairing off $\dfrac{1}{r} + \dfrac{1}{t(r)}, \dfrac{1}{r+1} + \dfrac{1}{t(r)-1}$, etc. **26.** 999 and 9999 are not autosquares since $9^2 + 9^2 + 9^2$ and $9^2 + 9^2 + 9^2 + 9^2$ are too small. Likewise for *all* integers above 243. Keep arguing this way until you hit 99. $(9^2 + 9^2 > 99.)$ At least you have no more than 99 integers to examine. **27.** 26 shows that $sos(n) < n$ if $n \geqslant 100$. If necessary, examine the first 99 integers individually. **28.** Treat two cases: (i) $a = b$; (ii) $a \neq b$. **29.** If $n = rs$ is composite then $r < n - 1$ and $s < n - 1$. **30.** Add up the divisors of $p \cdot 2^{n-1}$ (except the biggest). **31.** Induction on n. **32.** Use (after proving!) $\dbinom{k}{i-1} + \dbinom{k}{i} = \dbinom{k+1}{i}$ for all integers $0 \leqslant i - 1 \leqslant k$. **33.** Use the fact that each odd prime $\geqslant 5$ is of the form $6k - 1$ or $6k + 1$. **34.** For the second part note that from $37a + 127b = 37c + 127d = 1$ we get $37(a - c) = -127(b - d)$. Hence $37|(b - d)$ and $127|(a - c)$. Why? Hence $a = c + 127m$ for some integer m. **35.** Fix $m (m \geqslant 1)$. For this m use mathematical induction on $n (n \geqslant 1)$. **36.** Use 35 with $m = tn$ and $n = n$ and mathematical induction on t. **37.** Calling the terms $g(1), g(2), \ldots, g(n), \ldots$ use mathematical induction on n. **38.** For large n note that $n! + 2, \ldots, n! + n$ will include many $6k - 1, 6k + 1$ pairs. **39.** Assume that $r - \sqrt{2} = \sqrt{3} + \sqrt{5}$ where r is rational. Square up. Rearrange. Square up. **40.** Assume $x = a/b$, $y = u/v$. Multiply up by bv in order to work with integers. Work mod 5. **41.** Call first number x, second number y. Then $10y - x = \ldots$? **42.** The sum of an even number of odds is …? But are you *sure* you have understood the hypotheses about the bags? (Russian dolls?) **43.** What about the numbers around the edge? Do they 'add up'? **44.** Who says prices can't be silly? **45.** Don't be surprised if you can't decide! **46.** What are you being asked?

Appendix: solutions to the problems[1]

PROBLEM 1   For $k = 2$ and $k = 8$, $\sum\limits_{n=1}^{k} n!$ has values 3 and 46 233. Since, for $n \geqslant 3$, $3|n!$, we see that, for every $k \geqslant 2$, $\sum\limits_{n=1}^{k} n!$ is a multiple of 3. Since $27 \nmid 46\,233$ and since, for $n \geqslant 9$, $27|n!$, we see that, for $k \geqslant 9$, $\sum\limits_{n=1}^{k} n!$ is a multiple of 3 but is not a multiple of 27. Hence, for $k \geqslant 9$, $\sum\limits_{n=1}^{k} n!$ cannot be a perfect cube nor, indeed, any higher power. Since we easily check, for $k = 2, \ldots, 8$, that none of the sums $\sum\limits_{n=1}^{k} n!$ is a tth power for $t \geqslant 3$, the proof is complete.

PROBLEM 2   For $n = 1, 2, 3, 4, \ldots$ the sum is $1, 4, 10, 20, \ldots$, i.e. $1, 1 + 3, 1 + 3 + 6$, $1 + 3 + 6 + 10, \ldots$. Now 1, 3, 6, 10 are the first four triangular numbers, that is

---

[1] Better solutions – from students – gratefully accepted.

integers of the form $k(k + 1)/2$. I therefore conjecture that

$$1 \cdot n + 2 \cdot (n - 1) + \ldots + n \cdot 1 = \sum_{k=1}^{n} \frac{k(k + 1)}{2} = \sum_{k=1}^{n} \frac{k^2}{2} + \sum_{k=1}^{n} \frac{k}{2}$$

$$= \frac{1}{2} \left\{ \frac{n(n + 1)(2n + 1)}{6} + \frac{n(n + 1)}{2} \right\} = \frac{n(n + 1)(2n + 4)}{12}$$

(The proof, by mathematical induction, we leave to the reader.)

PROBLEM 3   To test, say, $12\,111\,110\,101\,099\,998\ldots22\,111\,111\,111\,111$ for divisibility by 9 we need only test $1 \cdot 12 + 2 \cdot 11 + 3 \cdot 10 + \ldots + 11 \cdot 2 + 12 \cdot 1$ rather than $1 + 2 + 1 + 1 + 1 + 1 + 1 + 0 + 1 + 0 + \ldots + 2 + 2 + 1 + 1 + 1 + 1 + 1 + 1 + 1 + 1 + 1 + 1 + 1 + 1$ since, modulo 9, $12 \equiv 1 + 2$, $11 \equiv 1 + 1$, $10 \equiv 1 + 0$, etc. It is easy to check that the integer $2\,524\,242\,323\,232\,222\,222\,221\ldots$ $2\,221\,111\,111\,111\,111\,111\,111\,111\,111$ is a multiple of 9 since $1 \cdot 25 +$ $2 \cdot 24 + 3 \cdot 23 + \ldots + 24 \cdot 2 + 25 \cdot 1 = \dfrac{25 \cdot 26 \cdot 54}{12} (= 25 \cdot 13 \cdot 9)$ by Problem 2.

PROBLEM 4   If we write $10k = 15 + (10k - 15)$, $10k + 2 = 27 + (10k - 25)$, $10k + 4 = 9 + (10k - 5)$, $10k + 6 = 21 + (10k - 15)$, $10k + 8 = 33 + (10k - 25)$ we see that, if $k \geqslant 4$, then each bracketed integer is composite. Therefore all even integers $\geqslant 40$ are expressible as claimed. You can easily check that $38 = 37 + 1 = 35 + 3 = 33 + 5 = 31 + 7 = 29 + 9$, etc. is not so expressible.

PROBLEM 5   For the first inequality note that

$$\left(1 + \frac{1}{n}\right)^n = 1 + \frac{n}{1!} \cdot \frac{1}{n} + \frac{n(n - 1)}{2!} \cdot \left(\frac{1}{n}\right)^2 + \frac{n(n - 1)(n - 2)}{3!} \cdot \left(\frac{1}{n}\right)^3 + \ldots + \frac{n!}{n!}\left(\frac{1}{n}\right)^n$$

$$< 1 + \frac{1}{1!} + \frac{1}{2!} + \frac{1}{3!} + \ldots + \frac{1}{n!} < 1 + 1 + \left(\frac{1}{2}\right) + \left(\frac{1}{2}\right)^2 + \left(\frac{1}{2}\right)^3$$

$$+ \ldots + \left(\frac{1}{2}\right)^n = 3 - \left(\frac{1}{2}\right)^n < 3$$

For the second inequality we compare the above term by term with

$$\left(1 + \frac{1}{n + 1}\right)^{n+1} = 1 + \frac{n + 1}{1!} \cdot \frac{1}{n + 1} + \frac{(n + 1)n}{2!} \cdot \left(\frac{1}{n + 1}\right)^2 + \frac{(n + 1)n(n - 1)}{3!} \cdot \left(\frac{1}{n + 1}\right)^3$$

$$+ \ldots + \frac{(n + 1)!}{(n + 1)!}\left(\frac{1}{n + 1}\right)^n$$

Since $\dfrac{n - 1}{n} < \dfrac{n}{n + 1}$, $\dfrac{n - 2}{n} < \dfrac{n - 1}{n + 1}$, etc., each term of the former is less than the corresponding term of the latter – and the latter also has an extra term to boot. This proves the result claimed.

PROBLEM 6

(a) Put $y = x + t$. Then

$$\left(\frac{y}{x}\right)^x = \left(\frac{x+t}{x}\right)^x = \left(\frac{x+t}{x+t-1}\right)^x \left(\frac{x+t-1}{x+t-2}\right)^x \dots \left(\frac{x+1}{x}\right)^x$$

$$= \left(1 + \frac{1}{x+t-1}\right)^x \left(1 + \frac{1}{x+t-2}\right)^x \dots \left(1 + \frac{1}{x}\right)^x$$

$$\leqslant \left(1 + \frac{1}{x}\right)^x \left(1 + \frac{1}{x}\right)^x \dots \left(1 + \frac{1}{x}\right)^x$$

$$\leqslant 3 \cdot 3 \cdot \dots \cdot 3 = 3^{y-x} \text{ (by the first part of Problem 5)}$$

$$\leqslant x^{y-x} \text{ (since } 3 \leqslant x, \text{ given)}$$

Hence

$$y^x \leqslant x^{y-x} \cdot x^x = x^y$$

(b) By 6(a) we only need consider the cases $x = 1, 2$. Clearly $x = 1$ will yield no solution. So look at $x = 2$. Now for $n \geqslant 3$, $(n + 1)^2/n^2 = (1 + 1/n)^2 < 2$ (since $1 + 1/n < \sqrt{2}$). Hence, if $2^n > n^2$ (and $n \geqslant 3$) then $2^{n+1} > 2n^2 > (n + 1)^2$. But $2^5 > 5^2$. Hence, by mathematical induction, for each $n \geqslant 5$, $2^n > n^2$. Checking the other cases (i.e. $2^1, 2^2, 2^3, 2^4$) we see that only $2^4 = 4^2$ fulfils the conditions.

PROBLEM 7   For all values of x

$$|a_{n-1}x^{n-1} + \dots + a_1 x + a_0| \leqslant |a_{n-1}||x^{n-1}| + \dots + |a_1||x| + |a_0|$$

Hence, if $x \geqslant 1$, then

$$|a_{n-1}x^{n-1} + \dots + a_1 x + a_0| \leqslant Ax^{n-1}$$

where $A = |a_{n-1}| + \dots + |a_1| + |a_0|$. It follows that

$$p(x) \geqslant x^{n-1}(a_n x - A) \tag{I}$$

Since $a_n x^n$ is the dominant term of $p(x)$ it is clear that $a_n > 0$. Now, from (I), for all $x > A/a_n$, $p(x)$ steadily increases as x increases. If m is such a value of x and if $p(m) = t > 1$ then, using the binomial expansion on powers $(zt + m)^k$, we see that, for each integer z, $p(zt + m)$ is a multiple of t. Therefore $p(x)$ is composite for infinitely many (integer) values of x.

PROBLEM 8   Solving the equations as usual we obtain

$$(ad - bc)x = de - bf \quad \text{and} \quad (ad - bc)y = af - ce$$

Hence, if $ad - bc \neq 0$ then the equations have unique solution

$$x = \frac{de - bf}{ad - bc}$$

$$y = \frac{af - ce}{ad - bc}$$

with x and y clearly rational. If $ad - bc = 0$ then $de = bf$ and $ce = af$. If we suppose that $a \neq 0$ then $x = e/a$ and $y = 0$ is a rational solution of the equations – since $a(e/a) + b0 = e$ and $c(e/a) + d0 = f (a/a) + d0 = f$ (because $ce = fa$). The other cases where $b \neq 0$, $c \neq 0$ or $d \neq 0$ are similar. If $a = b = c = d = 0$ then $e = f = 0$ and $x = y = 0$ will be a rational solution.

PROBLEM 9

(a) $23 = 2^3 + 2^3 + 2^3 + (-1)^3$

(b) If $x = 3k$ (respectively, $3k + 1$, $3k - 1$) then $x^3 \equiv 0$ (respectively, 1, -1) modulo 9 (since $(3k \pm 1)^3 = 27k^3 \pm 27k^2 + 9k \pm 1$). Hence a sum $x^3 + y^3 + z^3$ of three cubes can be congruent to $0, 1, 2, 3, -1, -2, -3$ but not to 4 or -4 modulo 9. Consequently no integer of the form $9k \pm 4$ can be a sum of three cubes. But 23 is of form $9k - 4$.

PROBLEM 10 $(u^2 + 2v^2)^2 - 2(2uv)^2 = (u^2 - 2v^2)^2 = 1^2 = 1$. Starting from $u = 3$, $v = 2$ we obtain

$$U = u^2 + 2v^2 = 17$$

$$V = 2uv = 12$$

with

$$U^2 - 2V^2 = 1$$

Then, from $(U^2 + 2V^2)^2 - 2(2UV)^2 = (U^2 - 2V^2)^2 = 1$ we obtain

$$X^2 - 2Y^2 = 1$$

where

$$X = U^2 + 2V^2 = 577$$

$$Y = 2UV = 408 \text{ etc.}$$

In particular, $\sqrt{2}$ is better and better approximated by $\frac{3}{2}, \frac{17}{12}, \frac{577}{408}, \ldots$.

PROBLEM 11

(a) $\sum_{k=1}^{n} k = n(n + 1)/2$ will be a square if we can find integers u and v such that $n = 2v^2$ and $n + 1 = u^2$. This requires $u^2 - 2v^2 = 1$. Solutions, as above, include $u = 3, v = 2; u = 17, v = 12; u = 577, v = 408;$ etc. Thus $\sum_{k=1}^{n} k = n(n + 1)/2$ is a square for infinitely many n.

(b) $\sum_{k=1}^{n} k^2$ is a square iff $n = 1$ or $n = 24$. But there is no equally easy proof that these are the only two possible values of n. (See Guy (1994, p. 147).)

PROBLEM 12   *Assuming* that there is an integer $\ldots d \cdot 10^3 + c \cdot 10^2 + b \cdot 10 + a$ whose cube ends in $\ldots 111$ we see that $a^3 \equiv 1 \pmod{10}$. Hence $a = 1$. But then

$(b \cdot 10 + 1)^3$ $(= b^3 \cdot 10^3 + 3b^2 \cdot 10^2 + 3b \cdot 10 + 1)$ ends in $\ldots 11$. So $30b + 1 \equiv 11$ (mod 100). Hence $30b \equiv 10$ (mod 100) so consequently $3b \equiv 1$ (mod 10). Hence $b = 7$. Next

$$(100c + 71)^3 = \ldots + 300c \cdot 71^2 + 71^3 \equiv 111 \text{ (mod 1000)}.$$

We find

$$300c \cdot 71^2 \equiv -357\,800 \text{ (mod 1000)}, \quad 3c \cdot 71^2 \equiv -3578 \text{ (mod 10)}$$

so that

$$3c \cdot 1^2 \equiv -8 \text{ (mod 10) from which } c = 4$$

The required integer is 471.

PROBLEM 13   Each integer k which is odd and a sum of two squares is necessarily the sum of an odd square and an even square. Thus k is of the form

$$(2m + 1)^2 + (2n)^2 = 4(m^2 + m + n^2) + 1$$

Since $1\,234\,567$ is of the form $4t + 3$ (just look at the last two digits, 67) we see that $1\,234\,567$ is not the sum of two integer squares.

PROBLEM 14 $(n - 1) + n + (n + 1) + (n + 2) = 4n + 2$ is not a square (since each integer square is of the form $4k$ or $4k + 1$).

PROBLEM 15

(a) If $(m - 1)m(m + 1) = (m^2 - 1)m$ is a square then m and $m^2 - 1$ are both squares since they are coprime. But if $m > 1, m^2 - 1 = t^2$ is impossible (since successive positive integers cannot both be squares. Hence $n(n + 1)(n + 2)$ is not a square if $n \geqslant 1$.

(b) $n(n + 1)(n + 2)(n + 3) = (n^2 + 3n + 1)^2 - 1$. (Why? For $n = 1, 2, 3, 4$ the values of $n(n + 1)(n + 2)(n + 3)$ are 24, 120, 360, 840 – always 1 short of a square. So I *guessed* that $n(n + 1)(n + 2)(n + 3) = (n^2 + an + b)^2 - 1$ for suitable integers a and b which are easily found to be 3 and 1.) Hence $n(n + 1)(n + 2)(n + 1)$ is not a square since it is a positive integer 1 less than a square.

PROBLEM 16   Let $f(x) = \log_e x - (x - 1)$. Then $f(1) = 0 - 0 = 0$. Also

$$f'(x) = \frac{1}{x} - 1$$

which is negative if $x > 1$. Consequently $f(x)$ decreases if $x > 1$. Hence $f(x)$ is negative if $x > 1$, i.e. $\log_e x < x - 1$ if $x > 1$.

PROBLEM 17   Suppose a is an integer such that $a^2 \equiv -1$ (mod p). Then $(a, p) = 1$ {since if $p|a$ then $p|-1$}. Hence, if p is of the form $4k + 3$ then

$$a^{p-1} = a^{4k+2} = (a^2)^{2k+1} \equiv (-1)^{2k+1} \equiv -1 \text{ (mod } p)$$

But $a^{p-1} \equiv 1 \pmod{p}$, by Fermat's Little Theorem, a manifest contradiction. Hence p *cannot* be of the form $4k + 3$. Hence p must be of the form $4k + 1$. In particular, if p divides $2^2(5 \cdot 13 \cdot 17...)^2 + 1$, then

$$(2 \cdot 5 \cdot 13 \cdot 17...)^2 \equiv -1 \pmod{p}$$

so that p must be of the form $4k + 1$.

Now suppose that there are only t primes $p_1 = 5$, $p_2 = 13$, $p_3 = 17,...$ of the form $4k + 1$. Let

$$N = 4 \cdot (5 \cdot 13 \cdot 17 \cdot 29...)^2 + 1 = q_1 \cdot q_2 \cdot ... \cdot q_t$$

where the q_i are primes. By the first paragraph, each q_i is of the form $4k + 1$. This means that each q_i is one of the p_j. But, then, $q_i | N$ and $p_j(=q_i)|(N - 1)$. Hence $q_i | 1$, which is a contradiction. There are, therefore, infinitely many primes of the form $4k + 1$.

PROBLEM 18 $38^2 = 1444$. Now if $(100a + b)^2 = 10\,000a^2 + 200ab + b^2$ (where $0 \leqslant b \leqslant 99$) ends in four 4s then $b^2 = 12^2 (=144)$, $38^2 (=1444)$, $62^2 (=3844)$, or 88^2 $(=7744)$ (since it must end in 44). We therefore need $2ab$ to end in 43, 30, 06, 67 respectively. The first and last are impossible (since $2ab$ is even). Thus we need $2 \cdot a \cdot 38$ to end in 30 or $2 \cdot a \cdot 62$ to end in 06. In other words, working (mod 100) we want $76a \equiv 30 \pmod{100}$ or $124a \equiv 6 \pmod{100}$. Each of these is impossible since 100, 76 and 124 are multiples of 4 whereas 30 and 6 are not.

PROBLEM 19   For $n \in \mathbb{Z}^+$, $n \geqslant 2$, and $h > 0$ we have

$$(1 + h)^n = 1 + nh + \frac{n(n-1)}{2}h^2 + ... \geqslant \frac{n(n-1)}{2}h^2$$

Now, for $n \geqslant 2$, we have $n^{1/n} > 1$. Hence $n^{1/n} = 1 + k(n)$ for some positive $k(n)$ depending on n. Since $k > 0$ (and assuming $n \geqslant 2$) we have, from the above,

$$n \geqslant \frac{n(n-1)}{2!} \cdot k^2$$

Hence

$$k^2 \leqslant \frac{2}{n-1}$$

Therefore, as n increases indefinitely (slightly more formally: as n tends to infinity) we see that $k(n)$ (which is positive) must tend towards 0. Consequently $n^{1/n} = 1 + k(n)$ can be made as near to 1 as we wish.

PROBLEM 20   By Fermat's Little Theorem, $10^{30} \equiv 1 \pmod{31}$. Hence 9_{30} (the integer comprising 30 nines) is divisible by 31. Since $(9, 31) = 1$ the integer 3_{30} is divisible by 31. Therefore the integer $3_{30} \cdot 10^2 + 31$ is divisible by 31. Hence, in particular, the 1st and the 31st members in the sequence are not coprime.

PROBLEM 21   Given k, let p be a prime dividing $k + 2$. Since k is odd, p is odd. Hence by Fermat's Little Theorem

$$2^p \equiv 2 \pmod{p}$$

Therefore $p|(k + 2) + 2^p - 2$, that is $p|(k + 2^p)$.

PROBLEM 22 Factorize n as $n = rs$ where r is odd and s is a power of 2. If $r > 1$ we put $y = 2^s$ in

$$y^r + 1 = (y + 1)(y^{r-1} - y^{r-2} + y^{r-3} - \dots - y + 1)$$

thus giving a factorization of $2^n + 1$. Hence, if $2^n + 1$ is prime then $r = 1$ and n is a power of 2.

PROBLEM 23 We know (from arithmetic mean $\geqslant$ geometric mean) that

$$a_1 a_2 \dots a_n \leqslant \{(a_1 + a_2 + \dots + a_n)/n\}^n = \left(\frac{t}{n}\right)^n$$

which is a fixed number since t and n are fixed. Thus $a_1 a_2 \dots a_n$ can never exceed $(t/n)^n$. And yet $a_1 a_2 \dots a_n$ can equal $(t/n)^n$ when each a_i is equal to t/n.

PROBLEM 24 Consider the set P of all ordered pairs (x, y) where x and y are people at the party who have shaken hands. Then (y, x) also belongs to P; indeed each handshake generates two ordered pairs. Thus P contains an even number of elements. If $\langle x \rangle$ denotes the number of ordered pairs with first member x (that is, the number of people with whom x has shaken hands) then the sum

$$\langle x \rangle + \langle y \rangle + \dots \text{ (taken over all people } x, y, \dots \text{ at the party)} = |P|$$

is even. Now some of $\langle x \rangle$, $\langle y \rangle, \dots$ may be odd – but there cannot be an odd *number* of these since odd·odd + even is not even.

PROBLEM 25

$$\frac{1}{r} + \frac{1}{r + 1} + \dots + \frac{1}{3r - 1} + \frac{1}{3r} = \left(\frac{1}{r} + \frac{1}{3r}\right) + \left(\frac{1}{r + 1} + \frac{1}{3r - 1}\right)$$

$$+ \dots + \left(\frac{1}{2r - 1} + \frac{1}{2r + 1}\right) + \frac{1}{2r}$$

Now, for each k with $0 \leqslant k \leqslant r - 1$

$$\frac{1}{r + k} + \frac{1}{3r - k} = \frac{4r}{(r + k)(3r - k)} \geqslant \frac{1}{r}$$

because $(r - k)^2 \geqslant 0$ implies $4r^2 \geqslant 3r^2 + 2rk - k^2$. Since the sum

$$\left(\frac{1}{r} + \frac{1}{3r}\right) + \left(\frac{1}{r + 1} + \frac{1}{3r - 1}\right) + \dots + \left(\frac{1}{2r - 1} + \frac{1}{2r + 1}\right) + \frac{1}{2r}$$

contains r 'pairs', the sum itself exceeds $r \cdot (1/r) = 1$. Consequently the partial sums of $\frac{1}{1} + \frac{1}{2} + \frac{1}{3} + \frac{1}{4} + \dots$ have at least the values 1, 2, 3, 4, 5, 6, $\dots$, certainly by the 1st, 6th, 21st, 66th, 201st, 606th, $\dots$ terms compared with the 1st, 4th, 16th, 64th, 256th, 1024th, $\dots$ terms in Oresme's method. (In fact, from the above, one can *immediately* replace $t(r) = 3r$ by $t(r) = 3r - 1$. Why?)

PROBLEM 26 Let $k \geq 3$ be an integer. Let n be any integer such that $10^k \leq n < 10^{k+1}$. Then n has $k + 1$ digits. The sum of the squares of these digits is no greater than $9^2 \cdot (k + 1)$ which, for $k \geq 3$, is less than 10^k. (Proof by mathematical induction.) This shows that there is no autosquare greater than 999. Since $9^2 + 9^2 + 9^2 = 243$, no integer in the range 244 to 999 can be an autosquare. Since $2^2 + 9^2 + 9^2 = 166$, no integer in the range 167 to 299 is an autosquare. Continuing in this way the reader can easily check that for each integer $n \geq 100$ the sum of the squares of the digits of n is less than n. So we only have to check that no integer in the range 2 to 99 is an autosquare. These cases are easily checked.

PROBLEM 27 Since, for every integer $n \geq 100$, we have (by the solution to Problem 26) $sos(n) < n$, we see that we need only check the claim made in the problem for integers ≤ 99. This is easily checked.

PROBLEM 28 If $a = b$ then

$$\{\{a\}, \{a, b\}\} = \{\{a\}, \{a\}\} = \{\{a\}\}$$

Hence $\{\{c\}, \{c, d\}\}$ has just one element. This implies that $\{c\} = \{c, d\}$ whence $c = d$. From this we deduce that $\{\{a\}\} = \{\{c\}\}$, hence that $\{a\} = \{c\}$ and, consequently, $a = c$. It follows that $b = d$ (since $b = a = c = d$). If $a \neq b$ then $\{a\} \neq \{a, b\}$. Hence $\{\{a\}, \{a, b\}\}$ is a set with two elements. Therefore so too is $\{\{c\}, \{c, d\}\}$. Consequently $\{c\} \neq \{c, d\}$ so that $\{c, d\}$ has two distinct elements. From

$$\{\{a\}, \{a, b\}\} = \{\{c\}, \{c, d\}\}$$

we see that $\{a\} = \{c\}$ or $\{a\} = \{c, d\}$. But this latter equality cannot hold since $\{a\}$ contains only one element whereas $\{c, d\}$ contains two. Therefore $\{a\} = \{c\}$ and $a = c$ follows. Also $\{a, b\} = \{c, d\}$. So, since $a = c$ and since $\{a, b\}$ and $\{c, d\}$ each have two elements, we deduce that $b = d$.

PROBLEM 29 Let n be an integer > 1. Suppose $n = rs$ (with $1 < r < n$ and $1 < s < n$). Then r divides n and, consequently, divides $(n - 1)! + 1$. However, r divides $(n - 1)!$ (since $1 < r \leq n - 1$). Hence $r \mid 1$: an impossibility since $r > 1$. Hence n must be prime.

PROBLEM 30 Let p denote $2^k - 1$. The divisors of $2^{k-1} \cdot p$ are the integers 1, 2, $2^2, \ldots, 2^{k-1}$ and $1 \cdot p, 2 \cdot p, 2^2 \cdot p, \ldots, 2^{k-1} \cdot p$. Their sum is

$$(p + 1)(1 + 2 + 2^2 + \ldots + 2^{k-1}) = (p + 1)(2^k - 1) = 2^k \cdot p$$

Hence the total sum *excluding* $2^{k-1} \cdot p$ itself is

$$2^k \cdot p - 2^{k-1} \cdot p = 2^{k-1} \cdot p$$

that is, the given integer.

The first five perfect numbers correspond to $p = 2, 3, 5, 7, 13$. ($2^{11} - 1$ is not prime.)

PROBLEM 31 Let $S(n)$ be the statement: each integer from 1 to n inclusive is expressible as a sum of distinct powers of 2.

Clearly $S(1)$ is true, since $1 = 2^0$. Now suppose that $S(k)$ is true and consider the integer $k + 1$. Let t be such that $2^{t+1} > k + 1 \geq 2^t$. Now consider the integer $K = k + 1 - 2^t$ (so $K < 2^t$, by choice of t). By the induction assumption, K is expressible as a sum of distinct powers of 2, none exceeding 2^{t-1} (since $K < 2^t$). Hence $k + 1 \, (= 2^t + K)$ is expressible as a sum of 2^t and distinct powers of 2 none of which exceeds 2^{t-1}. Hence $S(k + 1)$ is true, and the claimed result is true by mathematical induction.

PROBLEM 32   We prove

$$(x + y)^n = \binom{n}{0} x^n + \binom{n}{1} x^{n-1} y + \dots + \binom{n}{i} x^{n-i} y^i + \dots + \binom{n}{n-1} x y^{n-1} + \binom{n}{n} y^n$$

by induction on n. For $n = 1$ the claimed result is trivially true. Now suppose that the claimed result is true for $n = k$ and consider

$$(x + y)^{k+1} = (x + y)^k \cdot (x + y)$$

The 'general term' $c \cdot x^{(k+1)-i} y^i$ in $(x + y)^{k+1}$ is obtained from the two terms

$$\binom{k}{i} x^{k-i} y^i \cdot x \quad \text{and} \quad \binom{k}{i-1} x^{(k+1)-i} y^{i-1} \cdot y$$

of the product $(x + y)^k \cdot (x + y)$. This means that

$$c = \binom{k}{i} + \binom{k}{i-1}$$

$$= \frac{k!}{i!(k-i)!} + \frac{k!}{(i-1)!(k-i+1)!}$$

$$= \frac{k!}{(k-i)!(i-1)!} \left\{ \frac{1}{i} + \frac{1}{k-i+1} \right\} = \frac{k!}{(k-i)!(i-1)!} \left\{ \frac{k+1}{i(k-i+1)} \right\}$$

$$= \frac{(k+1)!}{i!(k-i+1)!}$$

$$= \binom{k+1}{i}$$

This shows that the formula is true for the case $n = k + 1$ and completes the proof by induction.

PROBLEM 33   If $x = 2$ then 204 is not prime, and if $x = 3$ then $209 = 11 \cdot 19$ is not prime. If x is a prime greater than 3, then x is of the form $6k - 1$ or $6k + 1$. Then

$$x^2 + 200 = 200 + 6(6k^2 \pm 2k) + 1$$

Since $201 = 3 \cdot 67$ we see that $x^2 + 200$ is then a multiple of 3.

PROBLEM 34

$$25\,273 = 3\cdot 7363 + 3184$$
$$7363 = 2\cdot 3184 + 995$$
$$3184 = 3\cdot 995 + 199$$
$$995 = 5\cdot 199 + 0 \quad \text{Hence } (7363, 25\,273) = 199$$

We may then write

$$
\begin{aligned}
199 &= 3184 - 3\cdot 995\\
&= 3184 - 3\cdot(7363 - 2\cdot 3184) = 7\cdot 3184 - 3\cdot 7363\\
&= 7\cdot(25\,273 - 3\cdot 7363) - 3\cdot 7363 = 7\cdot 25\,273 - 24\cdot 7363 \tag{II}
\end{aligned}
$$

Here $s = 7$ and $t = -24$. Subtracting and then adding $7363\cdot 25\,273$ to (II) gives

$$199 = -7356\cdot 25\,273 + 25\,249\cdot 7363$$

with $t = 25\,249$ being positive, as required.

To show that 199 *cannot* be written in the form $25\,273s + 7363t$ with $0 < t < 100$ suppose

$$199 = 25\,273s + 7363t = 7\cdot 25\,273 - 24\cdot 7363$$

We deduce

$$7363\cdot (t + 24) = 25\,273\cdot (7 - s)$$

Dividing through by 199 gives

$$37\cdot (t + 24) = 127\cdot (7 - s)$$

Since 127 is prime and $127 \nmid 37$ we have $127 \mid (t + 24)$. Hence $t + 24 = 127m$, that is $t = 127m - 24$, m being an integer. Putting $m = 0$ and $m = 1$ gives $t = -24$ and $t = 103$ respectively. Thus we cannot write 199 in the form $25\,273s + 7363t$ where $0 \leqslant t \leqslant 100$.

PROBLEM 35   We prove the result for each integer m by using mathematical induction on n. (Recall $m, n \in \mathbb{Z}^+$.) Given m we must first check that

$$f(m + 1) = f(m - 1)\cdot f(1) + f(m)\cdot f(2)$$

which is correct since

$$f(1) = f(2) = 1 \quad \text{and} \quad f(m + 1) = f(m) + f(m - 1)$$

Now, by definition,

$$f(m + \{k + 1\}) = f(m + k) + f(m + \{k - 1\})$$

and by the induction assumption we have

$$f(m + k) = f(m - 1)f(k) + f(m)f(k + 1)$$

and

$$f(m + \{k - 1\}) = f(m - 1)f(k - 1) + f(m)f(k)$$

Adding these we obtain

$$f(m-1)[f(k)+f(k-1)]+f(m)[f(k+1)+f(k)]$$
$$= f(m-1)f(k+1)+f(m)f(k+2)$$

This completes the proof.

PROBLEM 36 For each $n \in \mathbb{Z}^+$ we prove, by mathematical induction on t, that, for $t \in \mathbb{Z}^+$, $f(n)|f(tn)$. Clearly this result holds if $t = 1$. Suppose the result proved for $t = k$. Then, by Problem 35

$$f(kn+n) = f(kn-1)f(n)+f(kn)f(n+1)$$

This shows immediately that $f(n)|f(kn+n)$ since $f(n)|f(n)$ and $f(n)|f(kn)$.

PROBLEM 37 Let the nth term of the sequence be denoted by $g(n)$. Let $S(n)$ be the statement: $g(n)$ is not a multiple of 9. Then $S(1)$ is clearly true. Suppose that $S(k)$ is true. Working modulo 9 we see that the equality $g(k+1) = g(k) + \{$sum of digits of $g(k)\}$ implies the congruence $g(k+1) \equiv 2g(k) \pmod 9$ (since $g(k) \equiv \{$sum of digits of $g(k)\} \pmod 9$). Hence $g(k+1) \not\equiv 0 \pmod 9$ since $g(k) \not\equiv 0 \pmod 9$. In other words $9 \nmid g(k+1)$.

PROBLEM 38 Every sequence of eight successive integers includes exactly one pair of integers of the form $6k-1$, $6k+1$. Hence for each $n \in \mathbb{Z}^+$ with $n \geqslant 2$, the sequence $n!+2, n!+3, \ldots, n!+9$ includes exactly one pair of integers of the required form both of which are composite.

PROBLEM 39 Let $r = \sqrt{2} + \sqrt{3} + \sqrt{5}$. Then $r - \sqrt{2} = \sqrt{3} + \sqrt{5}$. Consequently

$$r^2 - 2\sqrt{2}r + 2 = 3 + 5 + 2\sqrt{15}$$

Hence $r^2 - 6 - 2\sqrt{2}r = 2\sqrt{15}$ and so, squaring up

$$(r^2-6)^2 - 4\sqrt{2}r(r^2-6) + 8r^2 = 60$$

If r *were* rational then the previous equality would show (in the usual way) that $\sqrt{2}$ is rational. This contradiction shows that r is *not* rational.

PROBLEM 40 A solution $(a/b)^2 - 2(u/v)^2 = 5$ in $\mathbb{Q}$ would tidy up to

$$(av)^2 - 2(ub)^2 = 5(bv)^2$$

Working (mod 5) and trying all cases $av \equiv 0, 1, 2, 3, 4$ with all cases $ub \equiv 0, 1, 2, 3, 4$, we find that only the case $av \equiv 0$ *and* $ub \equiv 0$ are possibilities. Hence 5 divides a or v and 5 divides u or b. In both cases 5^2 divides $(av)^2$ *and* $(ub)^2$ and so $5|(bv)^2$. Therefore $5|bv$ – and hence $5|b$ or $5|v$. Now if, say, $5|b$ then we may assume that $5 \nmid a$ (why?). Hence $5|v$. (Likewise $5|v$ implies that $5|b$.) Writing $V = v/5$ and $B = b/5$ we find,

$$\left(\frac{a}{B}\right)^2 - 2\left(\frac{u}{V}\right)^2 = 5^3$$

An identical argument now proves that B and V are multiples of 5 and that b and v are multiples of 25. Continuing in this manner we see that b and v are divisible by arbitrarily large powers of 5, an obvious absurdity. Hence the given equation can have no rational solution.

PROBLEM 41   Let $x = 2.345\,678\,910\,111\,213...$ and $y = 1.234\,567\,891\,011\,121....$
Then

$$10 = 10y - x = 9y + (y - x)$$

Hence if $y - x$ were rational so too would be $9y$ and hence y itself. Since this is clearly not the case, we infer that $y - x$ must be irrational. If $x + y$ were rational then so would be

$$(x + y) + 10 = (x + y) + (10y - x) = 11y$$

This, again, is impossible.

The next five questions were included to see if you spotted that the hypotheses are too few or too many or just too unclear to solve the problem completely.

PROBLEM 42   An odd number of balls in each of 12 bags totals 45? Impossible because a sum of 12 odd numbers is even. However, who said that the bags are all separate? You can solve the problem if you put 45 balls in bag 1, balls and bag 1 inside bag 2, balls, bag 1 and bag 2 inside bag 3, etc. Surely there is an odd number of balls in each bag?

PROBLEM 43   The sum of the row totals must be equal to the sum of the column totals, since each is just the sum of all the entries in the 4×5 box. Since these two totals are *not* the same, the problem cannot be solved.

PROBLEM 44 $7807 = 37 \cdot 211$ so one naturally answers that he had 211 which he sold for 37 pence each. However, he might have had 780 700 which he sold at '100 for one penny'.

PROBLEM 45   Draw any triangle ABD with AD = BD. Now draw AC so that angle DAC = 39°. (The newspaper apologized the following week for the insufficient hypotheses.)

PROBLEM 46   White sheep *do* eat more than black ones. After all, there are many more of them!

For your computer

PROBLEM 47 $169 = 13^2 = 5^2 + 12^2 = 3^2 + 4^2 + 12^2 = 10^2 + 8^2 + 2^2 + 1^2 = 11^2$
$+ 6^2 + 2^2 + 2^2 + 2^2$. Now if $n \in \mathbb{Z}^+$ and $n > 169$, we may write $n - 169$ as a sum of one, or two, or three or four non-zero squares. Then n is expressible as a sum of *exactly* five non-zero squares by using the appropriate sum of squares for 169. You

can check that 33 is not so expressible and that all integers from 34 to 169 *are* so expressible (by computer if you wish).

PROBLEM 48   You *should* find a counterexample.

PROBLEM 49   There is an example with $t < 100$.

PROBLEM 50   No: 905 is a counterexample. Is it the first?

CHAPTER 15

Hints/Answers to the Exercises

Chapter 1

EXERCISE 1

(a) *Conjecture:* For each (odd) positive integer n, $1 + 3 + 5 + ... + n = \{(n + 1)/2\}^2$.
(c) *Conjecture:* Each even integer $\geqslant 4$ can be written as a sum of two primes.
(e) *Conjecture:* $2^n - 2$ is divisible by n when and only when n is a prime.

Conjecture (a) is true, (e) is false (since $2^{341} - 2$ is a multiple of 341 but 341 is not prime). (c) is *Goldbach's conjecture.*[1] No-one yet knows whether it is true or false (see Guy (1994, p. 105)).

EXERCISE 2   (b) $n = 26$;   (c) f may not even have a derivative at $x = 3$, for example: $f(x) = 0$ if $x = 3$, $f(x) = 1$ if $x \neq 3$.

EXERCISE 4   We do not yet have a proof of $A \rightarrow B$ (unless from some outside source we know I is true) because, to get B we need P (and N) and hence (L and) M and hence I. Since we can deduce L from A, $L \rightarrow I$ will allow us to deduce I from A. To get P we need to know I. Thus, to use P to derive I (in order to deduce P) is engaging in *unacceptable circular argument.*

DOUBTFUL PROOFS

3. If an assumption (here that $-1 = 1$) leads to a correct conclusion (here $1 = 1$) it does not necessarily mean that the assumption is correct. (See the IF/THEN truth table in Chapter 2 (page 27).

[1] Christian Goldbach (18 March 1690–20 November 1764).

5. The logarithm of a negative number can be defined. It turns out to be an infinite collection of complex numbers. Assuming a to be positive, $\log(-a)^2$ would be $2\log(-a)$ *if* $\log(-a)$ *were real*. The 'equality' $2\log(a) = 2\log(-a)$ does *not* show that $\log(-a)$ is real because it depends upon the assumption that $\log(-a^2) = 2\log(-a)$ which requires the logs to be real. A classic piece of circular reasoning!
6. In attempting to give meaning to a sum of *infinitely many numbers* one has to take extreme care. (Courses in mathematical analysis examine this problem.)

FINAL EXERCISES

1. (a) Use
$$-1 \leqslant \sin x \leqslant 1 \quad \text{and} \quad -1 \leqslant \cos x \leqslant 1$$

3. (a), (e) are true; (b), (c), (f), (g) are false; (d) is an unsolved problem.
4. There is a value of x such that all three lines are concurrent. (When the shaded triangle 'vanishes'.)
5. Pythagoras' theorem.
6. They (and several other angles) are equal from the *symmetry of the figure*.
7. Here's a starter: there are infinitely many primes ending in a '1'.

Chapter 2

EXERCISE 1   (a), (b)   F   T
                                              F   T

EXERCISE 2   You cannot tell who is oldest. All you know is that $w = 34$ OR $b = 35$ OR $n = 36$.

EXERCISE 5   (a)

P	Q	$\neg(P \vee Q)$	$\neg P \wedge \neg Q$
T	T	F	F
T	F	F	F
F	T	F	F
F	F	T	T

EXERCISE 6   True (since $2 = 2$ is true, even though $2 < 2$ is false).

EXERCISE 7

(a) *If* 1997 is prime *then* 1997 is not divisible by any prime less than 44 AND *if* 1997 is not divisible by any prime less than 44 *then* 1997 is prime.

EXERCISE 8

(a) True, since the antecedent is false.
(d) False, since $3^2|n^2$ does not necessarily imply $3^2|n$. (For notation $x|y$ see Definition 1.)

EXERCISE 12

(a) Take P to be any true statement, Q to be any false one.

EXERCISE 13

(a) P, Q and R must each be true.
(b) S must be T. It follows that $[\{P \vee Q\} \wedge \{Q \rightarrow R\} \wedge \{(R \vee P) \rightarrow S\}] \rightarrow S$ is a tautology.

EXERCISE 14 $\neg P \rightarrow \neg Q$ is F only for P being F and Q being T. Column (c) is (a) $\wedge$ (b).

EXERCISE 15   (b) Only if. (The father *meant* to say 'If you *don't* finish ... by 6 o'clock (then) you *can't* go.')

EXERCISE 16   (a) is necessary; (c) is necessary and sufficient; (d) is necessary but not sufficient (36 is divisible by 12 but 6 *isn't*).

EXERCISE 17

(a) is correct. (Suppose that XYZ is a triangle with side lengths a, b, c such that $a^2 + b^2 = c^2$ and let UVW be a triangle with sides UV and UW of length a and b meeting in a right-angle at U. Then, by Pythagoras' theorem, VW has length c. Consequently XYZ and UVW are congruent triangles. Hence XYZ is right-angled.)

EXERCISE 19   Each is a pair of equivalent statement forms.

EXERCISE 20   (a) $\vee$; (b) $\wedge$. Exercises 19(a) and 20(a) are sort of 'dual'. So are 19(b) and 20(b).

EXERCISE 22   An integer is *rectangular* iff it is the product of two distinct primes. Another (very silly) definition also fitting the information given would be: an integer is *rectangular* iff it is equal to 6 or to 34 or to 143.

EXERCISE 23   An integer n is an *autocube* iff it is equal to the sum of the cubes of its digits.

EXERCISE 24   An integer is *square-free* iff it is not divisible by the square of any prime.

EXERCISE 25   Of course π is delicious!

EXERCISE 26   What do 'no part', 'length', 'breadth' mean?

EXERCISE 27   See page 268.

EXERCISE 29

(a) 7 divides 3311 and 8526, and, hence their sum.
(b) $m = 2$, $a = b = 1$.

EXERCISE 31   ...*neither is* $2^{11\,213} - 1$.

EXERCISE 33   Yes: In line 4 we are forced to deduce P is T and Q is F. But, then, the second F in line 2 is contradicted.

EXERCISE 34   1. A→B   Given
              2. A∨C   Given
              3. C→¬D   Given
              4. D   Assumed
              5. ¬C   (3), (4), *modus tollens*
              6. A   (5), (2), *modus tollens* equivalence
              7. B   (6), (1), *modus ponens*

EXERCISE 35   1. A→B   Given
              2. B→C   Given
              3. D→¬C   Given
              4. A∧D   Given
              5. A   (4), conjunctive simplification
              6. B   (5), (1), *modus ponens*
              7. C   (6), (2), *modus ponens*
              8. D   (4), conjunctive simplification
              9. ¬C   (8), (3), *modus ponens*
             10. C∧¬C   (7), (9)

EXERCISE 36   Let T, S, I, U, L stand for government reduces Taxes, government increases Spending, Inflation will get worse, Unemployment increases, government will Lose the election. The argument given is

$$(T \wedge S) \to I, \ \neg S \to U, \ (I \vee U) \to L$$

Therefore $(T \vee L)$. Now, if L is false, then so are both I and U. Hence ¬S is false and T∧S is false. Together these latter two imply that S is true, and hence T is false. Thus the assumed falsity of L does *not* imply the truth of T. Consequently the argument is unsound.

EXERCISE 37   If P is F, Q is T and R is F then $(P \to Q) \to R$ is F whilst $P \to (Q \to R)$ is T. Consequently $P \to Q \to R$ is ambiguous – unless, perhaps, you *define* $P \to Q \to R$ to *mean* $(P \to Q) \to R$ (as some people do) – or as $P \to Q$ AND $Q \to R$, as many undergraduates seem to bring with them from school!

Chapter 3

EXERCISE 2

(a) Each digit can be the final digit of a cube.
(b) No.

EXERCISE 3

(a) Squaring $00, 01, 02, ..., 09, 10, 11, ..., 49$ produces integers whose last pair of digits is one of the list $00, 01, 04, 09, 16, 25, 36, 49, 64, 81, 21, 44, 69, 96, 56, 89, 24, 61, 41, 84, 29, 76$. Since each integer n is expressible as $X \cdot 100 \pm a$, where X is an integer and $0 \leqslant a \leqslant 50$, each n^2 will have last two digits as above.

EXERCISE 4   Take n to be one of the forms $15k \pm r$ where $r = 0, 1, 2, ..., 7$. (It is even quicker to use $5k$, $5k \pm 1$ and $5k \pm 2$ to show that $n^2 - 2$ is not divisible by 5 – and hence not by 15.)

EXERCISE 5 15 divides $n^2 + 11$ iff n is of the form $15k \pm 2$ or $15k \pm 7$.

EXERCISE 6   Equally 'clearly' $\sqrt{(181k - 4)}$ is not an integer. And yet, if $k = 8...$.

EXERCISE 7   How do you know that π has the value claimed for it?

EXERCISE 9   The proof only shows that, for each circle, c/d lies somewhere between 3 and 4. For all we know this ratio may vary from circle to circle.

EXERCISE 10   The point O *cannot* lie inside the triangle. (Of course, careful drawings cannot 'prove' this since they may not be careful enough and there are infinitely many of them and lines should have no width, etc.)

EXERCISE 13   Both have solutions since (a) the graphs of $y = \cos x$ and $y = x$ cross somewhere between $x = 0$ and $x = \pi/2$; (b) as t changes from 0 to $\pi/4$, the graph of $y = \cos t - \sin t$ moves from value 1 to 0 and hence crosses the line $y = \frac{1}{2}$.

EXERCISE 14   The primes are $101, 103, 107, 109, 421, 1997$.

EXERCISE 15 $2^{8.5}$ (which is, approximately, 362) and $2^{9.5}$.

EXERCISE 17   For 3(b), suppose that the sum of the primes is at least $125\,000$. The first 28 primes (25 up to 100 and the next three, $101, 103, 107$) add up to at most $25 \cdot 100 + 311$. If n more primes (each less than 998) are required to get the total past $125\,000$ then we clearly need at least $(125\,000 - 2811)/997 \cong 122.6$ of them. Thus we need at least 151 primes.

EXERCISE 18   If n is *not* greater than or equal to 4 then $n = 1, 2$ or 3. Now $f(1) = -102$, $f(2) = -43$, $f(3) = -14$. This does it.

EXERCISE 20   If $2^{67} + 1$ is *not* divisible by 3 then $2^{67} - 1$ is not prime.

EXERCISE 22   You only need to observe that
$$\frac{364 \cdot 363 \cdot ... \cdot 343}{365 \cdot 365 \cdot ... \cdot 365} < \frac{1}{2}$$

Twins may spoil the converse.

EXERCISE 23 ...iff $4|94$.

EXERCISE 25 The system $x + y = 1$ and $2x + 2y = 2$ does not satisfy $ad - bc \neq 0$ and yet it clearly has a solution – indeed infinitely many!

EXERCISE 26 See Figure 15.1 in which the dotted lines represent red string and the solid lines blue.

Chapter 4

EXERCISE 1 Each is false.

EXERCISE 3

(a) $1, 2, 3, \{4, 5\}, \{6, \{7, 8\}\}$;
(d) This set is the empty set $\emptyset$ (cf. part (b)).

EXERCISE 5 For P(x): $x \in \mathbb{Z}^+$ and $x|1234$ and $x|4321$. For Q(x): x is prime and $887 < x < 907$.

EXERCISE 6 For Q(x): x is a set $\{a\}$ comprising exactly one integer. (Note that each such set x is certainly *not* an integer.)

EXERCISE 7 The candidate is claiming (i) the set A is greater than the integer 0 (!!), and (ii) the set A is a real number (!!). (He perhaps *meant* to say that each member of A is greater than zero and $A \subset \mathbb{R}$.)

EXERCISE 9 Yes, it is equal to $\emptyset$.

EXERCISE 10 (b) *Conjecture:* A set with n elements has 2^n subsets.

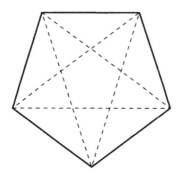

Figure 15.1

EXERCISE 11

(b) $c \notin S, \{c,d\} \in S, \varnothing \notin S, S \notin S, \{c,d\} \not\subseteq S, \{\{c,d\}\} \subset S, \{b,47\} \subset S, \{c,d,47\} \not\subseteq S, S \subseteq S.$

EXERCISE 12

(a) $\varnothing \neq \{\varnothing\}$ and $\varnothing \neq \{\{\varnothing\}\}$ since neither $\{\varnothing\}$ nor $\{\{\varnothing\}\}$ is the empty set. $\{\varnothing\} \neq \{\{\varnothing\}\}$ since these sets have different elements, namely $\varnothing$ and $\{\varnothing\}$.

EXERCISE 13   No! $\{\varnothing\} \in \{\{\varnothing\}\}$ but $\{\varnothing\} \not\subseteq \{\{\varnothing\}\}$.

EXERCISE 14   (a) $\{17 + 20k: k \in \mathbb{Z}\}$; (b) 5, 11, 17, 23, 29, ...

EXERCISE 15

(b) No such sets exist. For all sets A, B we have $A \cap B \subseteq A \cup B$, and yet here $n \in C \cap D$ but $n \notin C \cup D$.

EXERCISE 16   A picture soon convinces you that the equality doesn't *always* hold. A specific example is $A = B = \varnothing$, $C = \mathbb{Z}$ (indeed, any non-empty set).

EXERCISE 17   *Proof*: let $x \in (A \cap B) \cap C$. Then $x \in A \cap B$ AND $x \in C$. Hence $(x \in A$ AND $x \in B)$ AND $x \in C$. We deduce that $x \in A$ AND $(x \in B$ AND $x \in C)$, that is $x \in A \cap (B \cap C)$. This shows that $(A \cap B) \cap C \subseteq A \cap (B \cap C)$.

EXERCISE 19   We prove half of (i) of Exercise 18(b). *Proof*: Let $x \in (A \cup B)^c$. Then $x \notin (A \cup B)$. Hence $x \notin A$ AND $x \notin B$. Hence $x \in A^c$ AND $x \in B^c$. That is, $x \in A^c \cap B^c$. Therefore $(A \cup B)^c \subseteq A^c \cap B^c$.

EXERCISE 20   Figure 4.7 may persuade you that the 'equality' is correct but the diagram is not as general as it should be. In fact, if you relabel C and D as D and C you will see that the 'equality' fails. Indeed, taking $A = \{1\}$; $B = \varnothing$; $C = \varnothing$; $D = \{1\}$ shows the 'equality' to be false in general.

EXERCISE 21

(a) No!
(b) We deduce $A \subseteq B \subseteq C \subseteq A$. (Does this help?)

EXERCISE 22   Yes. $((A \cup B) \cup C)^c = (A \cup B)^c \cap C^c = A^c \cap B^c \cap C^c$ (using Exercise 18(b)(i) twice.

EXERCISE 23

(a) For all real numbers x, $x^2 - 1 = (x - 1)(x + 1)$.

EXERCISE 24 $(\forall x \in \mathbb{Z})(\exists y \in \mathbb{Z})[x < y]$.

EXERCISE 26   There exists an integer y which is greater than every integer x. (This is obviously false.) *Conclusion*: $\forall \exists$ may be different from $\exists \forall$.

EXERCISE 27 (b) $(\forall x \in A)[x \in B] \wedge (\exists y \in B)[y \notin A]$.

EXERCISE 28 $P(x, y)$ is $x + y = x$ will do.

EXERCISE 29 If an integer x is odd then its square is odd. (That is, the square of an odd integer is odd.)

EXERCISE 31

$$(\forall x \in \mathbb{Z})[\{(\exists a \in \mathbb{Z})(\exists b \in \mathbb{Z}) \rightarrow x = a^2 + b^2\} \rightarrow \neg\{(\exists k \in \mathbb{Z})[x = 4k + 3]\}]$$

EXERCISE 32 For every positive real number r there exists a positive integer N such that for every integer n larger than N, $1/n$ is smaller than r. (This is true.)

EXERCISE 33 There exists a prime greater than $2^{1\,257\,787} - 1$.

EXERCISE 34 If, given integers x and y such that $x \geqslant 3$ and $1 < y < x$, you find that $x/y \notin \mathbb{Z}$, then x is prime. (This is false: take $x = 6$ and $y = 4$.)

EXERCISE 36

(a) $(\exists x \in \mathbb{Z})[x^2 - 10x + 26 \leqslant 0$ or $x^4 - x + 2$ is odd]
(c) $(\exists x \in \mathbb{Z}^+)[x$ is prime AND $4^x - 3$ is not prime]. (recall: $\neg(P \rightarrow Q) \equiv \neg(\neg P \vee Q)$
$\equiv P \wedge \neg Q$.)

EXERCISE 38 *Negation:* $(\forall x \in \mathbb{Z}^+)[x = 1$ or x is composite or $x^2 + 200$ is composite]. (Note that '$x = 1$' is redundant here since 201 is composite.)

EXERCISE 41 Suppose $|x| < a$. Two cases: (i) if $x \geqslant 0$ then $(-a <)0 \leqslant x = |x| < a$; (ii) if $x < 0$ then $-a < -|x| = x < 0 < a$. Hence, in both cases, $-a < x < a$. Conversely ... left to you.

EXERCISE 42 Suppose that $5^\alpha | n!$ but $5^{\alpha+1} \nmid n!$. Now show that $\alpha = [n/5] + [n/25] + [n/125] + [n/625] + [n/3125] + \dots$ (Of course, as soon as the denominator exceeds n, the corresponding bracket has value 0.) Then, since $2 < 5$, $2^\alpha | n!$ and so $n!$ will end in exactly α 10s.

EXERCISE 43 (a) Not onto, not $1-1$; (b) $1-1$ but not onto; (c) onto but not $1-1$; (d) onto and $1-1$.

EXERCISE 44 f is onto: for $a \in \mathbb{Z}$, $f(2a + 2) = a$ if $a \geqslant 0$; $f(-2a - 1) = a$ if $a < 0$. f is $1-1$: for if $f(2s) = f(2t)$ then $s - 1 = t - 1$, hence $s = t$. Likewise $f(2s - 1) = f(2t - 1)$ implies $s = t$. Finally $f(2s) = f(2t - 1)$ is impossible (why?).

EXERCISE 45 Just check $f(1) = g(1)$, $f(2) = g(2)$ and $f(3) = g(3)$.

EXERCISE 46 (a)(i) $7 \cdot 7 \cdot 7 \cdot 7$; (b) $7 \cdot 6 \cdot 5 \cdot 4$; (c) 8400 (see Bogart (1983, p. 50)).

EXERCISE 47

(b) Let $(u, v) \in (A \times B) \cap (C \times D)$. Then (i) $u \in A$ and $u \in C$; (ii) $v \in B$ and $v \in D$. Hence $(u, v) \in (A \cap C) \times (B \cap D)$. This argument also 'goes backwards' to prove that, if $(u, v) \in (A \cap C) \times (B \cap D)$ then $(u, v) \in (A \times B) \cap (C \times D)$.

(c) is false. Figure 15.2 gives an idea why.

EXERCISE 48   Five students must study Russian.

Chapter 5

EXERCISE 2   For each $n \in \mathbb{Z}^+$, $n(n + 1) + 2$ is even and greater than 2, hence not prime.

EXERCISE 4 $10\,009$ is prime.

EXERCISE 5

(b) Truth tables show that $\{(A_1 \rightarrow B_1) \wedge (A_2 \rightarrow B_2)\} \rightarrow \{(A_1 \wedge A_2) \rightarrow (B_1 \wedge B_2)\}$ but not conversely.

EXERCISE 6   Look up Fermat's Little Theorem.

EXERCISE 7   Let x and y be real numbers. If $x \neq y$ AND (if) $x \neq -y$ then $x^2 \neq y^2$. *Proof*: Since $x - y \neq 0$ and $x + y \neq 0$ we have (if we assume that the product of two non-zero real numbers is non-zero)

$$x^2 - y^2 = (x - y)(x + y) \neq 0$$

that is $x^2 \neq y^2$.

EXERCISE 9   If $x^2 < 4$ then $-2 < x$ AND $x < 2$ (in brief $-2 < x < 2$). (Note the AND – rather than OR – as in Statement 9(b).)

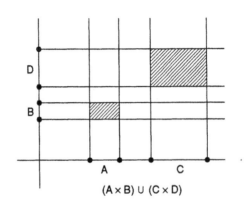

$(A \times B) \cup (C \times D)$

Figure 15.2

EXERCISE 10 (a) Let x be even. Then $x = 2m$ for some suitable integer m. It follows that $x^2 = 4m^2$ which is clearly even.

EXERCISE 11 The statement 'if $x = y$ then $x^2 = y^2$' seems to say that if two numbers are equal then so are their squares. So to 'prove' it by saying 'if you square two equal numbers you get equal numbers' seems to get us nowhere.

EXERCISE 12 Put $y = x(x - 1)$. Then y is even (since one of x, $x - 1$ is). But $x = x^2 - y$. Hence x is odd – since x^2 is odd and y is even.

EXERCISE 13 Odd integers are certainly of the form $2k + 1$. But there is no justification for assuming the odd number x^2 is of the form $4m^2 + 4m + 1$.

EXERCISE 14

(a) The contrapositive version of the statement is: if a and b are both odd then ab is odd.
(c) Use the contrapositive: if x is odd then x^2 is odd.

EXERCISE 15 The assertion about 6571 is *not* obvious, mainly because it is false.

Chapter 6

EXERCISE 1 If $x = 1 + \sqrt{2}$ were rational so, too, would $x - 1$ be rational. But $x - 1 = \sqrt{2}$ and we know that $\sqrt{2}$ is *not* rational.

EXERCISE 2 From $x + \sqrt{2} = \sqrt{3}$ we obtain $(x + \sqrt{2})^2 = 3$, and hence $2\sqrt{(2)}x = 3 - x^2 - 2$ so that

$$\sqrt{2} = \frac{3 - x^2 - 2}{2x}$$

Hence, if x were rational so too would be $\sqrt{2}$. This contradiction shows that x cannot be rational.

EXERCISE 3 ... provided

$$n > \frac{2m + 1}{2 - m^2}$$

EXERCISE 4

(a) The proof that $\sqrt{3}$ is irrational is a word-for-word copy, except that, from $a^2 = 3b^2$ we need to deduce $a = 3m$ and from $3m^2 = b^2$ we need to deduce $3|b$. (To prove that $a^2 = 3b^2$ implies that $3|a$ just show that, if $a = 3k + 1$ or if $a = 3k - 1$, then a^2 is *not* a multiple of 3.)

EXERCISE 5 Suppose $b \neq d$. From $a - c = (d - b)\sqrt{2}$ we obtain $\sqrt{2} = (a - c)/(d - b)$. But this is impossible since $\sqrt{2} \notin \mathbb{Q}$. Hence $b = d$, and $a = c$ follows.

EXERCISE 6 $p(-1) = -14$, $p(0) = 3$, $p(2) = -5$, $p(4) = 11$. Hence there are real roots between -1 and 0, between 0 and 2 and between 2 and 4.

EXERCISE 8 Now

$$h(x) = \cos x + \cos\left(x - \frac{\pi}{2}\right) - \frac{1}{2}$$

$$= 2\cos\left(x - \frac{\pi}{4}\right)\cos\frac{\pi}{4} - \frac{1}{2}$$

But in the range $0 \leqslant x \leqslant \pi/2$

$$\cos\left(x - \frac{\pi}{4}\right) \geqslant \frac{1}{\sqrt{2}}$$

Hence

$$h(x) \geqslant 2 \cdot \frac{1}{\sqrt{2}} \cdot \frac{1}{\sqrt{2}} - \frac{1}{2} = \frac{1}{2} > 0$$

EXERCISE 9 Define $f: \mathbb{R} \to \mathbb{R}$ by $f(x) = -1$ if $x < 0$; $f(x) = 1$ if $x \geqslant 0$. Draw its graph.

EXERCISE 11 No: $2 \cdot 3 \cdot 5 \cdot 7 \cdot 11 \cdot 13 + 1 = 59 \cdot 509$.

EXERCISE 12

(a) None! (For they are, respectively, divisible by $2, 3, 4, \ldots, 100$.) (b) Take $a = (10^6)! + 1$. Then x is composite if $a < x < a + (10^6 + 1)$.

EXERCISE 13 Let p_k be prime and let $p_1, p_2, \ldots, p_k$ be the primes $\leqslant p_k$. Set $N = p_1 \cdot p_2 \cdot \ldots \cdot p_k + 1$. Then N is divisible by a prime distinct from (and hence greater than) p_k.

EXERCISE 14 Now $(6k \pm 1)^2 + 2 = 3(12k^2 \pm 4k + 1)$ is divisible by 3. Thus only $3^2 + 2$ is a *candidate* for primeness – and, of course, it is prime!

EXERCISE 15 $n!$ has n digits if $n = 22, 23, 24$. $25!$ has 26 digits. Clearly, for all larger n, $n!$ has at least $n + 1$ digits (since $n \geqslant 10$).

EXERCISE 16 $t = 91$. It is not the only triple. Indeed, $t = 213$ starts a sequence of five such successive odd numbers.

Chapter 7

Note: After the first solution the proofs are written out quite briefly.

EXERCISE 1

(a) (Cf. Exercise 1, Chapter 1) *Conjecture:* S(n): For each positive integer n, $A(n) = n^2$. *Proof:* $A(1) = 1 = 1^2$. Thus S(1) holds. Now suppose that S(k) holds for some positive integer k. That is, we assume that $A(k) = k^2$. Now

$$A(k + 1) = A(k) + (2\{k + 1\} - 1)$$

Hence

$$A(k + 1) = k^2 + 2k + 1 = (k + 1)^2$$

Thus, if S(k) is true so is S(k + 1). Consequently, by mathematical induction, S(n) is true for all $n \in \mathbb{Z}^+$.

EXERCISE 2 The 'proof' says: 'Assuming that S(k) and S(k + 1) are true then, by combining them, we find

$$\frac{2}{k} \leqslant \frac{2}{2^{k-1}}$$

This is S(k), which we are assuming to be true, therefore S(k + 1) is true.' (But {X AND Y} will *always* imply Y – whether Y is true or not. So we may *not* infer the truth of Y.)

EXERCISE 3

(b) *Fact:* The formula is

$$\left(\frac{2n(2n + 1)}{2}\right)^2 - 2^3\left(\frac{n(n + 1)}{2}\right)^2 = n^2(2n^2 - 1)$$

(Can you see why? No mathematical induction needed! Use the formula of Statement 16.1.)

EXERCISE 5 $A(1) = \frac{1}{2}$, $A(2) = \frac{5}{6}$, $A(3) = \frac{23}{24}$. So I guess that

$$A(n) = 1 - \frac{1}{(n + 1)!}$$

In the proof use

$$A(k) + \frac{k + 1}{(k + 2)!} = \left\{1 - \frac{1}{(k + 1)!}\right\} + \left\{\frac{k + 2}{(k + 2)!} - \frac{1}{(k + 2)!}\right\}$$

EXERCISE 7 One checks easily that $n! > n^{10}$ if $n = 15$. Supposing $k! \geqslant k^{10}$ (and $k \geqslant 15$) we deduce that

$$\left(\frac{k + 1}{k}\right)^{10} \leqslant (1 + \tfrac{1}{15})^{10} < (1.07)^{10} < 2$$

Thus, for $k \geqslant 15$, $(k + 1)! = (k + 1)k! > 2k^{10}$ (why?) $> (k + 1)^{10}$ (why?).

EXERCISE 8   Let $S(n)$ be: $C(n) < 2^{n-1}$. $S(6)$ is easily checked to be true. We prove $C(k + 1) \leqslant 2C(k)$. Now

$$C(k + 1) = \frac{(k + 1)k(k^2 - 3k + 14)}{24} + 1$$

Further

$$(k + 1)k(k^2 - 3k + 14) \leqslant 2k(k - 1)(k^2 - 5k + 18)$$

if (cancelling the common k)

$$k^3 - 2k^2 + 11k + 14 \leqslant 2(k^3 - 6k^2 + 23k - 18)$$

that is, if

$$10k^2 + 50 \leqslant k^3 + 35k$$

Now if $k \geqslant 10$ we have $10k^2 \leqslant k^3$ and $50 < 35k$. Hence the truth of $C(k)$ implies that of $C(k + 1)$ *provided* $k \geqslant 10$. Checking $C(7), C(8), C(9)$ are correct we have our result.

EXERCISE 10   Let $S(n)$ be: no integer > 1 divides both $f(n)$ and $f(n + 1)$. Clearly $S(1)$ is true. Suppose $S(k)$ is true. If the (positive) integer c divides both $f(k + 1)$ and $f(k)$ then c divides

$$f(k + 1) - f(k) = f(k - 1)$$

Thus c divides both $f(k)$ and $f(k - 1)$. The truth of $S(k)$ implies $c = 1$.

EXERCISE 12   Use m.i. Let $S(n)$ be $\sum_{t=1}^{n} f(t)^2 = f(n)f(n + 1)$. Since $f(1)^2 = f(1)f(2)$ we see that $S(1)$ is true. Now suppose $S(k)$ is true and consider

$$\sum_{t=1}^{k+1} f(t)^2 = \sum_{t=1}^{k} f(t) + \{f(k + 1)^2\} = f(k)f(k + 1) + \{f(k + 1)^2\}$$

$$= f(k + 1) \cdot \{f(k) + f(k + 1)\} = f(k + 1) \cdot f(k + 2)$$

EXERCISE 14   Let $h > -1$. Let $S(n)$ be: for $n \in \mathbb{Z}^+$, $(1 + h)^n \geqslant 1 + nh$. Clearly $S(1)$ is true. Suppose $S(k)$ is true and consider $(1 + h)^{k+1}$. Now

$$(1 + h)^{k+1} = (1 + h)^k (1 + h)$$
$$\geqslant (1 + nh)(1 + h)$$
$$= 1 + \{n + 1\}h + nh^2$$
$$\geqslant 1 + \{n + 1\}h \quad \text{(why?)}$$

This suffices. (Where do we use the condition $h > -1$?)

EXERCISE 17   We use m.i., but note the different style. The result claimed holds for $n = 1$ since $1 \leqslant h(1) = 1 < 2$. Now suppose $1 \leqslant h(k) < 2$. Then $\frac{1}{2} < 1/h(k) \leqslant 1$. Now, from the given formula

$$h(k + 1) = \tfrac{1}{2}h(k) + \frac{1}{h(k)}$$

Hence $h(k + 1)$ lies between $\frac{1}{2} \cdot 1 + \frac{1}{2}$ and $\frac{1}{2} \cdot 2 + 1$, that is, $1 < h(k + 1) < 2$. In particular $1 \leqslant h(k + 1) < 2$.

EXERCISE 19   In each case it seems difficult to perform the induction step (II).

THE INVALID ARGUMENT

$S(1)$ is indeed true. But the 'proof' of the induction step $S(k) \rightarrow S(k + 1)$ fails in the case $k = 1$ (that is when there are two people in the room). There is no-one who remains in the room when both persons 1 and $k + 1$ are asked in turn to leave, and so no-one with whom 1 and $k + 1$ share a common height.

Chapter 8

EXERCISE 1   7. The section Going Backwards in this chapter will help you to see why.

EXERCISE 2 $\frac{3}{2} = \frac{6}{5} \cdot \frac{5}{4} = \frac{9}{8} \cdot \frac{8}{7} \cdot \frac{7}{6} = \frac{12}{11} \cdot \frac{11}{10} \cdot \frac{10}{9} \cdot \frac{9}{8} = \cdots$

EXERCISE 5   From $(a =) mb + r = m_1 b + r_1$ we obtain

$$(r_1 - r) = b(m - m_1)$$

which is a multiple of b. Show that

$$-|b| < r_1 - r < |b|$$

Deduce $r_1 - r = 0 \cdot |b|$. Thus $r = r_1$ and hence $m = m_1$.

EXERCISE 7   If $u = (a, b)$ and $v = (-a, b)$ then $v | a$ (since $v | -a$) and $v | b$. Hence v is a common divisor of a and b and, consequently, $v \leqslant u$. Similarly, one easily proves that $u \leqslant v$. Hence $u = v$. All the other equalities are proved likewise.

EXERCISE 8

(b) In the range $1, 2, \ldots, p^2$ the integers *not* coprime to p are precisely $p, 2p, \ldots, (p - 1)p$, p^2, that is, p of them in total. Hence $\phi(p) = p^2 - p$.

EXERCISE 9 $(312\,497, 453\,919) = 2281$.

EXERCISE 10   (a) No; (b) No. (Find your own counterexamples.)

EXERCISE 11

(a) $5 \cdot 649 - 21 \cdot 154 = 11$. Hence $10 \cdot 649 - 42 \cdot 154 = 22$.
(b) If there were such integers z and t then we should have

$$649z + 154t = 11 \cdot (59z + 14t)$$

which cannot be equal to 23 (why not?).

EXERCISE 12

(b)

$$43 = 38\,008 \cdot 39\,001 - 38\,865 \cdot 38\,141$$

is one way.

EXERCISE 13   If $(a/d, b/d) = x$ then $a/d = mx$ and $b/d = nx$ for suitable $m, n \in \mathbb{Z}$. Multiplying up we see that this implies that xd divides both a and b. But this is impossible unless $x = 1$. However $(a, b/d) > 1$ is possible. Find an example.

EXERCISE 14   Let $(a, b) = d$. If ($\leftarrow$) assuming $dr = c$, $as + bt = d$ implies that

$$a(sr) + b(tr) = c$$

only if ($\rightarrow$): from $d|a$ and $d|b$ we deduce $d|(ax + by)$, i.e. $d|c$.

EXERCISE 15   If d is positive and a *multiple* of every common divisor of a and b then clearly d is numerically the greatest of all the common divisors of a and b. Conversely, if d is *numerically* the gcd of a and b, then d is a *multiple* of every common divisor c, say. For, by Theorem 5, $d = sa + tb$ for suitable $s, t \in \mathbb{Z}$. But then, since $c|a$ and $c|b$ we see, immediately, that $c|d$, as claimed.

EXERCISE 16   If $p|nn$ then from Theorem 6, either $p|n$ or $p|n$.

EXERCISE 17

(b) Given that $(n, a) = 1$ there exist $r, s \in \mathbb{Z}$ such that

$$rn + sa = 1$$

Hence

$$rnb + sab = b$$

Since $n|n$ and $n|ab$ we deduce that $n|b$.

EXERCISE 18   If p (prime) divides c and ab, then either $p|(c, a)$ or $p|(c, b)$.

EXERCISE 20   For $n = 1033$, $n(n + 1)(n + 2)$ is divisible by 5, but not by 5^2.

EXERCISE 21   From $p|a^2 = aa$ we deduce that either $p|a$ or $p|a$. Then, trivially, $p^2|a^2$. Hence $p^2|pb^2$ from which we deduce $p|b^2$ and – as above – $p|b$. The proof that $\sqrt{p}$ is not rational follows that of $\sqrt{2} \notin \mathbb{Q}$ with each 2 being replaced by p.

EXERCISE 22   Let p^α ($\alpha > 0$) be the greatest power of the prime p which divides a. Then $p|u$ and so, if p^β is the greatest power of p dividing u then $p^{2\beta}$ is the exact power of p dividing u^2. Since $p \nmid b$, we deduce that $p^\alpha = p^{2\beta}$. Thus $\alpha = 2\beta$ is even. As this is true for each prime dividing a we deduce that a is a perfect square – and likewise for b.

EXERCISE 23

(a) Factorizing a as $r_1^{\alpha_1} r_2^{\alpha_2} \ldots r_k^{\alpha_k}$ where the r_i are distinct primes we see that one of the r_i must be p and a different r_i must be q. Hence $pq|a$.
(b) (i) Take $p = 4$, $q = 6$ and $a = 12$. (ii) Take $p = q = 2$ and $a = 2$.

EXERCISE 24   The error is in assuming the *existence* of p. For example, if $a = 1$ and $b = 0$, then $a + b|ab$ but there is no prime dividing $a + b$.

EXERCISE 25   By Theorem 9 each *potential* rational solution comes from the set $\{1, -1, 2, -2\}$.

EXERCISE 26   49 is an H-prime because, apart from $a = 1$, $b = 49$ or $a = 49$, $b = 1$, we cannot express 49 as a product ab with a and b in H. 45 is *not* an H-prime because $45 = 5 \cdot 9$ (and 5 and 9 *are* in H). For the first factorization: take $a = 9$, $b = 49$, $c = 21$.

EXERCISE 27   One of two or three 'correct' orderings is (e), (f), (g), (d), (b), (i), (j), (a), (h), (c).

EXERCISE 28   (a) $x = 5$;   (b) $\{5 + 7k: k \in \mathbb{Z}\}$

EXERCISE 29

(a) $5 \cdot 9 \equiv 1$
(b) $5 \cdot (3 \cdot 9) \equiv 3$ (using (a))
(e) No solution. (For any such x we would have $6x = 7 + 8m$ for suitable $m \in \mathbb{Z}$. But 7 is not even.)
(g) Requires us to find x and m such that $17x - 73m = 1$. (If the x you obtain is outside the range 0–72 you may adjust it by adding/subtracting multiples of 73.)
(h) From $51x \equiv 3$ (mod 219) we deduce that $17x \equiv 1$ (mod 73) (since $51x = 3 + 219m \longleftrightarrow 17x = 1 + 73m$). By (g), $x = 43$. Hence the full set of solutions is $\{43 + 73k: k \in \mathbb{Z}\}$, and so the solutions we want are 43, 116, 189. (Note that there are three of them – and we divided by 3. Is this a coincidence?)

EXERCISE 30   'modulus'.

EXERCISE 33   Note that $10^1 \equiv -1$ (mod 11), $10^2 \equiv +1$ (mod 11), $10^3 \equiv -1$ (mod 11).... This explains the presence of the alternating $+/-$ signs.

EXERCISE 34   Since $N = 1000 \cdot s + a_2 \cdot 10^2 + a_1 \cdot 10 + a_0$ (for suitable integer s) and since $8|1000$, therefore $8|N$ iff $8|a_2 \cdot 10^2 + a_1 \cdot 10 + a_0$.

EXERCISE 35   Working mod 10, the product is congruent to $7 \cdot 7$ (i.e. $\equiv 9$).

EXERCISE 36   Any such sum as the one given, must, modulo 9, be congruent to $0 + 1 + 2 + \ldots + 9$ which is congruent to 0 (mod 9). But $100 \not\equiv 0$ (mod 9).

EXERCISE 37 (b) $2^{322} = 2^{256} \cdot 2^{64} \cdot 2^2$. Working modulo 323 we find $2^4 \equiv 16$, $2^8 \equiv (16)^2 \equiv -67$, $2^{16} \equiv (-67)^2 \equiv -33$, and, likewise, $2^{64} \equiv 188$, $2^{256} \equiv 35$. Hence $2^{322} \equiv 35 \cdot 188 \cdot 4 \equiv 157 \not\equiv 1$. Hence 323 is not prime.

Chapter 9

EXERCISE 1

(a) My gas-driven hand-calculator says $\frac{1}{17}$'$=$'$0.058\,823\,5$ and that $0.058\,823\,5 \cdot 17 = 0.999\,999\,5$. We therefore now need to know the result of dividing 17 into $0.000\,000\,5$, so find $\frac{5}{17} = 0.294\,117\,6$ and then $0.294\,117\,6 \cdot 17 = 4.999\,999\,2$. Next find $\frac{8}{17}$.

(b) The repeating length of $1/n$ is at most $n - 1$ since on division of an integer by n only $n - 1$ non-zero remainders are possible.

(d) $x = 167\,341\,645\,157/2\,283\,105\,000$ (in lowest terms).

EXERCISE 3

(a) Let a denote some rational number and b and c irrational numbers. Then $a + b$ must be irrational since, if $a + b = d$ were rational, so too would be $b = (a - d)$, a contradiction.

(b) $b + c$ may well be rational, as, for example, when b is irrational and $c = -b$.

(c) I used *reductio*.

EXERCISE 5 $0 < \frac{-1}{-2} < \frac{3}{4}$ but $\frac{2}{2}$ doesn't lie between $\frac{-1}{-2}$ and $\frac{3}{4}$.

EXERCISE 7

(a) Since $0 < 1/n \leqslant 1$ and $0 \leqslant 1 - 1/n < 1$, then

$$0 \leqslant \left(\frac{1}{n}\right)\left(1 - \frac{1}{n}\right) \leqslant 1$$

Hence 0 (and therefore every negative real number) is a lower bound; 1 and every greater real number is an upper bound.

EXERCISE 8 $\mathbb{Z}$.

EXERCISE 9   (b) 1 is lub, -1 is glb.

EXERCISE 10   (b) no lub; glb 1 (in); (c) lub $2/\sqrt{2}$ (in); glb -1 (not in).

EXERCISE 11 $\{x \in \mathbb{R}; \pi < x < \frac{355}{113}\}$.[2]

EXERCISE 12   Suppose X and Y are lubs for the set S. Then (since X is an upper bound and since Y is a least upper bound) we have $Y \leqslant X$. A symmetrical argument shows that $X \leqslant Y$. From these two inequalities $X = Y$ follows.

---

[2] $355/113$ was a remarkable approximation to π discovered by the Chinese Tsu Chung-Chi (430–501).

EXERCISE 13   lub A.

EXERCISE 14   Let $-B$ denote the set $\{-b: b \in B\}$. Then $-B$ is bounded *above* (since, if $M < b$ for each $b \in B$ then $-M > -b$ for each $-b \in -B$). But, by the completeness axiom, $-B$ has a lub L, say. Hence $-L$ is a glb for B. (This seems reasonable – can you supply a proof for the doubters?)

EXERCISE 15   In the proof of Theorem 2, replace the 2 under the $\sqrt{2}$ by 3, 5, 6, 7, etc.

EXERCISE 16 $(\forall \alpha \in \mathbb{R}^+)(\exists n \in \mathbb{Z}^+)[0 < 1/n < \alpha]$.

EXERCISE 17 $AP \rightarrow AP'$: Given $\alpha > 0$ the AP tells us that $\exists n \in \mathbb{Z}^+$ such that $1/n < 1/\alpha$ (since $1/\alpha \in \mathbb{R}^+$). From this follows $\alpha < n$.
   $AP' \rightarrow AP$: Given $\alpha > 0$, AP' tells us that $\exists n \in \mathbb{Z}^+$ such that $1/\alpha < n$ (since $1/\alpha \in \mathbb{R}^+$). From this follows $(0 <) 1/n < \alpha$.

EXERCISE 18

(a) Choose $n \in \mathbb{Z}^+$ such that $1/n < b - a$ as in Theorem 3(i). Then replace $1/n$ by $1/n\sqrt{2}$.

EXERCISE 19   To show $\mathbb{Q}$ fails to satisfy CP, use $\sqrt{2}$.

EXERCISE 20

(a) Think of:

$$\begin{array}{ccccccc} 1 & 2 & 3 & 4 & 5 & 6... \\ 1 & -1 & 3 & -3 & 5 & -5... \end{array}$$

   i.e. $f(2n - 1) = 2n - 1$, $f(2n) = -(2n - 1)$.
(b) The pairing

$$\begin{array}{cccccc} 1 & 2 & 3 & 4 & 5 & 6... \\ 1 & 2 & 4 & 5 & 7 & 8... \end{array}$$

   suggests each $2n$ 'pairs with' $3n - 1$. Take f to be defined by: for all $n \in \mathbb{Z}^+$: $f(2n) = 3n - 1$, $f(2n - 1) = 3n - 2$.

EXERCISE 23

(a) Let r be a real number. By Theorem 3(i) there is a rational number s, say, between r and $r - h$. (Indeed there are infinitely many!) Then r lies inside the circle of radius h centred on s.
(c) You can 'see' to infinity along every line out of the origin which has irrational slope. (Why does no such line meet an 'integer point' (a, b)?) Now make sure that the trunk of each tree is so small that the trunk does not meet this line. (The idea is similar to that in (b). If all the trees have one and the same thickness then, no matter how small this thickness, you cannot see to infinity. This seems to be quite hard. Send me your solutions, if you wish!)

Chapter 10

EXERCISE 2 $\dfrac{-15 + 23i}{13}$. *Guess*: $\dfrac{-15 - 23i}{13}$ (Change 'all' i to $(-i)$.)

EXERCISE 3   Once you have decided which x root to take, the y root is forced on you since you know that $2xy = 1$.

EXERCISE 4 $2 - i$ is one (what is the other?).

EXERCISE 5   The polynomial factorizes as $(z^2 - 14z + 53)(z^2 - 7z - 78)$.

EXERCISE 6   Nowhere! The proof tells us that if 3 is a root then 3 is a root – which is *not* a surprise. This doesn't imply $(x - 3)^2$ is a factor.

EXERCISE 7   Because Theorem 1 *depends* on the Fundamental Theorem of Algebra. So any use of Theorem 1 to prove FTA would be circular.

EXERCISE 8   *Note*: In the polar form my choice of angle is only one of infinitely many possibilities.   (b) 3, π, $3(\operatorname{cis}\pi)$;   (c) 1, $\vartheta + 2n\pi$, $\operatorname{cis}\vartheta$ (where n is chosen so that $-\pi < \vartheta + 2n\pi \leqslant \pi$);   (e) 1, $\frac{\pi}{2}$, $1(\operatorname{cis}\frac{\pi}{2})$;   (h) $\sqrt{2}$, $-\frac{\pi}{4}$, $\sqrt{2}(\operatorname{cis}(-\frac{\pi}{4}))$;   (i) 5, ϑ (where $\cos\vartheta = \frac{3}{5}$, $\sin\vartheta = -\frac{4}{5}$), $\operatorname{cis}\vartheta$;   (k) the given complex number must first be put in $\cos\vartheta + i\sin\vartheta$ form.

EXERCISE 9   (a) $\frac{\sqrt{3}}{2} + \frac{1}{2}i$;   (c) -4.

EXERCISE 10

(f) holds because $x \leqslant +\sqrt{(x^2 + y^2)}$.

EXERCISE 11

(b) Set $X = \alpha - \beta$ and $Y = \beta - \gamma$ and use $|X + Y| \leqslant |X| + |Y|$ again.

EXERCISE 13

(a) $(\cos\frac{3\pi}{28} + i\sin\frac{3\pi}{28})^7 = \cos\frac{3\pi}{4} + i\sin\frac{3\pi}{4} = \frac{-1}{\sqrt{2}} + \frac{i}{\sqrt{2}}$
(c) The answer is *not* $\sin\frac{11\pi}{4} + i\cos\frac{11\pi}{4}$ (why not?)
(d) Write $\cos\frac{2\pi}{3} - i\sin\frac{2\pi}{3}$ as $\cos(-\frac{2\pi}{3}) + i\sin(-\frac{2\pi}{3})$
(e) $(\sqrt{3} + i)^{201} = \{2(\cos\frac{\pi}{6} + i\sin\frac{\pi}{6})\}^{201} = 2^{201}(-i)$

EXERCISE 14

(c) $z^3 = 2i = 2(\operatorname{cis}\frac{\pi}{2})$. Hence $z = \sqrt[3]{2}\operatorname{cis}\left\{\dfrac{\pi/2 + 2k\pi}{3}\right\}$ $(k = 0, 1, 2)$
(f) $(z^2)^2 + 2z^2 + 4$ has roots

$$z^2 = \frac{-2 \pm \sqrt{(4 - 16)}}{2} = -1 \pm i\sqrt{3}$$

EXERCISE 15

(c) -1 has argument π so the six roots are the vertices of a regular hexagon, centre the origin and radius 1, one vertex having polar coordinates $1 \cdot \text{cis}\,\frac{\pi}{6}$ (that is, Cartesian coordinates $(\cos\frac{\pi}{6}, \sin\frac{\pi}{6})$).

EXERCISE 17

(a) $\sin 3\vartheta$ is the imaginary part of

$$(\cos\vartheta + i\sin\vartheta)^3 = c^3 + 3c^2 is + 3c(is)^2 + (is)^3$$

This imaginary part is

$$3c^2 s - s^3 = 3(1 - s^2)s - s^3 = 3s - 4s^3$$

(c) No, since for all ϑ, $\sin 4\vartheta = -\sin(-4\vartheta)$, but $\cos m\vartheta = \cos(-m\vartheta)$

EXERCISE 18   If $\vartheta = 36°$ then $5c^4 - 10c^2 s^2 + s^4 = 0$ (since $\sin 5\vartheta = 0$). Hence

$$16c^4 - 12c^2 + 1 = 0$$

giving

$$c^2 = \frac{3 \pm \sqrt{5}}{8}$$

Now, $\cos 36° > \frac{1}{2}$ and so $c^2 = (3 + \sqrt{5})/8$. Hence

$$\cos 72° = 2\cos^2 36° - 1 = \frac{-1 + \sqrt{5}}{4}$$

Likewise, $\cos 144° = (-1 - \sqrt{5})/4$. The sine values can be worked out from the identity $\cos^2\vartheta + \sin^2\vartheta = 1$.

EXERCISE 19

(a)

$$\sin^4\vartheta = \left\{\frac{1}{2i}(v - v^{-1})\right\}^4 = \tfrac{1}{16}\{(v^4 + v^{-4}) - 4(v^2 + v^{-2}) + 6\}$$

$$= \tfrac{1}{8}\cos 4\vartheta - \tfrac{1}{2}\cos 2\vartheta + \tfrac{3}{8}$$

Chapter 11

EXERCISE 1   *Intelligent(?) guess*: The men were loaned £900 and £600 respectively.

EXERCISE 2   (b) (v) [(iv) isn't good enough. Try it!]; (c) None: but (iii) + (v) will do nicely.

EXERCISE 3   *By false position*: Assume A gets $2 (rather than $1 so as to avoid fractions if possible). Then B gets $9 and C receives $11. Therefore $2 for A means $22 for A, B, C together. Since $1100/22 = 50$, A must receive $50 \cdot (\$2)$, i.e. $100.

EXERCISE 4   For large n, n^3 is approximately $125\,000\,000 = 500^3$. (Also $n^3 + 125n$ ends in a 6 so n may end in a 1 or a 6 but not a $2, 3, 4, 5, 7, 8, 9, 0$.)

EXERCISE 5   Set $f(x) = x^3 - 9x + 9$. Then $f(2) = -1$ and $f(3) = 9$. (Since $f(2)$ is nearer to 0 than is $f(3)$ you might risk trying $f(2.2)$.) If $f(2.2) > 0$ then the root lies between 2 and 2.2; if $f(2.2) < 0$ the root will lie between 2.2 and 3. Continuing refining in this way you should find a root near 2.227.

EXERCISE 6   A first guess (it seems that the number of zeros in $n!$ is roughly $n/4$) shows[3]

$$\left[\frac{492}{5}\right] + \left[\frac{492}{25}\right] + \left[\frac{492}{125}\right] + \left[\frac{492}{625}\right] + \ldots = 98 + 19 + 3 + 0 + \ldots = 120$$

We then quickly see that 495! will end in 121 zeros and 500! will end in 124.

EXERCISE 7   My conjecture, based on a few 'small' trials, would be 199 tears. Each tear reduces a rectangular sheet to a greater number of *smaller rectangular sheets*. So we try the principle of mathematical induction. I claim that the following statement is true. S(n): If n stamps form a rectangular block then the number of tears required to separate the stamps is $n - 1$.

Proof: S(2) is clearly true. Suppose k stamps are given in a rectangular block, $r \cdot s$, say. Tearing the given block into two blocks of size $t \cdot s$ and $(r - t) \cdot s$ the induction hypothesis tells us that $(t \cdot s) - 1 + \{(r - t) \cdot s\} - 1$ further tears will separate all the stamps. Therefore $(t \cdot s) - 1 + \{(r - t) \cdot s\} - 1 + 1$ tears (that is $n - 1$) are necessary to separate the stamps. (Notice that this is independent of the way of tearing.)

EXERCISE 8   They remind me of the coefficients in the binomial expansion of $(1 + x)^n$. So my *guess* is

$$\frac{d^5 f}{dx^5} = \frac{d^5 g}{dx^5} + 5\frac{d^4 g}{dx^4} \cdot \frac{dh}{dx} + 10\frac{d^3 g}{dx^3} \cdot \frac{d^2 h}{dx^2} + 10\frac{d^2 g}{dx^2} \cdot \frac{d^3 h}{dx^3} + 5\frac{dg}{dx} \cdot \frac{d^4 h}{dx^4} + \frac{d^5 h}{dx^5}$$

EXERCISE 9   I'd guess 'no harm', but I'd expect to deduce that $a = 0$.

EXERCISE 10   Since the relationship $L_n = L_{n-1} + L_{n-2}$ is exactly that of the Fibonacci sequence we have

$$L_n = c_1 \alpha_1^n + c_2 \alpha_2^n$$

again and we only have to find c_1 and c_2 from the initial conditions, namely $L_1 = 1$ $L_2 = 3$.

EXERCISE 11   (i) can't be right. (Try $a = b = c = 1$.)

EXERCISE 12   *Method?* Mathematical induction (on the number of factors). Let S(n) be: If n integers, each of the form $4t + 1$ are multiplied together, then the result is (again) an integer of the form $4t + 1$.

---

[3] We choose 492 since $492 = 4 \cdot 123$.

EXERCISE 13 The main reason for failure is that the analogue of '$(4u + 1)(4v + 1)$ is of $4t + 1$ form', namely '$(4u + 3)(4v + 3)$ is of the form $4t + 3$' is *false*.

EXERCISE 16 Because the degree of $f(x)$ may be less than $m - 1$.

EXERCISE 17 $q(x) = 3x^3 - 2x^2 + \frac{1}{12}x + \frac{4}{9}$, $r(x) = \frac{-47}{432}x + \frac{1}{6}$

EXERCISE 20 See Allenby (1995, pp. 11, 12).

EXERCISE 21 Drop the perpendiculars OL, OM, ON from O onto the sides BC, CA, AB respectively. Then the lengths of OL and OM are equal (since OC bisects angle BCA) and the lengths of OL and ON are equal (similar reasons). Thus the lengths of OM and ON are equal. It follows that OA bisects angle CAB.

EXERCISE 23 The middle triangle is equilateral. (This is called Morley's theorem: it is hard to prove.)

EXERCISE 24 No, it is $\sqrt{65}$ m.

EXERCISE 26 (a) $\begin{pmatrix} 103 \\ 4 \end{pmatrix}$; (b) $\begin{pmatrix} 23 \\ 3 \end{pmatrix}$.

Chapter 12

EXERCISE 1 $1! + 2! + 3! + 4! = 33$. Subsequently each new factorial added is a multiple of 10. Hence, *every* sum of the form $1! + 2! + ... + n!$ with $n \geqslant 4$ ends in a 3.

EXERCISE 2 At least one of $a, -a, b, -b, c, -c$ is positive.

EXERCISE 3 $(66, 110, 165) = 1 \cdot 165 + 7 \cdot 110 - 14 \cdot 66$

EXERCISE 5

(a) In $(x + y + z)/3 \geqslant \sqrt[3]{(xyz)}$ replace x by ab, y by bc and z by ...*guess what!*
(b) In Theorem 5 replace each a_i by a_i^m.

EXERCISE 6 The second solution, with $a = t$, is the first solution with $a = t + 1$.

EXERCISE 7 Given n, set $m = 3^2n$ at the outset. Eliminate maximum powers 2^a and 5^b from m to obtain the integer $N = m/2^a5^b$. Then $(N, 10) = 1$. Therefore $N | (10\phi N - 1)$. Hence $\frac{N}{9} | 111...1$, an integer with $\phi(N)$ digits, all ones. Then $\frac{N}{9} \cdot 2^c5^c$, where c is the maximum of a, and b is a multiple of n whose only digits are zeros and ones.

EXERCISE 8 When h is very very small the cone is nearly the disc of area πr^2. But the formula given in Example 1 would have us believe that the area ought to be $2\pi r$ {very very small}. Thus the formula in Example 1 is wrong.

EXERCISE 9   More. To convince yourself of this, without actually doing the calculation, let the wind speed increase to 400 kph. Then, with respect to the ground, the plane flies from A to B at $400 + 400$ kph. From B to A it flies at $400 - 400$ kph, that is, it *never* gets back. (Extreme cases, even if silly, can often point the way to the correct answer. It is a common misconception that the total journey time will be independent of the wind speed because 'what the plane gains on the way from A to B it loses on the way from B to A.')

EXERCISE 11 $\frac{1}{378}(x^3 - 11x^2 + 416x - 40)$. The difference of two such polynomials would be a polynomial of degree at most 3 but possessing at least 4 roots.

Chapter 13

EXERCISE 1   The minute hand 'catches up' the hour hand by 55 minutes in every 60 minutes. It will therefore 'catch up' the initial 5-minute difference in $60/11$ minutes.

EXERCISE 2

$$10 + \tfrac{1}{10} + \tfrac{1}{1000} + \tfrac{1}{100000} + \dots = 10.101\,010\dots \text{ metres}$$

that is $10\frac{10}{99}$ metres.

Example 1
Mrs Achilles thought, incorrectly, that a sum of an 'infinite number' of non-zero time intervals must add up to an infinite length of time. This is not the case.

EXERCISE 3   As x changes from -1 to 1 it passes through a point ($x = 0$) where the integrand is not defined. (The integral is an example of an 'improper infinite integral'. Sense can be made of such objects if care is taken.)

EXERCISE 4   The arbitrary constant c has gone missing.

Example 2
The substitution $u = 1/x$ cannot be allowed over the range from -1 to 1. (When $x = 0$, u is not defined.)

EXERCISE 5   It is possible (sometimes) to define the sum of infinitely many numbers. It needs care, and that care would exclude the infinite sum here from having a meaning.

Example 3
This example emphasizes that you cannot take liberties with infinite sums.

Example 4
As the jagged line gets nearer to AC its 'jaggedness' increases just sufficiently to allow its total length to remain the same.

Example 5
If you imagine the outer edge of the flange running on a parallel rail you may more readily see that, with respect to that rail, the outer flange is *slipping*.

EXERCISE 6   Flatten the cone out. It becomes a sector of a circle of radius H. The perimeter of this sector is $2\pi r$. The area of this sector is therefore

$$\frac{2\pi r}{2\pi H}\cdot \pi H^2\ (=\pi rH)$$

Example 6
For ease of explanation, look at the corresponding two-dimensional *analogue*; the area and side-lengths of the triangle in Figure 15.3. First, as more and more thinner rectangles are used, the difference between their total area and the area of the triangle, being twice the sum of the areas of all the triangles $Y_iX_{i+1}Y_{i+1}$, clearly approaches zero. On the other hand, the sum of the heights X_iY_i of the rectangles is just the vertical height of the triangle, no matter how many rectangles are employed (and is not, therefore, equal to the side length AB of the triangle.)

Example 7
Solution 1 shows that *if* x satisfies (III) and (IV) then x *must* equal 4 or -1. Solution 2 shows similarly that x *must* equal $-\frac{3}{4}$. The correct inference is that *no value of* x *satisfies both (III) and (IV) simultaneously.*

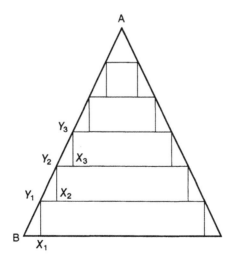

Figure 15.3

Example 8
To be sure that $p|\binom{p}{r}$ we should need to know that

$$\frac{(p-1)\cdot\ldots\cdot(p-r+1)}{2\ \cdot\ldots\cdot\ \ r}$$

is an *integer*. Is that obvious?

Example 9
The argument assumes that $A \neq \emptyset$. In fact, if you choose $A = \emptyset$, $D = \emptyset$ and B and C arbitrary but non-empty, you will obtain a counterexample.

Example 10
The inequality after the word 'But' uses the result that the example is trying to establish. So the argument is circular.

Example 11
The proof says 'If x were odd... therefore x is odd'. (What is to be proved has been assumed.)

Example 12
A single case cannot establish the general assertion.

Example 13
The second sentence is wrong. (If $n = 14$, say, so that $r = 4$ then $n^2 + 4$ does not 'end' in 16. Furthermore, a number may end in 16 and still be divisible by 7. Indeed one of $16, 116, 216, 316, 416, 516, 616$ *must* be divisible by 7.

Example 14
No, this argument is correct. (However, we do claim that $\sqrt{b} \in \mathbb{Z}$ even though it only seems – for certain – to belong to $\mathbb{Q}$. Is that fair?)

Example 15
(i) Who says $a \cdot b$ is 1 less than a perfect square?
(ii) From $a \cdot b = c \cdot d$ one may not, in general, infer that $a = c$ and $b = d$ (e.g. $5 \cdot 12 = 4 \cdot 15$).

Example 16
It may not be known whether or not $2^{(2^{127} - 1)} - 1$ is prime but you certainly cannot claim that all the integers arising in that sequence are prime on the basis that the first five are.

Example 17
The statement is false but the 'proof' does not prove it to be false. The example claims there is *at least one* such y. To disprove the claim, the student must show there is *no* such y. He has (only) shown that y cannot be equal to 1. He has not yet shown that it is not some other integer.

Example 18
It is correct to say that $1 + 1/n$ approaches the value n, but it is *incorrect* to infer that $(1 + 1/n)^n$ approaches the value 1^n. It is unfair to allow the n inside the bracket to trundle off 'to infinity' whilst allowing the (same) n in the power to go there afterwards.

Example 19
This argument, involving letting A 'tend to infinity' is perfectly acceptable and the result is correct.

Example 20
Before the final examination, T-D had scored 1069 marks on 15 papers, average 71.27% : E had scored 563 marks on 8 papers, average 70.38%. However, you will find $1167/16 < 658/9$. (The reason for this is that a high mark has more effect on the average of a small number of weaker marks than on the average of a large number of weaker marks.)

Example 21
The proof claims to show that $(a + b + c)/3 \geqslant \sqrt[3]{(abc)}$ but it actually *calls upon that very result* in the final inequality. So this is another example of circular reasoning.

Example 22
The ploy of proving x to be odd by comparing it with the positive integer $x - 2$ fails if we try to apply it in the case $x = 2$. (But for this anomaly all integers could be proved odd – by mathematical induction.)

Example 23
Suppose you choose a 4. There are 6^3 ways of throwing 3 dice. Of these, 5^3 will not show a 4, $3 \cdot 1 \cdot 5^2$ will show one 4, $3 \cdot 1 \cdot 1 \cdot 5$ will show two 4s and one will show three 4s. (In particular the claim that there is an even chance of winning and losing is incorrect. The student has counted some winning throws twice [which?].) Consequently, for each set of 216 different throws the stallholder will collect $125 \cdot 10\text{p}$ and pay out $75 \cdot 10 + 15 \cdot 20 + 1 \cdot 30 = 1080$ pence. Compared with a fair game, the stallholder is being overgenerous on the 'one-4' throw but not generous enough with the other three types of throw, ludicrously so in the case of 'three 4s'.

'I don't believe it'

1. None of them is true for *all* pairs a, b. (i), (ii), (iii) are false for *all* pairs of positive reals. (iv) is true iff a and b are *both* non-negative or *both* non-positive. (v), (vi), (vii) can be true for special pairs of a, b (for example, (v) is true if $a > 1$ and $b = a/(a - 1)$).
2. No! f may be the function given by

$$f(x) = e^{-\frac{1}{x^2}}$$

   if $x \neq 0$ and $f(0) = 0$. (This property of f is proved in certain books on mathematical analysis.)

3. You may, of course *define* the zero polynomial to have degree 0, but if you do, you can no longer have the formula that, for all polynomials $p(x)$, $q(x)$, degree $\{p(x) \cdot q(x)\} = $ degree $\{p(x)\} + $ degree $\{q(x)\}$. For this formula would fail if you take $p(x) = 0$ to have degree 0 and $q(x)$ to be a non-zero polynomial. (For example, with $p(x) = 0$ and $q(x) = x$ we get: $\deg\{p(x) \cdot q(x)\} = 0$ whereas $\deg\{p(x)\} + \deg\{q(x)\} = \deg\{q(x)\} \neq 0$.) Hence the correct version of the Division Theorem for polynomials with coefficients in $\mathbb{R}$ say (or $\mathbb{Q}$ or $\mathbb{C}$) is: *Given polynomials $a(x)$ and $b(x)$ there exist polynomials $m(x)$ and $r(x)$ such that $a(x) = m(x)b(x) + r(x)$ where either (i) $r(x) = 0$, the zero polynomial or (ii) $r(x) \neq 0$ and $\deg\{r(x)\} < \deg\{b(x)\}$.*

4. Fil was of course speaking rubbish. If all numbers have an equal chance of winning then it is immaterial which 10 tickets you have.

5. Julie's twin sister, Kulie, runs at a steady 10 mph. Barbara alters her speed like a sine curve so that she is 1 yard behind Kulie after 5, 25, 45, ... yards, level at 10, 20, 30, ... yards and 1 yard ahead of Kulie after 15, 35, 55, ... yards. Because the race is over 46 415 yards, Kulie and Barbara are level at 46 140 yards but Barbara loses over the last five yards. Clearly, Barbara and Kulie run *every* distance of 1760 yards (1 mile) in the same time. Now, over the 26 miles 385 yards Julie runs uniformly but more slowly than her sister, so much so that she finishes 1 whole inch behind her. This means she runs *every* distance of 1 mile more slowly than her sister and hence more slowly than Barbara. However, at 5, 25, 45, ... yards Kulie is always 1 yard ahead of Barbara. So, at the finishing tape, Julie beats Barbara by almost 1 yard.

6. In fact the 'new' equations have solution $x = -8.347...$, $y = +21.826...$.

References

Allenby, R.B.J.T. (1991) *Rings, Fields and Groups: An Introduction to Abstract Algebra*, 2nd edition. London: Edward Arnold.

Allenby, R.B.J.T. (1995) *Linear Algebra*. London: Edward Arnold.

Allenby, R.B.J.T. and Redfern, E.J. (1989) *Introduction to Number Theory – with Computing*. London: Edward Arnold.

Bogart, Kenneth P. (1983) *Introductory Combinatorics*. Boston: Pitman.

Boyer, Carl B. (1968) *A History of Mathemctics*. New York: Wiley.

Bunt, Lucas N.H., Jones, Phillip S. and Bedient, Jack D. (1988) *The Historical Roots of Elementary Mathematics*. New York: Dover Publications.

Burton, David M. (1995) *History of Mathematics*, 3rd edition. Dubuque, IA: William C. Brown.

Dictionary of Scientific Biography. New York: Charles Scribner's Sons (1970–1980).

Durell, Clement V. (1950) *General Arithmetic for Schools*. London: G. Bell.

Eves, Howard (1976) *An Introduction to the History of Mathematics*. New York: Holt, Rinehart & Winston.

Gardiner, Tony (1995) *The Mathematical Gazette*, vol. 79, pp. 526–32.

Guy, Richard K. (1994) *Unsolved Problems in Number Theory*, 2nd edition. New York: Springer-Verlag.

Hall, H.S. and Knight, S.R. (1887) *Higher Algebra*. New York: Macmillan.

Heath, Sir Thomas L. (1897) *The Works of Archimedes*. Cambridge: Cambridge University Press. (Reprinted Dover Publications, New York, 1955.)

Hirst, K.E. (1995) *Numbers, Sequences and Series*. London: Edward Arnold.

Just, Erwin and Schaumberger, Norman (1973) 'A curious property of the integer 38', *Math. Magazine*, vol. 46, p. 221.

Kline, M. (1972) *Mathematical Thought from Ancient to Modern Times*. New York: Oxford University Press.

Liu, C.L. (1968) *Introduction to Combinatorial Mathematics*. New York: McGraw-Hill.

Phillips, E.G. (1950) *Analysis*. Cambridge: Cambridge University Press.

Riesel, Hans (1985) *Prime Numbers and Computer Methods for Factorization*. Boston: Birkhauser.

Salomaa, Arto (1990) *Public Key Cryptography*. Berlin: Springer-Verlag.

Slomson, Alan (1991) *An Introduction to Combinatorics*. London: Chapman & Hall.

Index

Printed and bound by CPI Group (UK) Ltd, Croydon, CR0 4YY

03/10/2024

01040331-0008